# Phytochemicals
# in Health
# and Disease

# OXIDATIVE STRESS AND DISEASE

Series Editors

**LESTER PACKER, PH.D.**
**ENRIQUE CADENAS, M.D., PH.D.**
University of Southern California School of Pharmacy
Los Angeles, California

1. Oxidative Stress in Cancer, AIDS, and Neurodegenerative Diseases, *edited by Luc Montagnier, René Olivier, and Catherine Pasquier*
2. Understanding the Process of Aging: The Roles of Mitochondria, Free Radicals, and Antioxidants, *edited by Enrique Cadenas and Lester Packer*
3. Redox Regulation of Cell Signaling and Its Clinical Application, *edited by Lester Packer and Junji Yodoi*
4. Antioxidants in Diabetes Management, *edited by Lester Packer, Peter Rösen, Hans J. Tritschler, George L. King, and Angelo Azzi*
5. Free Radicals in Brain Pathophysiology, *edited by Giuseppe Poli, Enrique Cadenas, and Lester Packer*
6. Nutraceuticals in Health and Disease Prevention, *edited by Klaus Krämer, Peter-Paul Hoppe, and Lester Packer*
7. Environmental Stressors in Health and Disease, *edited by Jürgen Fuchs and Lester Packer*
8. Handbook of Antioxidants: Second Edition, Revised and Expanded, *edited by Enrique Cadenas and Lester Packer*
9. Flavonoids in Health and Disease: Second Edition, Revised and Expanded, *edited by Catherine A. Rice-Evans and Lester Packer*
10. Redox–Genome Interactions in Health and Disease, *edited by Jürgen Fuchs, Maurizio Podda, and Lester Packer*
11. Thiamine: Catalytic Mechanisms in Normal and Disease States, *edited by Frank Jordan and Mulchand S. Patel*
12. Phytochemicals in Health and Disease, *edited by Yongping Bao and Roger Fenwick*

*Related Volumes*

Vitamin E in Health and Disease: Biochemistry and Clinical Applications, *edited by Lester Packer and Jürgen Fuchs*

Vitamin A in Health and Disease, *edited by Rune Blomhoff*

Free Radicals and Oxidation Phenomena in Biological Systems, *edited by Marcel Roberfroid and Pedro Buc Calderon*

Biothiols in Health and Disease, *edited by Lester Packer and Enrique Cadenas*

Handbook of Antioxidants, *edited by Enrique Cadenas and Lester Packer*

Handbook of Synthetic Antioxidants, *edited by Lester Packer and Enrique Cadenas*

Vitamin C in Health and Disease, *edited by Lester Packer and Jürgen Fuchs*

Lipoic Acid in Health and Disease, *edited by Jürgen Fuchs, Lester Packer, and Guido Zimmer*

*Additional Volumes in Preparation*

Carotenoids in Health and Disease, *edited by Norman I. Krinsky, Susan T. Mayne, and Helmut Sies*

Herbal Medicine, *edited by Lester Packer, Choon Nam Ong, and Barry Halliwell*

# Phytochemicals in Health and Disease

edited by
## YONGPING BAO
## ROGER FENWICK
*Institute of Food Research*
*Norwich, England*

## CRC Press
Taylor & Francis Group
Boca Raton London New York

CRC Press is an imprint of the
Taylor & Francis Group, an **informa** business

CRC Press
Taylor & Francis Group
6000 Broken Sound Parkway NW, Suite 300
Boca Raton, FL 33487-2742

First issued in paperback 2019

ISBN-13: 978-0-8247-4023-8 (hbk)
ISBN-13: 978-0-367-39428-8 (pbk)

**Library of Congress Cataloging-in-Publication Data**
A catalog record for this book is available from the Library of Congress.

Visit the Taylor & Francis Web site at
http://www.taylorandfrancis.com

and the CRC Press Web site at
http://www.crcpress.com

# Series Introduction

Oxygen is a dangerous friend. Overwhelming evidence indicates that oxidative stress can lead to cell and tissue injury. However, the same free radicals that are generated during oxidative stress are produced during normal metabolism and thus are involved in both human health and disease.

Free radicals are molecules with an odd number of electrons. The odd, or unpaired, electron is highly reactive as it seeks to pair with another free electron.

Free radicals are generated during oxidative metabolism and energy production in the body.

Free radicals are involved in:
Enzyme-catalyzed reactions
Electron transport in mitochondria
Signal transduction and gene expression
Activation of nuclear transcription factors
Oxidative damage to molecules, cells, and tissues
Antimicrobial action of neutrophils and macrophages
Aging and disease

Normal metabolism is dependent on oxygen, a free radical. Through evolution, oxygen was chosen as the terminal electron acceptor for respiration. The two unpaired electrons of oxygen spin in the same direction; thus, oxygen is a biradical, but is not a very dangerous free radical. Other oxygen-derived free radical species, such as superoxide or hydroxyl radicals, formed during metabolism or by ionizing radiation are stronger oxidants and are therefore more dangerous.

In addition to research on the biological effects of these reactive oxygen species, research on reactive nitrogen species has been gathering momentum. NO, or nitrogen monoxide (nitric oxide), is a free radical generated by NO synthase (NOS). This enzyme modulates physiological responses such as vasodilation or signaling in the brain. However, during inflammation, synthesis of NOS (iNOS) is induced. This iNOS can result in the overproduction of NO, causing damage. More worrisome, however, is the fact that excess NO can react with superoxide to produce the very toxic product peroxynitrite. Oxidation of lipids, proteins, and DNA can result, thereby increasing the likelihood of tissue injury.

Both reactive oxygen and nitrogen species are involved in normal cell regulation in which oxidants and redox status are important in signal transduction. Oxidative stress is increasingly seen as a major upstream component in the signaling cascade involved in inflammatory responses, stimulating adhesion molecule and chemoattractant production. Hydrogen peroxide, which breaks down to produce hydroxyl radicals, can also activate NF-(B, a transcription factor involved in stimulating inflammatory responses. Excess production of these reactive species is toxic, exerting cytostatic effects, causing membrane damage, and activating pathways of cell death (apoptosis and/or necrosis).

Virtually all diseases thus far examined involve free radicals. In most cases, free radicals are secondary to the disease process, but in some instances free radicals are causal. Thus, there is a delicate balance between oxidants and antioxidants in health and disease. Their proper balance is essential for ensuring healthy aging.

The term oxidative stress indicates that the antioxidant status of cells and tissues is altered by exposure to oxidants. The redox status is thus dependent on the degree to which a cell's components are in the oxidized state. In general, the reducing environment inside cells helps to prevent oxidative damage. In this reducing environment, disulfide bonds (S-S) do not spontaneously form because sulfhydryl groups kept in the reduced state (SH) prevent protein misfolding or aggregation. This reducing environment is maintained by oxidative metabolism and by the action of antioxidant enzymes and substances, such as glutathione, thioredoxin, vitamins E and C, and enzymes such as superoxide dismutase (SOD), catalase, and the selenium-dependent glutathione and thioredoxin hydroperoxidases, which serve to remove reactive oxygen species.

Changes in the redox status and depletion of antioxidants occur during oxidative stress. The thiol redox status is a useful index of oxidative stress mainly because metabolism and NADPH-dependent enzymes maintain cell glutathione (GSH) almost completely in its reduced state. Oxidized glutathione (glutathione disulfide, GSSG) accumulates under conditions of oxidant exposure, and this changes the ratio of oxidized to reduced glutathione; an increased ratio indicates oxidative stress. Many tissues contain large amounts of glutathione, 2-4 mM in erythrocytes or neural tissues and up to 8 mM in hepatic tissues. Reactive oxygen and nitrogen species can directly react with glutathione to lower the levels of this substance, the cell's primary preventative antioxidant.

Current hypotheses favor the idea that lowering oxidative stress can have a clinical benefit. Free radicals can be overproduced or the natural antioxidant system defenses weakened, first resulting in oxidative stress, and then leading to oxidative injury and disease. Examples of this process include heart disease and cancer. Oxidation of human low-density lipoproteins is considered the first step in the progression and eventual development of atherosclerosis, leading to cardiovascular disease. Oxidative DNA damage initiates carcinogenesis.

Compelling support for the involvement of free radicals in disease development comes from epidemiological studies showing that an enhanced antioxidant status is associated with reduced risk of several diseases. Vitamin E and prevention of cardiovascular disease is a notable example. Elevated antioxidant status is also associated with decreased incidence of cataracts and cancer, and some recent reports have suggested an inverse correlation between antioxidant status and occurrence of rheumatoid arthritis and diabetes mellitus. Indeed, the number of indications in which antioxidants may be useful in the prevention and/or the treatment of disease is increasing.

Oxidative stress, rather than being the primary cause of disease, is more often a secondary complication in many disorders. Oxidative stress diseases include inflammatory bowel diseases, retinal ischemia, cardiovascular disease and restenosis, AIDS, ARDS, and neurodegenerative diseases such as stroke, Parkinson's disease, and Alzheimer's disease. Such indications may prove amenable to antioxidant treatment because there is a clear involvement of oxidative injury in these disorders.

In this series of books, the importance of oxidative stress in diseases associated with organ systems of the body is highlighted by exploring the scientific evidence and the medical applications of this knowledge. The series also highlights the major natural antioxidant enzymes and antioxidant substances such as vitamins E, A, and C, flavonoids, polyphenols, carotenoids, lipoic acid, and other nutrients present in food and beverages.

Oxidative stress is an underlying factor in health and disease. More and more evidence indicates that a proper balance between oxidants and antioxidants is involved in maintaining health and longevity and that altering this balance in favor of oxidants may result in pathological responses causing functional disorders and disease. This series is intended for researchers in the basic biomedical sciences and clinicians. The potential for healthy aging and disease prevention necessitates gaining further knowledge about how oxidants and antioxidants affect biological systems.

*Lester Packer*
*Enrique Cadenas*

# Preface

Research into the biological activity of phytochemicals has a long history. Initially, the focus was on natural product chemistry, with degradation studies providing a basis for structure determination that was confirmed by elegant multistage synthesis. Many key reactions of organic chemistry were first developed out of the specific synthetic requirements of this area.

At this early stage, synthesis and isolation afforded significant amounts of individual compounds that could be used for biological investigation. Such studies addressed the effects of a single compound and provided a context for future research that studied effects produced by concentrations vastly in excess of those normally encountered.

Much of this early research centered on pharmacologically active compounds, in which the biological effects were clear and measurable. In a minority of cases, the biological effects were antinutritional. Texts on "natural toxicants" published since the 1970s reveal greater knowledge within the chemical sector than within the biological, veterinary, and clinical areas.

For example, many thousands of papers were published between 1970 and 1990 addressing the antinutritional and "toxic" effects of rapeseed meals on farm livestock and poultry. The vast majority of these were focused on the effects of glucosinolates and their breakdown products. In retrospect one can see the great divergence between "chemical" and "biological" investigations. The former, centered on development of analytical approaches, became focused on increasingly rapid, sensitive, and specific methods, in almost total isolation from the biological studies.

One consequence was that considerable human and financial resources were wasted and fundamental knowledge denied. An overall lack of agreement on analytical protocols and best practice also bedeviled attempts to build on data produced in different laboratories. This is not to overly criticize the activities conducted, rather it represented the barriers to effective international cooperation and funding available.

The next phase of research was conducted with animal models and, later, cell systems. This approach examined single compounds and was frequently not able to be translated to the human situation. Experimental doses were administered in ways that were inconsistent with the natures, times, and levels associated with human exposure. The failure to identify and agree on standardized experimental protocols and methodologies made it difficult to compare data from different laboratories and to send a clear message to the consumer, over and above the necessary but unexciting exhortation to "eat more fruit and vegetables."

We are in a new era of phytochemical and health research where past knowledge and experience of chemical, biochemical, nutritional, and clinical research is allied to the new generation of molecular biology and trinomic research. Over the next decade we will witness scientific advances as well as raise new questions and challenges for future scientists to answer.

The beginning chapters of this volume are devoted to aspects of phytochemical research that are fundamental to this new era of activity. Chapter 1 introduces the fundamentals of nutritional genomics (nutrigenomics), notably DNA arrays, proteomics, and metabolomics, while Chapter 2 indicates the manifold factors affecting bioavailability, critically compares various methods for its measurement, and highlights crucial differences between studies on new drug formulations and dietary phytochemicals. The fundamental importance of considering metabolic processes is addressed in Chapter 3, with particular attention to those associated with the ubiquitous polyphenols.

Genistein has recently received much scientific and unscientific publicity, and Chapter 4 presents current knowledge of genistein-induced gene expression and its role in regulating proliferation of cancer, especially that of the bladder. Chapter 5 offers an illustration of how the effects and the mechanisms of action of *Ginkgo biloba* extracts have been studied, while Chapter 6 centers on another widely studied phytochemical, the isothiocyanate sulforaphane. Chapter 7 is concerned with demonstrating how heterocyclic amine–induced DNA adduct formation, a biomarker of DNA damage and risk of cancer, can be affected by phytochemicals.

Chapters 8–11 are devoted to consideration of two main classes of phytochemicals; Chapter 8 centers on organosulfur compounds from the *Allium* genus, and their role in cancer prevention, while subsequent chapters are devoted to polyphenols and their wide-ranging effects against cancers, cataracts, and cardiovascular disease. Chapter 9 examines the cancer-protective and therapeutic potential of polymethylated flavonoids such as occur in citrus, while Chapter 10 summarizes recent cancer chemoprevention research in an area particularly relevant to the British and Chinese backgrounds of the editors, notably tea. A much less well-studied area, that of cataracts and their prevention by dietary flavonoids, is presented in Chapter 11.

Chapter 12 summarizes the role of dietary phytochemicals, especially those from fruits, herbs, and spices, in the prevention of cardiovascular disease. Chapter 13 is devoted to the effects and mechanisms of action of resveratrol, a polyphenolic phytoalexin that has received considerable publicity. The same is true of lycopene, the topic of Chapter 14, which occurs in especially high levels in fresh and processed tomatoes. Consideration is given in Chapter 15 to the effects of oltipraz on phase 1 and 2 xenobiotic-metabolizing enzymes; this draws together nutritional, clinical, and pharmacological investigation and provides some pointers for the future. These are elaborated upon in Chapter 16, which addresses research directions, challenges, and opportunities.

This book presents illustrations on a broad range of approaches and methodologies, some new, others less so, and brings them together to address some of the fundamental questions of phytochemicals and health. In this context it is important to remember that the questions posed by consumers are simple to present, but highly complex to address: What should I eat? How often? How much?

One important area that demands further study relates to the ability of plant scientists and agronomists to modify the composition of fruits and vegetables to enhance levels of phytochemicals or to introduce relevant genes into other species in order to broaden the occurrence of these compounds in the diet, as presented in Chapter 6, on sulforaphane. In this area, as in many others, the public's attitude to future genetically manipulated plants and foods will be crucial.

The point has been made in many of these chapters that methodological and scientific development over recent years is such that very considerable advances will be made over the next decade; indeed, one suspects that such is the pace of advance that the information presented in this volume will before too long be considered dated. This is to be welcomed and, in our opinion, is greatly preferable to the former situation, in which texts on natural toxicants/phytochemicals/secondary metabolites often contained much the same general coverage over many years.

This volume is written for the research and scientific community. However, it must be remembered that if the full health benefits of phytochemicals are to be delivered to society, it will be necessary for researchers and scientists to explain their interests and activities to the general public. This, perhaps, is an equally important challenge as that addressed in the laboratory. If we, as scientists, fail in this area, then members of the public will seek information and assurance from less well-informed and less independent sources. Since many of these individuals will have personal or family reasons for seeking such information, the consequences could be both serious and dangerous.

We are grateful to the contributors (from 16 different organizations in 8 countries) for their enthusiasm, professionalism, and attention to detail. All are internationally recognized experts in their respective fields and it was a pleasure to work with them. We would also like to thank Susan Lee, Elyce Misher, and Richard Johnson at Marcel Dekker, Inc. for their support, encouragement, and assistance throughout this project.

*Yongping Bao*
*Roger Fenwick*

# Contents

# Contributors

**James R. Bacon, M.R.S.C., M.I.F.S.T.** Department of Nutrition, Institute of Food Research, Colney, Norwich, United Kingdom

**Yongping Bao, Ph.D.** Department of Nutrition, Institute of Food Research, Colney, Norwich, United Kingdom

**Ann M. Bode, Ph.D.** The Hormel Institute, University of Minnesota, Austin, Minnesota, U.S.A.

**Rainer Cermak, D.V.M.** Department of Agricultural and Nutritional Sciences, Christian-Albrechts-University of Kiel, Kiel, Germany

**Andrea J. Day, B.Sc., Ph.D.** The Procter Department of Food Science, University of Leeds, Leeds, United Kingdom

**Zigang Dong, M.D.** The Hormel Institute, University of Minnesota, Austin, Minnesota, U.S.A.

**Ruan M. Elliott, M.A., Ph.D.** Department of Nutrition, Institute of Food Research, Colney, Norwich, United Kingdom

**Roger Fenwick, Ph.D.** International Coordination, Institute of Food Research, Colney, Norwich, United Kingdom

**Regina Goralczyk, D.V.M.** Department of Human Nutrition and Health, DSM Nutritional Products, Basel, Switzerland

**André Guillouzo, Ph.D.**  Department of Toxicology, University of Rennes, Rennes, France

**Birgit Holst, Ph.D.**  Department of Nutrition, Nestlé Research Center, Nestec Ltd., Lausanne, Switzerland

**Sophie Langouët, Ph.D.**  University of Rennes, Rennes, France

**Ching Li, Ph.D.**  Department of Microbiology and Immunology, Chung Shan Medical University, Taichung, Taiwan

**Jen-Kun Lin, Ph.D.**  Institute of Biochemistry, College of Medicine, National Taiwan University, Taipei, Taiwan

**W. Russell McLauchlan, Ph.D.**  Department of Nutrition, Institute of Food Research, Colney, Norwich, United Kingdom

**John Milner, Ph.D.**  Division of Cancer Prevention, National Cancer Institute, National Institutes of Health, Rockville, Maryland, U.S.A.

**Fabrice Morel, Ph.D.**  University of Rennes, Rennes, France

**Michael R. A. Morgan, B.Sc., M.Sc., Ph.D.**  The Procter Department of Food Science, University of Leeds, Leeds, United Kingdom

**Akira Murakami, Ph.D.**  Division of Food Science and Biotechnology, Kyoto University, Kyoto, Japan

**Hajime Ohigashi, Ph.D.**  Division of Food Science and Biotechnology, Kyoto University, Kyoto, Japan

**Gerald Rimbach, Ph.D.**  School of Food Biosciences, University of Reading, Reading, United Kingdom

**Joseph A. Rothwell, B.Sc., M.Sc.**  The Procter Department of Food Science, University of Leeds, Leeds, United Kingdom

**Samir Samman, Ph.D.**  University of Sydney, Sydney, New South Wales, Australia

**Julie Sanderson, B.Sc., Ph.D.**  School of Biological Sciences, University of East Anglia, Norwich, Norfolk, United Kingdom

**Biehuoy Shieh, Ph.D.** Department of Biochemistry, Chung Shan Medical University, Taichung, Taiwan

**Ulrich Siler, Ph.D.** Department of Human Nutrition and Health, DSM Nutritional Products, Basel, Switzerland

**Gary Williamson, Ph.D.** Department of Nutrition, Nestlé Research Center, Nestec Ltd., Lausanne, Switzerland

**Siegfried Wolffram, D.V.M.** Department of Agricultural and Nutritional Sciences, Christian-Albrechts-University of Kiel, Kiel, Germany

**Yuesheng Zhang, M.D., Ph.D.** Department of Cancer Prevention, Roswell Park Cancer Institute, Buffalo, New York, U.S.A.

# 1
# Nutritional Genomics

**Ruan M. Elliott, James R. Bacon, and Yongping Bao**
*Institute of Food Research, Colney, Norwich, United Kingdom*

## I. INTRODUCTION

It seems almost a given now that phytochemicals in our diet have health-promoting effects. The body of experimental evidence supporting this hypothesis (some of which is described elsewhere in this book) is substantial and continues to grow. However, many key questions remain. Which phytochemicals are the most efficacious? How do they work? Do they act independently or synergistically? How much should we consume, and in what form, to achieve optimal health benefits? How does this vary from person to person?

Until very recently, most of the research targeted at addressing these questions has had to focus upon single (or at most a few) genes, signaling pathways, or cellular processes and, most commonly, upon individual compounds. Over the last few years, molecular biological approaches, such as competitive and real-time RT-PCR and gene knockout models, have started to prove their worth in such studies [1–3]. These techniques are extremely valuable because they enable close scrutiny of possible modes of action. With sufficient such studies, it will be possible to begin to draw together the many different facets of the biological activities of phytochemicals to form a truly coherent view of how they work in vivo as agents present in real diets. But this goal is still some way off.

The use of the new functional genomic approaches in nutrition research, so-called nutritional genomics, presents the possibility for an entirely new approach that seeks to develop a global view of the biological effects of food [4,5]. This approach, in which the ultimate aim is to examine all genes and cellular pathways simultaneously in complex organisms, is sometimes termed systems biology.

If used sensibly, and in conjunction with the full range of nutrition research tools already available, nutritional genomics will significantly enhance the speed of movement towards a comprehensive understanding of how different phytochemicals promote health and how much we should aim to consume for maximum benefits. In conjunction with the rapidly developing knowledge of interindividual genetic variation, there is the very real opportunity to start to include in the overall process an assessment of the differences in dietary requirements for optimal health between individuals dependent on their age, gender, lifestyle, and genetic makeup.

Functional genomic techniques have another vital advantage over more traditional methods. They all enable the simultaneous examination of the regulation of many thousands of different cellular processes, meaning that unexpected effects are far less likely to be overlooked. This makes them ideal techniques for developing new hypotheses for mechanisms of action, for determining how different processes may interact together, and for identifying potential risks as well as defining the mechanisms that underlie health promotion. If this succeeds in reducing the number of health scares that result from dietary advice issued with an incomplete understanding of the full scope for beneficial and adverse biological effects, it can only be good for nutrition research, policy makers, and the general public.

In this chapter, the fundamentals of the two current mainstays of functional genomic technology, DNA arrays and proteomics, will be detailed, together the complementary technology of real-time RT-PCR (TaqMan® assay). Developments in a third and newest key area of functional genomics, namely metabolomics, will also be described. Finally, the challenges that remain to be addressed will be discussed and some of the potential pitfalls that need to be recognized and avoided if the opportunities presented by these new technologies are to be fully realized will be highlighted.

## II. TERMINOLOGY AND DEFINITIONS

Molecular biology and functional genomics, in particular, are jargon-rich disciplines. This can be daunting and confusing to those unfamiliar with the terminology. To assist the general reader and help avoid confusion, all the key terms used throughout this chapter are defined below:

> A *genome* is the complete complement of nucleotide sequences, including structural genes, regulatory sequences, and noncoding DNA (or RNA) segments, in the chromosomes of an organism.
> *Genomics* is, therefore, the study of all of the nucleotide sequences, including structural genes, regulatory sequences, and noncoding DNA segments, in the chromosomes of an organism. Genomics is sometimes divided into two separate components: structural and functional genomics.

*Structural genomics* is the DNA sequence analysis and construction of high-resolution genetic, physical, and transcript maps of an organism's genes and genome.

*Functional genomics* is the application of global (genome-wide or system-wide) experimental approaches to assess gene function.

*Nutritional genomics* is defined here as the application of all or any structural and functional genomic information/techniques within nutrition research.

A *transcriptome* is the complete complement of transcription products (i.e., RNA species) that can be produced by transcription of any components of an organism's genome. In many cases one gene can give rise to multiple different transcripts through, for example, tissue-specific variations in transcription start sites and posttranscriptional processing. All such variants should be considered as part of the transcriptome.

*Transcriptomics* is the study of the transcriptome and its regulation.

A *proteome* is the complete complement of translation products (i.e., polypeptides and proteins) that can be produced from the code held in an organism's genome. As for a transcriptome, a full proteome will include protein variants produced from individual genes by alternative splicing of RNA transcripts and alternative posttranslational modifications.

*Proteomics* is the study of a proteome and its expression.

*Metabolomics/Metabonomics* is the application of systems-wide techniques (normally based on NMR) for metabolic profiling. Some scientists use the term "metabolomics" to describe such analyses in both simple (cellular) and complex (tissue or whole body) systems. Others distinguish between "metabolomics" studies in simple systems only (e.g., cell culture) and "metabonomics" studies performed in complex systems (e.g., the human body).

*DNA arrays* are analytical tools for measuring the relative levels of thousands or tens of thousands of RNA species within cellular or tissue samples. Use of DNA arrays is sometimes referred to as "transcriptomics," although few if any current arrays for mammalian systems can truly be said to cover the entire transcriptome for the species being analysed.

*Genotyping* is the determination of interindividual variations in gene sequence.

*Reverse transcription polymerase chain reaction (RT-PCR)* is an experimental method for exponential amplification of a defined segment of complementary DNA (cDNA) produced by reverse transcription of RNA.

*Real-time RT-PCR* is a method based on RT-PCR for real-time detection of the products generated during each cycle of the PCR process, which are directly in proportion to the amount of template prior to the start of the PCR process.

There is an ongoing proliferation of ''-omic'' terminology in use in many different research areas (e.g., pharmacogenomics), reflecting the widely held view that the ''all-encompassing'' approaches made possible by structural and functional genomic technologies have vast potential to advance research in nearly every aspect of the life sciences. However, in most cases the technologies used are essentially the same as those described here, and it is really only the application to the different research specialities that differs. Since these are not of direct relevance to this chapter, they will not be discussed further here.

## III. REGULATION OF GENE EXPRESSION AND THE APPLICATION OF FUNCTIONAL GENOMICS

Gene expression is a multistep process for which many different possible modes of regulation have been recognized. It is this complexity that provides ample scope for the differential regulation of gene expression (e.g., developmental and tissue-specific switching on and off of defined genes) necessary for common precursor cells to differentiate into all the different cell types present in complex multicellular organisms, such as humans. Equally importantly, it enables each different cell type to respond in unique and defined ways to changes in their environment. Diet is a key factor that affects, both directly and indirectly, the environments to which cells in our tissues are exposed. As such, it should be no surprise that dietary components and their metabolites can have profound effects on gene expression patterns that will alter cellular function and ultimately impact upon health.

The gene expression process starts with transcription of a gene to produce RNA. If the RNA produced is a messenger RNA, it may then be translated into a protein. The protein product may be modified during or after synthesis, for example, by proteolytic cleavage, phosphorylation, or glycation. The amount of any given gene product and its activity (its level of expression) in a cell is determined by (1) the rate and efficiency of each of these processes, (2) the location of the product, and intermediates of its synthesis, within the cell, and (3) the rates at which the product and intermediates are degraded.

Changes in each of these processes can potentially be detected using DNA microarrays, proteomics, and metabolomics. Other technologies, such as real-time RT-PCR, complement these functional genomic techniques by providing more focused (i.e., small numbers of genes) but also more sensitive and accurate data to confirm and elaborate upon the broad spectrum data generated through functional genomic approaches.

A host of well-established molecular biological methods can be used to provide still more in-depth analysis; for example, determining whether an increase in the quantity of a specific RNA species is due to an increase in its rate of transcription or to a reduction in its rate of degradation. Descriptions of these approaches are beyond the scope of this chapter but can be found in many molecular biology texts (see, e.g., Ref. 6).

## IV. TRANSCRIPTOMICS AND THE APPLICATION OF MACRO- AND MICROARRAYS

DNA arrays, like more traditional hybridization techniques such as Southern and Northern blotting, make use of the fact that single-stranded nucleic acid species (DNA and RNA) that possess sequences complementary to each other will hybridize together with exquisite specificity to form double-stranded complexes. In the traditional approaches, the samples to be analyzed are distributed according to size by gel electrophoresis, transferred to and immobilized on solid membrane supports, and probed with specific, labeled complementary nucleic acid sequences.

With DNA arrays, this approach is turned on its head [7]. Multiple, individual and known DNA species are arranged as spots (elements) in a grid (array) pattern on the solid substrate (Fig. 1). It is the test sample that is fluorescently or radioactively labeled and hybridized with the DNA on the solid support. Wherever nucleic acids present in the test sample encounter complementary sequences on the array, they will be captured. After washing away nonhybridized sample, the array can be imaged using appropriate techniques (phosphor imaging for radiolabeled samples or laser scanning for fluorescently labeled samples). The signal intensity returned from each element on the array is indicative of the relative abundance of the corresponding nucleic acid in the test sample.

Arrays can be employed for the analysis of both DNA and RNA samples. Analysis of genomic DNA samples has been used to determine whether specific genes are present or absent in, for example, DNA samples from different strains of yeast or bacteria. This technique (termed genomotyping) is proving very valuable in microbiological research. It is of little importance to human nutrition research, but the potential use of arrays to determine the genotype of an individual for many thousands of defined genes has very exciting potential.

By far the most widespread use of arrays to date has been with RNA samples for the analysis of gene expression at the level of transcription. It is this application that will be described here. Arrays are available in an ever-increasing number of different formats; currently, three predominate: membrane (macro)arrays, microarrays produced on specially coated glass microscope slides, and Affymetrix GeneChip® arrays.

## A. Membrane Macroarrays

These are manufactured, using robotic systems, by spotting or spraying (using ink-jet technology) nucleic acid solutions (normally segments of genomic DNA

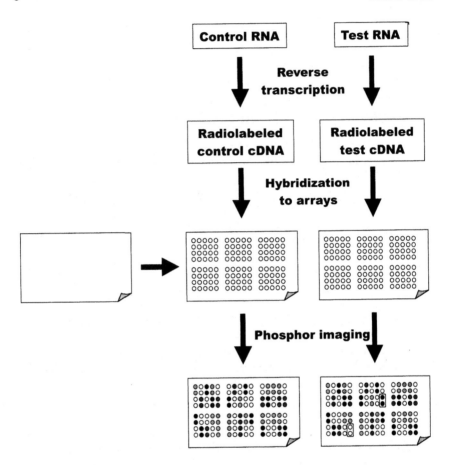

**Figure 1** Membrane array analysis. DNA or oligonucleotides are printed onto nylon membranes and covalently bound. RNA samples are reverse transcribed and radiolabeled nucleotides incorporated into the cDNA products. Hybridization is performed under standard conditions. The arrays are washed and radioactivity detected by phosphor-imaging. The patterns of spot intensities for control and test samples are compared to identify up- and downregulated transcripts.

or complementary DNA) onto nylon membranes to produce a defined grid (or array) with each spot corresponding to DNA from one gene (see Fig. 1). The spots of nucleic acids are denatured, if necessary, and fixed to the membrane, which is then ready for use. Due to their relatively large size, ranging in size from perhaps 5 to 25 cm along each side, membrane arrays are sometimes referred to as macroarrays.

Each array is hybridized with one test RNA sample. As with Northern and Southern blots, it is possible for the labeled sample to be stripped from the membrane after analysis so that the array can be reused. However, each time this process is performed there is some reduction in the performance of both the array and the quality of the data generated on subsequent use.

The labeled "probe" for the array is cDNA produced from the test RNA sample by a reverse transcription reaction that includes one or more deoxynucleotide triphosphates radiolabeled with $^{32}P$ or $^{33}P$ in the alpha position. The reverse transcription process may be initiated by including either oligo dT or random hexamers to nonamers as the primer in the reaction mix; this use of random hexamers to nonamers is only appropriate when the PolyA-enriched RNA is used. Alternatively, a primer mix can be specially prepared that contains oligonucleotides with sequences complementary to portions of each of the RNAs represented on the array. This approach has the advantage that only RNAs of interest are reverse transcribed, thus giving stronger specific signals and lower backgrounds.

Following the reverse transcription reaction, the RNA is either degraded with RNase or hydrolyzed by alkali. The reverse transcribed, radiolabeled cDNA is purified using any one of a number of different commercial kits. Hybridization is performed using a standard hybridization buffer and a hybridization oven essentially as for any Northern or Southern blot [6]. Following a hybridization step, normally conducted overnight, unbound radiolabeled cDNA is removed using a series of high stringency washes. The amount of radiolabeled probe hybridized to each element on the array is determined by phosphor imaging. The pattern of the different signal intensities for each of the spots is indicative of the relative abundance of the RNA species in the original sample. Thus, each array analyzed provides a gene transcript profile for the sample analysed.

If a control RNA sample is analyzed on one array and a sample obtained following a specific treatment (such as treatment of cells in culture with a bioactive phytochemical) is analyzed on a second array, the gene transcript profiles can be compared. This approach is used to identify genes up- and downregulated at the level of transcription by the treatment. Better still, multiple arrays can be analyzed using RNA isolated from cell or tissue samples following exposure, for example, to varying concentrations of a bioactive compound. This will generate data that can be used to describe not only the range of biological effects that the compound under investigation can elicit but also the dose-dependency of each of these effects.

In such experiments, the data obtained from each individual array have to be normalized to enable meaningful comparison between the data sets. The ratio of specific signal (correct hybridization of radiolabeled cDNA to the corresponding DNA on the array) to noise (nonspecific hybridization to the spots and background binding of radiolabeled probe to the membrane) will vary significantly from membrane to membrane. Normalization can be achieved by correction to the total specific signal (average spot intensity minus global or local background intensity) for all the elements on each array. Alternatively, the specific signal obtained for a range of housekeeping genes present on each array can be used to define normalization factors. Uneven hybridization or variations in background signal across an array can pose a real problem for data analysis. It may be possible to correct for these problems by, for example, determining the local background around each spot on the array. However, such problems are a potential source of error that the researcher must always remain alert to.

## B. Microarrays

As with macroarrays, microarrays are manufactured using robotic setups to spot nucleic acid solutions (genomic DNA, cDNA, or synthetic oligonucleotides) onto a solid substrate. In this case, however, the substrate is normally a specially coated glass microscope slide. A number of different coatings may be employed to enable the nucleic acids to be bound covalently to the surface of the slide and provide low background binding of the test samples during subsequent hybridization.

Arrays produced in this way are typically at a much higher density than those on membrane arrays. It is possible to spot 30,000 or more elements on a single, standard format, glass microscope slide.

Microarrays are used to perform gene transcript profiling in much the same manner as the membrane arrays. The most notable difference is that the detection system used is based on fluorescence rather than radioactivity. Fluorescent dyes, (most commonly Cyanine 3 [Cy3™] or Cyanine 5 [Cy5™]) are incorporated into the cDNA reverse transcribed from the RNA sample to be analyzed. This can be achieved by either direct or indirect labeling. The direct labeling process makes use of deoxynucleotide triphosphates in the reverse transcription mix, which have been modified by covalent attachment of the appropriate Cyanine dye. The indirect labeling protocol uses amino allyl dUTP in the reverse transcription mix. This analogue of dTTP is incorporated into the cDNA product during reverse transcription. The fluorescent Cyanine dyes are subsequently attached via chemical reaction with the amino allyl groups.

Use of fluorescence detection methods has certain notable advantages over the radioactive detection commonly used with macroarrays. In addition to removing the hazards associated with researcher exposure to ionizing radiation, fluorescence detection also provides the scope to use a range of different fluorophores that can be distinguished based on the wavelengths of the light they emit. This means that distinct cDNA samples can be labeled with different fluorophores and then hybridized to the same array. The most widely used format for microarrays employs Cy3 and Cy5 dyes to provide a two-sample, two-color system for each array (Fig. 2). This means that a control and test sample can be hybridized to the same array at the same time, thus overcoming, for a simple control versus treat experiment, many of the normalization issues discussed above. This approach also simplifies the normalization of data generated from multiple arrays since one common sample can be used as an internal standard on each array. Dual color detection also resolves most problems of localized variations in hybridization efficiency and background across an array, since at any given point on the array these factors should be the same for both the Cy3- and Cy5-labeled samples.

The fluorescently labeled cDNAs are hybridized to the array by placing the hybridization solution upon the slide under a cover slip (Fig. 2). The small volume required for this means that concentrations of the cDNAs in the samples are kept as high as possible, thus enhancing the rate and efficiency of hybridization.

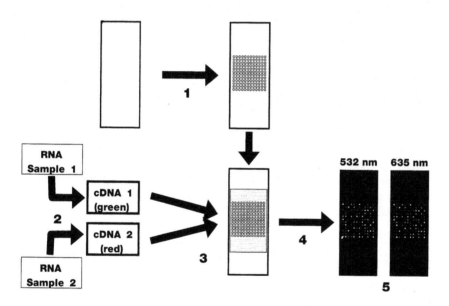

**Figure 2** Microarray analysis. DNA or oligonucleotides are printed onto coated slides (1) and covalently bound. Two RNA samples are reverse transcribed (2) and either green or red fluorescent dyes incorporated into the cDNA products. These cDNA samples are combined and hybridized to the array under a cover slip (3). The slide is washed and imaged using a microarray laser scanner at 532 and 635 nm (4). The fluorescence intensity signals from each of the two dyes are presented here in black and white (5) with cDNAs present in high abundance appearing white, while less abundant species appear in varying shades of grey. These images are often combined into a single false color image in which cDNAs present in only one of the two samples appear red or green and those present in both appear in varying shades of orange and yellow.

The slide is placed in a hybridization chamber specially designed to hold one slide. Hybridization is performed at an appropriate temperature (dependant on the hybridization buffer used and the type of DNA immobilized on the array) in the presence of a sufficient volume of buffer to maintain the humidity in the chamber and stop the hybridization solution from evaporating to dryness. After overnight incubation, the slide is put through a series of washes that remove the cover slip, the hybridization buffer, and nonhybridized, fluorescently labeled cDNA. Finally, the slide is imaged using a dual laser scanner to detect and quantify the fluorescence signals in each of the channels (532 nm for Cy3 and 635 nm for Cy5) from each element on the array. Two images are obtained for each dual-labeled array. These are often superimposed using a false color scheme (Fig. 2) to give the researcher a visual indication of the relative abundance of each the RNA species in the control and test samples. Dedicated computer software is used to analyze the images obtained; this locates and identifies each spot, then quantifies the signal intensities for each spot on the array in of the fluorescence channels, correcting for the local background signal.

Although dual color systems prevail at the moment in microarray analysis, the scope exists to introduce more fluorophores (such as other Cyanine dyes or Alexa Fluor™) so that three or even more samples can be analyzed simultaneously on one array. The use of multiple samples on a single array reduces still further the risk of introducing errors during normalization processes required for the comparative analysis of dual samples on multiple arrays. Nevertheless, in adopting such an approach, care would have to be taken to ensure that the signals from the different dyes could be distinguished with confidence. It will also be important to demonstrate that the arrays are not being swamped with cDNAs to such an extent that competition between the numerous samples interferes with the ratio of the signals obtained.

## C. Affymetrix Gene Chips

The third widely used format for microarrays is the gene chips produced by Affymetix. Their GeneChip® probe arrays are manufactured on quartz wafers using a process involving photolithography and combinatorial chemistry to synthesize oligonucleotides organised in very high-density arrays. Proprietary software is used to optimize the design of the oligonucleotides synthesized. Multiple probes are included for each gene. These arrays also make use of what the manufacturers term "the Perfect Match/Mismatch probe strategy." For each probe that is designed to be perfectly complementary to a target sequence, a second partner probe is generated that is identical except for a single base mismatch in its center. These probe pairs permit quantitation and subtraction of signals caused by nonspecific cross-hybridization.

The result is high-quality arrays that can be used with a fluorescence detection system in a manner similar to the glass slide microarrays, although only single color detection is used. cDNA is synthesized from a test RNA sample by reverse transcription followed by second strand synthesis. This is then used as a template for an in vitro transcription reaction to produce biotin-labeled cRNA. As well as introducing the label for detection, this step permits amplification of the RNA and thus enables less starting material to be used. The cRNA is hybridized to the GeneChip® and the biotin detected with an avidin-fluorophore conjugate. Hybridization of the sample to the chip is measured using a laser scanner.

## D. Analysis of Array Data

Data analysis for macro- and microarrays is no small undertaking. The many thousands of data points generated from a series arrays used in the execution of one experiment first need flagging and removal of any obviously erroneous data points and then careful normalization. Only after this has been completed can comparative analysis be started. The established dogma for microarray analysis states that fold changes in excess of two (i.e., >twofold increase or decrease in the level of an individual message) can be considered reliable indicators of real changes. In truth, the real cut-off value for identifying a genuine change may be smaller or greater depending on a number of factors. These include the system being used, the quality of the data generated, the signal strength, the number of biological replicates analyzed, and the reproducibility of the data obtained. The arbitrary twofold cut-off point should not be used as anything more than the most general rule of thumb, if that. Instead, the onus must be upon individual researchers to analyze the variability of their own data, generated with their own system, so as to determine an appropriate range of values for transcript level fold changes that will provide reliable indictors of genuine changes in the samples analyzed.

In addition to identifying those genes whose transcript levels change significantly as a result of treatment, with experiments that cover time courses, dose ranges, or exposure to a range of different agents, it is possible to attempt to cluster groups of genes according to common (or even opposing) patterns of response. This type of cluster analysis opens up the possibility of identifying families of genes that are regulated at the level of transcription through common mechanisms. Thereafter, it should be possible to develop and test new hypotheses for the modes of action of biological agents such as phytochemicals. For example, with the information available from the human genome mapping project, it is now possible to examine the sequences of all the genes in a cluster to look for common sequence motifs for known or previously unrecognized transcription or RNA stabilization/destabilization factors.

## V.  REAL-TIME RT-PCR (TaqMan®) FOR SENSITIVE AND ACCURATE QUANTIFICATION OF GENE EXPRESSION

The determination of changes in mRNA expression is key to the assessment of gene function and the biological activity of dietary phytochemicals. Frequently, the gene expression data obtained from microarrays may not agree (most often in terms of fold changes) with those from real-time RT-PCR, and, in most cases, the latter provides the benchmark for RNA quantification and confirmation [8]. Therefore, target genes identified from studies using microarray technology require further confirmation with quantitative real-time RT-PCR (TaqMan®) [9,10]. Real-time RT-PCR is the most sensitive technique for the determination of gene expression and has been applied to the detection of mRNA in single cells [11], and even at the single-copy level using a multiplex real-time RT-PCR [12]. RT-PCR with careful assay design and application of TaqMan® chemistry can be both highly specific and reproducible.

### A.  Basic Chemistry of the TaqMan® Assay

The TaqMan® real-time RT-PCR exploits the 5′ nuclease activity of AmpliTaq Gold DNA polymerase to cleave a TaqMan® probe during PCR. Such probes have a reporter dye (such as FAM) attached at the 5′ end and a quencher dye (for example, TAMRA) at the 3′ end. When the probe is intact, the quencher suppresses the intensity of the reporter fluorescence. During PCR, if the target of interest is present, the probe specifically anneals to its complementary sequence between the forward and reverse primer sites. The 5′–3′ nuclease activity of AmpliTaq Gold DNA polymerase only cleaves probes that are hybridized to the target releasing the reporter dye. The accumulation of the PCR product is detected directly by monitoring the increase in fluorescence of the reporter dye. The PCR reaction is performed in a 96- or 384-well plate, and a laser is directed to the sample wells through fiber optic cables to scan and excite the fluorescent dyes. The emission spectra from each well are collected by a charge-coupled device (CCD) camera and further processed by sequence detection software. Quantification is achieved by calculating threshold cycle (Ct) number.

### B.  Threshold Cycle

The concept of the threshold cycle (Ct) is the most important principle in TaqMan® assays. Values are calculated for the amount of reporter dye released during each PCR cycle, and this is representative of the amount of the product amplified. The Ct is the cycle number at which the fluorescent signal from the reporter dye is first detected at a statistically significant level above the background [13]. The more template mRNA present at the beginning of the reaction, the lower the Ct will be needed to reach this threshold level of fluorescence.

## C. Primer and Probe Design

The optimum amplicon length for TaqMan® real-time RT-PCR is 50–100 bp. For primers, the Tm should be 58–59°C and maximum of 2/5 G or C at the 3′ end.

For probes, the Tm should be 10°C higher than the primer Tm; there should be no G on the 5′ end because guanine is a natural quencher for FAM, and there must not be more Gs than Cs. Where possible, primers or probes should be designed to cross intron-exon boundaries so as to avoid false-positive results arising from contamination with genomic DNA [14].

## D. Relative Quantification and Normalization

### 1. Standard Curve Method

It is easy to prepare a standard curve for relative quantification because quantity is expressed relative to control samples. Normally, a RNA sample is diluted in a twofold series, e.g., 5, 10, 20, 40, and 80 ng, to construct a standard curve. For all experimental samples, the quantity of a target mRNA is determined from the standard curve and divided by the amount of untreated control. Thus, the control becomes 1 and all other samples are expressed as $n$-fold difference to the control. For quantification normalized to an endogenous control (housekeeping gene), standard curves are prepared for both the target and endogenous reference. For each sample, the amount of target and endogenous reference is determined from the appropriate curve. The target amount is then divided by the endogenous reference to obtain a normalized target value. Each of the normalized target values is further divided by the control normalized target value to generate the relative expression levels. However, if the absolute quantity of a standard (e.g., mRNA copy number) is known, the quantification can be absolute using an absolute standard curve method. Plasmid DNA or in vitro transcribed RNA are commonly used to prepare absolute standards. One limitation in using DNA as a standard, however, is the lack of a control for the efficiency of the reverse transcription step.

When a TaqMan® assay is validated (the efficiency of replication is close to 1), the relative quantity of the target gene can be calculated using a $\Delta$Ct method using the relationship: fold of induction $= 2^{Ct(control)-Ct(treated)}$.

When an endogenous reference is used, the amount of target mRNA is normalized to an endogenous reference and also relative to a control and calculated using $2^{-\Delta\Delta Ct}$ method as described in Applied Biosytems User Bulletin #2 (http://docs.appliedbiosystems.com/pebiodocs/04303859.pdf).

## 2. Multiplex PCR

Multiplex PCR can be used in relative quantification of a target and an endogenous reference in the same tube. The availability of multiple reporter dyes for TaqMan probes (FAM, VIC, JOE) makes it possible to detect amplification of more than one target in the same tube. Two important factors for multiplex PCR are (1) to limit the concentration of primers for the most abundant gene and (2) the deconvolution of the collected spectra to distinguish between reporter dyes. Multiplex real-time RT-PCR has been demonstrated in detection of single-copy mRNA of Xist gene in mouse embryos [12].

## 3. Housekeeping Genes

The ideal endogenous control or housekeeping gene should be expressed at a constant level within the cell types or tissue in an organ and should be unaffected by the experimental treatment. The three most commonly used housekeeping genes are glyceraldehydes-3-phosphate-dehydrogenase (GAPDH), β-actin, and 18S rRNA [15]. GAPDH is the most frequently used endogenous control for quantification of gene expression [16], though mRNA levels are not constant [17] with big interindividual variation. GAPDH was 40% upregulated by treatments with sulforaphane in Hep G2 cells. Goidin et al. [18] reported 18S rRNA to be a reliable internal standard with levels that are less likely to be modified than other housekeeping genes; its expression levels are, however, much higher than that of the target genes [18]. However, this research did not provide any information on the primer and probe sequence for 18S rRNA. β-Actin mRNA is expressed at moderately abundant levels in most cell types and encodes a cytoskeleton protein. It is one of the most commonly used internal standards though, unlike 18S rRNA, it is not suitable for serum-stimulating experiments [19].

In our laboratory, gene expression levels have been normalized against total RNA and a housekeeping gene, either 18S rRNA or β-actin. Total RNA from cultured cells, which contains a significant proportion of rRNA, is a good measure against which to express gene expression results, e.g., as copy number per ng of total RNA. Total RNA from cultured cells can be quantified from its absorbance at 260 nm and/or using a fluorescent labeling method such as the Molecular Probes Ribogreen assay. If more internal controls are needed, a human endogenous control plate with 11 genes is available from Applied Biosystems (http://docs.appliedbiosystems. com/pebiodocs/0438134.pdf).

## E. Application of TaqMan® Assay in Quantification of Phytochemicals and Gene Expression

Human hepatoma Hep G2 cells are a good model for the study of metabolism and induction of phase II enzymes such glutathione-*S*-transferase-α (GSTA1),

UDP-glucuronosyltransferase 1A1 (UGT1A1), and an antioxidant enzyme thiore-doxin reductase 1 (TR1) [2,20]. It has been shown that an isothiocyanate, sulfora-phane, can induce levels of GSTA1, UGT1A1, and TR1 mRNA in human HepG2 cells using TaqMan® assays (Fig. 3). Primers and probes are listed in Table 1. The assay condition used were TaqMan® one-step RT-PCR master mix Reagent kit in a total volume of 25 μL per well consisting of 100 nM probe, 200–300 nM forward and reverse primers, and 10 ng of total RNA. TaqMan® RT-PCR conditions are: 48°C 30 min, 95°C 10 min, then 40 cycles of 95°C for 15 s and 60°C for 1 min. Figure 3 shows that GSTA1, UGT1A1, and TR1 mRNA expres-sion were induced 2.4-, 5.6-, and 4-fold, respectively. The data were expressed against total RNA, since the expression of two housekeeping genes, 18S rRNA and β-actin, were unchanged after sulforaphane treatment. The expression of these two housekeeping genes has been shown in our laboratory to be relatively stable in most treatments with polyphenols and isothiocyanates in human Hep G2 cells (hepatocyte carcinoma), Caco2 (colon adenocarcinoma), and MCF 7 (breast adenocarcinoma) cells and may be used in most applications.

## VI. ANALYSIS OF GENE EXPRESSION AT THE LEVEL OF TRANSLATION AND POSTTRANSLATIONAL PROCESSING—PROTEOMICS

By far the most commonly used proteomic approaches exploit two-dimensional (2D) gel electrophoresis to separate proteins in a complex mixture isolated from

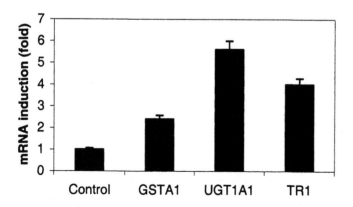

**Figure 3** Effect of sulforaphane on mRNA levels of GSTA1, UGT1A1, and TR1 in Hep G2 cells. For GSTA1 and UGT1A1, cells were exposed for 15 μM SFN for 18 h and 12 μM SFN for 24 h for TR1. (From Refs. 2 and 20.)

**Table 1**   Primer and Probe Sequences for GST, UGT, TR1, β-Actin, and 18S rRNA

| | | Sequence (5′ → 3′) |
|---|---|---|
| | Forward | CAG CAA GTG CCA ATG GTT GA |
| GSTA1 | Reverse | TAT TTG CTG GCA ATG TAG TTG AGA A |
| | Probe | FAM-TGG TCT GCA CCA GCT TCA TCC CAT C-TAMRA |
| | Forward | GGT GAC TGT CCA GGA CCT ATT GA |
| UGT1A1 | Reverse | TAG TGG ATT TTG GTG AAG GCA GTT |
| | Probe | FAM-ATT ACC CTA GGC CCA TCA TGC CCA ATA TG-TAMRA |
| | Forward | CCA CTG GTG AAA GAC CAC GTT |
| TR1 | Reverse | AGG AGA AAA GAT CAT CAC TGC TGA T |
| | Probe | FAM-CAG TAT TCT TTG TCA CCA GGG ATG CCC A-TAMRA |
| | Forward | CCT GGC ACC CAG CAC AAT |
| β-Actin | Reverse | GCC GAT CCA CAC GGA GTA CT |
| | Probe | FAM-ATC AAG ATC ATT GCT CCT CCT GAG CGC-TAMRA |
| | Forward | GGC TCA TTA AAT CAG TTA TGG TTC CT |
| 18S RNA | Reverse | GTA TTA GCT CTA GAA TTA CCA CAG TTA TCC A |
| | Probe | FAM-TGG TCG CTC GCT CCT CTC CCA C-TAMRA |

cells or tissue. 2D gel electrophoresis, which first separates proteins along a strip gel (first dimension) according to their isoeletric point and then by size in a slab gel (second dimension) run at right angles to the first dimension strip (Fig. 4), is by no means a new technology [21]. However, its use has been revolutionized by a number of recent technical developments that now make it possible to identify any protein that can be resolved on such a gel [22]. This is achieved through a series of steps performed subsequent to running the gel. First, the proteins in the gel are detected using a suitable stain, such as the fluorescent Sypro ruby. The image of the gel that is produced is analyzed using specialist software to identify particular proteins of interest. Here, at least two approaches can be adopted. The first is a ''shotgun'' approach in which the researcher sets out to characterize all the protein spots evident on the gel. The aim in this case is to develop a protein ''map'' for a particular cell type or tissue. Alternatively, a series of gels can be run to identify consistent differences in the protein patterns between control and treated samples. Special computer software is required to adjust all the gel images, taking into account the unavoidable variation in gel runs from day to day and batch to batch, so as to be able to align and overlay all the images accurately.

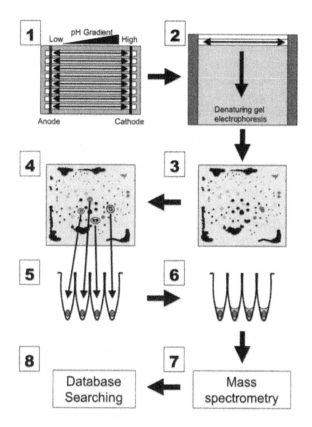

**Figure 4** Proteomics process. Protein extracts are subjected to isoelectric focusing on strip gel (1). Each strip gel is then placed on a slab gel and the sample subjected to denaturing gel electrophoresis at right angles (second dimension) to the direction of isoelectric focusing (2). The protein is visualized using a suitable staining method (3). Individual protein spots of interest are identified, excised from the gel (4), and transferred into a tube (5) for treatment with proteolytic enzymes (6). The peptide fragments produced are analyzed by mass spectrometry to determine amino acid content and, if necessary, sequence (7). This information is used to search protein databases (8) to identify each excised protein.

Differences in protein patterns may manifest as (1) the appearance or disappearance of a protein spot (switching on or off of one gene or a shift in subcellular location of the protein that causes it to appear in or disappear from the protein extract), (2) a change in the signal intensity of a spot relative to all the other spots on the gel (corresponding to an up- or downregulation of the expression of one gene), or (3) a shift in the relative position of a protein spot (perhaps indicating posttranslational modification).

In some respects, this comparative approach to hunting for changes that result from a given treatment is analogous to the common approach adopted for microarrays described above. However, one notable difference is that a change evident on an array can immediately be linked to one specific gene because the identity of each element on an array is predefined. With proteomics, unless a comprehensive cell or tissue protein "map" has already been prepared, the identity of most spots will not be known and characterization of the target proteins is still required.

Protein characterization is achieved via a protein spot of interest being excised from the gel. The protein in the excised gel segment is then subjected to proteolysis to yield peptide fragments. While both these steps potentially can be performed manually, robotic systems are now available that increase the precision, reliability, and throughput. The peptide fragments are then subjected to either MALDI-ToF (matrix-assisted laser desorption ionization time of flight) or Q-ToF (hybrid quadrupole time of flight) mass spectrometry; the former can be used to determine the molecular weight of each peptide fragment, from which it is possible to predict the amino acid content. This information describing the predicted amino acid content of all the fragments present in one protein spot from a gel is used to search databases of known protein sequences and putative protein sequences based on DNA sequence. In theory, since the entire human genome has now been sequenced, it should always be possible to find a good match and identify the protein with confidence. However, it is sometimes necessary to refine the mass spectrometric data used to search the databases. In this case, Q-ToF analysis can be employed to provide amino acid sequence data for the proteolytic fragments.

This technology enables researchers to isolate and identify individual proteins of interest present in highly complex mixtures. However, there are certain limitations; some proteins are highly abundant, giving rise to large spots on the gels that will mask the presence of other, less abundant proteins. On the other hand, the limitations of staining techniques means that some low-abundance proteins may not be detected at all (the detection limit being in the order of 0.5 ng). Simply because a protein is present in low abundance does not mean that it is not important. For example, many transcription factors are only ever present in low abundance, but a moderate change in the amount of one transcription factor, which could well be undetectable because of the low levels that it is present in, could have a profound effect on the expression of a number of other genes.

Some proteins will have sufficiently unusual isoelectric points and/or molecular weights that they run at the very periphery of the gel and are for this reason only poorly resolved. In particular, the hydrophobic nature of integral membrane and membrane-associated proteins pose a problem in that these proteins will often not be evident at all with most standard protein extraction and electrophoresis conditions. To obtain the maximum possible resolution of all proteins in a single tissue, extracts would have to be prepared and subjected to electrophoresis under a range of different conditions. This is rarely practical, and a number of research teams and other organizations are investigating alternatives to the 2D gel electrophoresis systems, including multidimensional chromatographic methods [23,24] and a number of different protein chip technologies [25]. However, while such approaches are still in development, 2D gel electrophoresis remains the current method of choice for most research groups.

## VII. METABOLOMICS

Metabolomics is the newest of the functional genomic technologies and also the least well defined in terms standard experimental approaches. A number of different spectroscopic methods can be used to obtain metabolite profiles from tissue, cellular, or extracellular fluid samples. These include NMR, mass spectroscopy, liquid chromatography, and optical spectroscopic techniques [26]. Of these, NMR approaches, such as magic angle spinning NMR, which can be used with intact cells and tissues, and mass spectrometric approaches appear to be the most promising.

As yet, very few groups are properly equipped to perform high-level, high-throughput metabolomic studies. This is in part because the technologies are still at a developmental stage but also because set-up costs involved are very substantial. In the long run, however, the importance of metabolomic studies should not be underestimated. These approaches will provide data on ultimate outcomes or endpoints of changes in gene expression. As such, they will form the vital link between genomic/transcriptomic/proteomic research and studies of the health-promoting properties of phytochemicals and other dietary components.

## VIII. DATA HANDLING/BIOINFORMATICS

All the functional genomic approaches described above rapidly yield vast data sets. The general approaches that can be used to analyze these data, described throughout this chapter, are really only a first step to interrogating the data obtained fully. One microarray dataset contains so much more information than simply which transcript levels go up and down. There is undoubtedly enormous significance in the actual transcript profile obtained, and as more advanced pattern recognition algorithms are developed it will be possible to extract more and more information from each data set obtained.

One particular challenge lies in integrating the data from the different functional genomic techniques. A change in transcript profile would be expected to elicit a change in protein expression that is almost guaranteed to alter the metabolic profile. But the interactions will not always be straightforward to predict. There will be a time offset between the changes and the possibility of feedback loops at each stage. These will vary from gene to gene and from cellular process to cellular process.

The rate of development of functional genomic techniques has, to some extent, outpaced the development of bioinformatic (computational, mathematical, and statistical) tools that are required to make the fullest use of the data they generate. However, the critical role of bioinformatics in the effective exploitation of functional genomic technologies is apparent. Currently, there is an influx of mathematicians into the field, and new degree and postgraduate courses in bioinformatics are being established. For many nutrition researchers, the exploitation of functional genomic techniques is a completely new possibility and acquiring bioinformatics skills is still another step away. It is, thus, essential to understand the crucial importance of developing at least a basic understanding of the principles underlying this discipline. The development of targeted training courses would be invaluable in this respect.

## IX. POTENTIAL PITFALLS, CHALLENGES, AND OPPORTUNITIES FOR NUTRIGENOMICS

The possibilities that functional genomic approaches present, outlined here, will radically change the way in which nutrition research is conducted in the immediate future. One potential pitfall is that every researcher could rush to exploit these technologies at the very first opportunity without appropriate pause for careful and considered experimental design. The result would be a mass of data that is of little, if any, physiological relevance (i.e., more phenomenology than valid nutrition research). Not only would such data be of questionable value, but the huge volume produced could also would waste enormous amounts of researcher time with extensive data analysis and experimental follow-up of initially promising avenues of research that ultimately prove to be false. Obviously, this pitfall can be easily avoided by adherence to well-established good scientific principles of study design, best practice, and application of common sense.

While functional genomic techniques are still in their current phase of rapid development, there are some legitimate concerns about the reliability and interoperability of data generated by different groups using different experimental systems. In nutritional genomics, this issue may be of particular significance since it is likely that many diet–gene interactions will influence health through relatively subtle but long-term effects. If the changes in gene expression provoked by specific dietary factors are indeed subtle, they will be more difficult to detect reliably than those more dramatic changes that might be associated with pharmaceutical agents. Early examples of the application of these techniques in nutrition research suggest that diet–gene interactions can be detected reliably [27], but extensive comparison, validation, and refinement of the different systems in use should be a high priority in nutrigenomics.

Provided these pitfalls and challenges are properly addressed, nutrition research faces a very exciting future in which, thanks to the use of functional genomic approaches, significant and far more rapid progress can be made towards the goal of defining optimal nutrition for populations, subpopulations, and individuals. More data still may be required to provide the conclusive proof of principle but with so many groups recognizing the opportunity, with increasingly effective interdisciplinary networking, and with significant funding being offered at national and regional levels, such developments should not be far away.

## REFERENCES

1. Chan MM-Y, Huang H-I, Fenton MR, Fong D. In vivo inhibition of nitric oxide synthase gene expression by curcumin, a cancer preventive natural product with anti-inflammatory properties. Biochem Pharmacol 1998; 55:1955–1962.
2. Basten GP, Bao YP, Williamson G. Sulforaphane and its glutathione conjugate but not sulforaphane nitrile induce UDP-glucuronosyl transferase (UGT1A1) and glutathione transferase (GSTA1) in cultured cells. Carcinogenesis 2002; 23:1399–1404.
3. Thomas SR, Leichtweis SB, Petterson K, Croft KD, Mori TA, Brown AJ, Stocker R. Dietary cosupplementation with vitamin E and coenzyme Q10 inhibits atherosclerosis in apolipoprotein E gene knockout mice. Arterioscler Thromb Vasc Biol 2001; 21:585–593.
4. Trayhurn O. Proteomics and nutrition—a science for the first decade of the new millennium. Br J Nutr 2000; 83:1–2.
5. Elliott R, Ong TJ. Nutritional genomics: exploiting the full potential of our diet. Br Med J 2002; 324:1438–1442.
6. Sambrook J, Russell DW. Molecular Cloning: A Laboratory Manual. 3rd ed. New York: Cold Spring Harbor Laboratory Press, 2001.
7. Eisen MB, Brown PO. DNA arrays for analysis of gene expression. Methods Enzymol 1999; 303:179–205.
8. Dong Y, Ganther HE, Stewart C, Ip C. Identification of molecular targets associated with selenium-induced growth inhibition in human breast cells using cDNA microarrays. Cancer Res 2002; 62:708–714.
9. Chuang Y-Y E, Chen Y, Chandramouli GVR, Cook JA, Coffin D, Tsai M-H, DeGraff W, Yan H, Zhao S, Russo A, Liu ET, Mitchell JB. Gene expression after treatment with hydrogen peroxide, menadione, or t-butyl hydroperoxide in breast cancer cells. Cancer Res 2002; 62:6246–6254.

10. Risinger JI, Maxwell GL, Chandramouli GVR, Jazaeri A, Aprelikova O, Patterson T, Berchuck A, Barrett JC. Microarray analysis reveals distinct gene expression profiles among different histologic types of endometrial cancer. Cancer Res 2003; 63:6–11.

11. Simone NL, Bonner RF, Gillespie JW, Emmert-Buck MR, Liotta LA. Laser-capture microdissection: opening the microscopic frontier to molecular analysis. Trends Genet 1998; 272:272–276.

12. Hartshorn C, Rice JE, Wangh LJ. Differential pattern of Xist RNA accumulation in single blastomeres isolated from 8-cell stage mouse embryos following laser zona drilling. Mol Reprod Dev 2003; 64:41–ɔ1.

13. Gibson UEM, Heid CA, Williams PM. A novel method for real time quantitative RT- PCR. Genome Res 1996; 6:995–1001.

14. Bustin SA. Absolute quantification of mRNA using real-time reverse transcription polymerase chain reaction assays. J Mol Endocrinol 2000; 25:169–193.

15. Bustin SA. Quantification of mRNA using real-time reverse transcription PCR (RT-PCR): trends and problems. J Mol Endocrinol 2002; 29:23–39.

16. Winer J, Jung CK, Shackel I, Williams PM. Development and validation of real-time quantitative reverse transcriptase–polymerase chain reaction for monitoring gene expression in cardiac myocytes in vitro. Anal Biochem 1999; 270:41–49.

17. Zhu G, Chang Y, Zuo J, Dong X, Zhang M, Hu G, Fang F. Fudenine, a C-terminal truncated rat homologue of mouse prominin, is blood glucose-regulated and can up-regulate the expression of GAPDH. Biochem Biophys Res Commun 2001; 281: 951–956.

18. Goidin D, Mamessier A, Staquet MJ, Schmitt D. Berthier-Vergnes O. Ribosomal 18S RNA prevails over glyceraldehyde-3-phosphate dehydrogenase and beta-actin genes as internal standard for quantitative comparison of mRNA levels in invasive and noninvasive human melanoma cell subpopulations. Anal Biochem 2001; 295: 17–21.

19. Schmittge TD, Zakrajsek BA. Effect of experimental treatment on housekeeping gene expression: validation by real-time, quantitative RT-PCR. J Biochem Biophys Methods 2000; 46:69–81.

20. Zhang JS, Svehlikova V, Bao YP, Howie AF, Beckett GJ, Williamson G. Synergy between sulforaphane and selenium in the induction of thioredoxin reductase 1 requires both transcriptional and translational modulation. Carcinogenesis 2003; 24: 497–503.

21. Farrell PH. High resolution two-dimensional electrophoresis of proteins. J Biol Chem 1975; 250:4007–4021.

22. Dongre AR, Opiteck G, Cosand WL, Hefta SA. Proteomics in the post-genome age. Biopolymers 2001; 60:206–211.

23. Link AJ, Eng J, Schieltz DM, Carmack E, Mize GJ, Morris DR, Barbera GM, Yates JR. Direct analysis of protein complexes using mass spectrometry. Nat Biotechnol 1999; 17:676–682.

24. Gygi SP, Rist B, Gerber SA, Turecek F, Gelb MH, Aebersold R. Quantitative analysis of complex protein mixtures using isotope-coded affinity tags. Nat Biotechnol 1999; 17:994–999.

25. Zhu H, Snyder M. Protein chip technology. Curr Opin Chem Biol 2003; 7:55–63.
26. Nicholson JK, Connelly J, Lindon JC, Holmes E. Metabonomics: a platform for studying drug toxicity and gene function. Nat Rev Drug Discov 2002; 1:153–161.
27. Moore JB, Blanchard RK, McCormack WT, Cousins RJ. cDNA array analysis identifies thymic LCK as upregulated in moderate murine zinc deficiency before T-lympohcyte population change. J Nutr 2001; 131:3189–3196.

# 2
# Methods to Study Bioavailability of Phytochemicals

**Birgit Holst and Gary Williamson**
*Nestlé Research Center, Nestec, Lausanne, Switzerland*

## I. INTRODUCTION

Phytochemicals have a profound influence on long-term health. However, their dietary intake does not necessarily reflect their bioavailability, i.e., the dose reaching the target tissue. Understanding the bioavailability of phytochemicals is critical and must always be considered. In comparison with that of macronutrients, the beneficial intake range of micronutrients and phytochemicals is relatively narrow and there is a possibility of both adverse effects at higher doses and no effect at very low doses. Isothiocyanates, derived from glucosinolates, may be cited as an example of a group of phytochemicals exhibiting such a small beneficial dose window. At low concentrations, these compounds activate mitogen-activated protein kinases (MAPK) such as ERK2, JNK1, or p38, leading to expression of survival genes (*c-Fos*, *c-Jun*) and defensive genes (phase 2 detoxifying enzymes, such as glutathione-S-transferase and quinone reductase) involved in the "homeostasis response." Increasing concentrations of isothiocyanates may additionally activate the caspase pathway, leading to apoptosis: finally, pharmacological and supra-pharmacological doses may lead to nonspecific necrotic cell death and genotoxicity [1].

Dietary supplements containing high concentrations of micronutrients and phytochemicals are becoming increasingly commonplace. However, because of the serious effects these elevated doses may provoke, such products may be better considered and studied as drugs rather than nutrients. In order for a bioactive compound to exert any activity, it must necessarily reach the target organ at a minimal concentration that determines both the biological effect and mechanism of action. The term bioavailability was introduced to describe the concentration of a given compound or its metabolite at the target organ; however, no single definition accurately encapsulating the multifactorial nature of the term (described in this chapter) is available. The U.S. Food and Drug Administration (FDA) defines bioavailability as *the rate and extent to which the therapeutic moiety is absorbed and becomes available to the site of drug action*. Because of difficulties in accessing organ sites in vivo in humans, attempts to use the term with quantitative precision or to calculate exact values for it are likely to be unsuccessful. As a consequence, the term "absolute bioavailability" is often used by clinical pharmacologists to describe the exact amount of a compound that reaches the systemic circulation; this is calculated as the fraction of the area under the curve (AUC) after oral ingestion compared with the AUC after intravenous administration. In the area of nutrition, however, "relative bioavailability" is commonly used to describe the bioavailability of a compound from one source compared with another. Given that once a compound is absorbed it is inevitably bioactive, Stahl et al. [2] have discussed whether bioactivity should not be included in the concept of bioavailability. Even if bioactivity is excluded, bioavailability already encompasses a very broad spectrum of processes, and as the bioactivity of phytochemicals is a diverse and extensive area itself, its inclusion would add an enormous amount of, probably indigestible, information. We have, therefore, confined ourselves to the "classical," pharmacological definition of bioavailability as the basis for this chapter; according to this, bioavailability consists of several linked and integrated processes—specifically, liberation, absorption, distribution, metabolism, and excretion (LADME).

This chapter focuses on the bioavailability of phytochemicals as a distinct group of xenobiotics;* rather than summarizing current knowledge for individual phytochemicals, general mechanisms of LADME, as applied to phytochemicals, are discussed. Examples will be presented of factors that are specific for dietary phytochemicals (as compared to drugs). The first part of this chapter provides an overview of the diversity of processes and parameters responsible for the enormous variation in phytochemical bioavailability. Given this information, and knowing that bioavailability directly determines the effect of a compound, it is unsurprising that many intervention trials lacking a biomarker of exposure (a measure of bioavailability) have failed to demonstrate the anticipated results.

There is a clear need for the development and exploitation of appropriate methods to study bioavailability and the individual processes contributing to the bioavailability of a compound (LADME). Currently used methods, mostly developed and applied in new drug development, are described in the second part of

---

* Throughout this chapter, the more general term *xenobiotic* will be applied whenever appropriate.

the chapter. Crucial differences between studying drugs and dietary phytochemicals are highlighted, and advantages and limitations of individual methods are presented that should facilitate critical discussion of experimental data.

## II. BIOAVAILABILITY OF PHYTOCHEMICALS

Many factors affect the bioavailability of a compound; these may be divided into exogenous factors (such as the complexity of the food matrix, the chemical form of the compound of interest and the structures and amounts of co-ingested compounds) and endogenous factors (including mucosal mass, intestinal transit time, rate of gastric emptying, and extent of conjugation). When taken together these factors can cause large interindividual and intraindividual variations in bioavailability, ranging from 0 to 100% of the ingested dose.

The pharmacokinetic processes following oral application of a compound (Fig. 1) are:

**L** = **Liberation**, processes involved in the release of a compound from its matrix.

**A** = **Absorption**, the diffusion or transport of a compound from the site of administration into the systemic circulation.

**D** = **Distribution**, the diffusion or transportation of a compound from intravascular space (systemic circulation) to extravascular space (body tissues).

**M** = **Metabolism**, the biochemical conversion or transformation of a compound.

**E** = **Excretion**, the elimination of a compound, or its metabolite, from the body via renal, biliary, or pulmonary processes.

Before considering the bioavailability of phytochemicals in a food matrix and relating these to biological effects in humans, it is important and necessary to understand that phytochemicals are mostly minor plant constituents whose concentration varies considerably according to, for example, seasonal and agronomic factors, the variety, age, and part of the plant examined. Such variability can lead to serious problems of interpretation of results from epidemiologal studies or human intervention trials if the dose of the phytochemical applied is not determined.

Most phytochemicals occur as glycosides in plant material and represent the precursors of a wide range of chemically diverse compounds that exhibit similarly diverse physiological activities. Deglycosylation of the parent compounds may occur during storage, food processing, mastication, digestion, and metabolism, where plant-derived enzymes (e.g., glucosinolate-degrading myrosi-

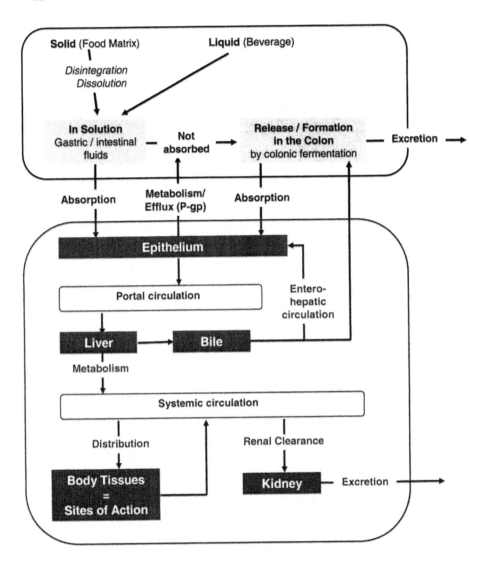

**Figure 1**  General scheme for uptake and pharmacokinetics of dietary phytochemicals.

nase) or human enzymes (e.g., lactase phlorizin hydrolase, LPH) catalyze the reaction [3].

The ubiquitous flavonoids may be cited as an example of the manifold factors affecting the bioavailability of phytochemicals; thus, the bioavailability of flavonoids differs between food sources, predominantly due to the nature of the attached sugar(s). The chemical structure (glycoside or aglycone) of a phytochemical will affect intestinal absorption and first-pass metabolism and will, therefore, modulate its pharmacokinetics [4,6].

## A. Liberation

Efficient oral absorption can only occur after the compound has been liberated from its food matrix and is present at the intestinal mucosal surface. After food processing, mastication is the first step that is likely to cause cell disruption and release of the phytochemical glycoside. Some phytochemicals, e.g., the glucosinolates, are not biologically active per se but serve as precursors for a range of bioactive compounds. Depending on the degree of cell disruption, glucosinolates are released and hydrolyzed by an enzyme, myrosinase, that co-occurs with its substrate. In the intact plant material, both enzyme and substrate are compartmentalized at cellular level; however, once cellular integrity is disrupted, enzyme and substrate interact, leading to formation of the aglycon—generally the active component. The nature of the reaction milieu can affect the formation of the aglycone and its subsequent chemical reaction. A number of different products possessing different activities may thus be expected.

Following their liberation from the food matrix, the compounds are dissolved into gastrointestinal fluids, complex media containing, e.g., bile salts, ions, lipids, cholesterol, as well as enzymes derived from secretions, shed enterocytes, and/or intestinal flora. A range of endogenous factors related to health status and dietary status determine the composition of this medium in which the food compound is presented to the mucosal surface [6].

Dissolution of the compound also depends on its physicochemical properties: notably, its aqueous solubility, ionizability (pKa), and lipophilicity (partition coefficient log $P$ for neutral species or log $D_{7.4}$ for partially ionized compounds). In addition to its effect on solubility, log $P$ is a crucial factor that governs passive membrane partitioning. However, while an increase in log $P$ enhances permeability, it indicates a reduced solubility and *vice versa*. It has been suggested that intestinal absorption may be optimal within a log $P$ range of 0.5–2.0 [7].

The digestion of other components of a meal will lead to the formation of peptides and smaller glycoproteins, thus providing substrates capable of associating and/or binding to phytochemicals. This may, at least in part, explain why the incubation of intestinal contents with intact glucosinolates gave an average recovery of only 76% of the initial doses, compared with the control [8,9].

Measurements of absolute solubility and dissolution are routinely performed during drug development, but knowledge of phytochemicals is limited. As described above, liberation of phytochemicals from the food matrix and subsequent dissolution into gastrointestinal fluids are both affected by a number of parameters. Thus, the bioaccessibility (i.e., the fraction of a substance available for absorption by an organism) of a compound in the small intestine may vary considerably, and a substantial proportion of food-derived phytochemicals may not be absorbed in the small intestine and reach the colon. In this case, colonic microflora will mainly contribute to their hydrolysis, but the precise nature and role of microbial hydrolysis remains controversial for many phytochemicals [10].

## B. Absorption

The rate and extent of absorption in the small intestine and presystemic metabolism represent the most important parameters that affect the early phases of the plasma concentration time curve. The intestinal epithelium is the major absorptive organ, which functions as a gatekeeper in controlling entry of nutrients and xenobiotics—drugs as well as phytochemicals. Absorption can occur at any place along the gastrointestinal tract; while the degree of absorption is strongly site-dependent, the duodenum and jejunum are the major sites of absorption. The intestinal mucosa has a large surface area (approximately 250 m$^2$ in a human) and is characterized by villi that constitute the anatomical and functional unit for the absorption of nutrients and xenobiotics [11]. Villi comprise an epithelial layer, lamina propria, a collagen matrix containing blood and lymphatic vessels, and the muscularis mucosa. Any xenobiotic entering the bloodstream has to pass through the epithelial layer, part of the lamina propria, and the wall of the respective vessel.

The cell membranes of the intestinal mucosa are phospholipid bilayers with polar head groups and inserted proteins that form aqueous pores; as a consequence, mass transport across the intestinal lining may occur by passive diffusion through the paracellular space and/or membranes (which requires no energy expenditure), vesicular uptake via endocytosis/pinocytosis, or active (energy-driven) transport. The exact mechanism of transport for a specific compound depends on its physical and chemical characteristics, such as partition coefficient, stereochemistry, molecular weight and size, pKa, solubility and charge distribution; it is possible to predict membrane permeability by passive diffusion with reasonable accuracy using these properties alone [12]. Depending on the size of the molecule—the smaller the molecule the more readily diffusion occurs—hydrophilic molecules permeate through aqueous pores while hydrophobic ones pass across the lipid domain of the membranes. A number of compounds, particularly nutrients, that are either too large or hydrophilic are actively transported, thereby allowing the compound to move against an electrochemical or concentration gradient. Such active transport systems are saturable and selective for certain structural features.

Physiological factors, including expression of transporters, gastric emptying, gastrointestinal motility, intestinal pH, blood flow, lymph flow, and pathological state, must also be taken into account when evaluating absorption [13,14].

Due to the lack of appropriate methodologies to study intestinal processes, there is only limited knowledge of in vivo intestinal absorption for most phytochemicals in humans. An intestinal perfusion technique (Loc-I-Gut®) has been developed as a "gold standard" for studying the effective intestinal permeability and metabolism of drugs [15–18]. Only recently has this method been applied to examine the intestinal absorption of phytochemicals, specifically sulforaphane, a glucosinolate-derived hydrolysis product (aglycone), and quercetin-3,4'-diglucoside, a flavonoid glycoside [19]. Because of its lipophilicity (log $P$ (octanol/water) $= 0.72$ [20]) and low molecular weight (177) sulforaphane was shown to rapidly diffuse into the enterocytes [19]. This diffusion was most likely driven by the rapid conjugation of sulforaphane with glutathione in enterocytes, leading to maintenance of the concentration gradient and an accumulation of intracellular sulforaphane-glutathione in the liver [21].

In contrast to sulforaphane, phytochemical glycosides are larger and hydrophilic molecules; as such they would be anticipated to have low membrane permeability and to be absorbed either through a paracellular route or by active transport. The presence of a glucose moiety in the molecule might suggest that active transport would occur via the glucose transporter route. However, quercetin-3,4'-diglucoside, unlike its monoglucosides, interacts very poorly with the sugar transporter SGLT1 [22], suggesting that carrier-mediated transport is unlikely. Quercetin-3,4'-diglucoside is a substrate for LPH, a membrane-associated β-glucosidase; this enzyme is expressed at the luminal site of enterocytes to afford direct contact with the luminal contents. LPH is, therefore, most likely responsible for the hydrolysis of quercetin-3,4'-diglucoside prior to absorption and the release of its highly permeable aglycone [23]. Petri et al. [19] confirmed that this, indeed, happens in vivo in the human jejunum and that the majority of the β-glucosidase activity is derived from intact enterocytes.

These are just some examples that demonstrate the complexity of the processes involved in intestinal absorption; other parameters that need to be taken into consideration when studying intestinal transport include transepithelial electrical resistance (TEER), active efflux pumps (such as P-gp), passive transcellular and paracellular diffusion.

## C. Distribution

Distribution describes the movement of a compound from its site of absorption to other areas of the body. When a compound is absorbed it passes through absorptive cells into the interstitial fluid of the organ; these body fluids (interstitial fluid, intracellular fluid, and blood plasma) are not isolated and separate, but represent one continuous pool. In contrast to fast-moving blood that allows mechanical transport to occur, interstitial- and intracellular fluids remain in place with a slow movement of components such as water and electrolytes into and out of cells. Any compound can leave the interstitial fluid by entering local tissue cells, the circulating blood, or the lymphatic system. After entry into blood plasma, the compound (in its free or bound form) may be transported to virtually all organs and tissues along with the blood.

   This distribution of a compound to cells and tissues within the body requires that the compound and/or its metabolite(s) penetrate the capillary cells and subsequently the cells of individual target organs. As has been mentioned above, the concentration gradient, molecular weight, lipid solubility, and polarity are all significant factors favoring the distribution of high concentrations of low molecular weight nonpolar phytochemicals.

   Distribution is also greatly affected by the ability of a given molecule to bind to plasma proteins; such compounds have a limited distribution throughout the body tissues but exhibit a prolonged half-life that will determine its efficacy.

   The concentration of a compound in the plasma is important since it generally reflects the level at the site of action. Once a compound has entered the bloodstream it may be excreted, stored, or metabolized; its metabolites may be excreted or stored; or the compound or its metabolites may reach target organs and interact with, or bind to, cellular components. Parameters that influence the distribution of a compound from the blood to the tissues of the body include blood flow, membrane barriers, ion trapping, plasma binding, and tissue affinity; most are well characterized and can be readily predicted.

## D. Metabolism

Even if their absorption is high, the bioavailability of many compounds may be limited by an extensive metabolism that can affect the in vivo activity profile irrespective of its route of administration. Metabolism is vital since it transforms absorbed nutrients into endogenous substances required to maintain body functions; for xenobiotics, including phytochemicals, metabolism represents the key body defence mechanism that converts them into less harmful, water-soluble, and thus excretable, compounds. Lipophilic, low molecular weight xenobiotics that are readily absorbed and distributed are difficult to eliminate and thus may accumulate to hazardous levels. Therefore, most lipophilic xenobiotics are metabolized into hydrophilic conjugates that are less likely to pass through membranes and, hence, can be more easily eliminated via the kidney.

Hepatic metabolism is the major factor determining the circulating concentration of xenobiotics, but significant metabolic potential also exists at extrahepatic sites, including the lung, skin, and gut. For many phytochemicals at dietary doses, the gut—and especially the small intestine—is the major metabolic organ. Accordingly, substantial metabolic activity was detected within the intestinal lumen or associated with the mucus layer where a number of enzymes are located within the absorptive cells. These enzymes, in combination with efflux pumps, such as P-gp, are responsible for the intestinal first-pass effect. The various enzymes involved in xenobiotic metabolism are divided into phase 1 (P450s) and phase 2 (including glutathione-S-transferases, sulfotransferases, N-acetyltransferases, and UDP-glucuronosyltransferases) enzymes. Both play a pivotal role in determining the bioavailability of a compound and, hence, the sensitivity of individual cells to environmental chemicals.

Phytochemicals not only are subject to xenobiotic metabolism, but can also modulate the latter. Knowledge on the modulation of phase 1/2 enzyme expression by phytochemicals is of crucial importance, since metabolism by P450 enzymes can not only lead to an increased rate of chemical deactivation (i.e., detoxification) but, under specific circumstances, also result in chemical activation and the generation of toxic intermediates. Thus, the induction of one particular P450 might be beneficial in protecting against one particular compound but might simultaneously potentiate the toxic effects of other compounds.

Because access to the small intestine is difficult, it is hard to distinguish between first-pass metabolism in the liver and intestine. However, using a human intestinal perfusion, Petri et al. [19] have studied the intestinal metabolism of phytochemicals and examined the effects of sulforaphane and quercetin on expression of xenobiotic-metabolizing enzymes in enterocytes in vivo; their results show that both compounds were extensively metabolized in the small intestine. A proportion of the metabolites, sulforaphane-glutathione conjugate and quercetin-3'-glucuronide, was effluxed back into the jejunal lumen, thereby providing support for a coordinated regulation of xenobiotic-metabolizing enzymes and transporters, such as the apical conjugate export pump MRP2 [19,24]. In the same publication [19], evidence was presented that the two compounds were able to induce enzymes responsible for their own metabolism. A number of phytochemicals have been shown to similarly induce their own metabolism and thus facilitate the parallel metabolism of toxins. This process, in fact, represents the major chemopreventive mechanism of many phytochemicals.

## E. Excretion

The terms excretion and elimination describe the processes by which a compound leaves the body; elimination is sometimes used in a broader sense that includes metabolic removal of the absorbed xenobiotic as well as its excretion.

Xenobiotics and their metabolites can be eliminated from the body by several routes; including urine, feces, and exhaled air. The primary organ systems involved in excretion are, therefore, the kidney and the gastrointestinal and respiratory systems. Xenobiotics are removed in the kidney by passive filtration of the blood or active secretion by carriers into the forming urine; selective reabsorption from the forming urine may also occur. Fat-soluble compounds are more likely to be reabsorbed than those that are water soluble. Measurements of xenobiotics and their metabolites in the urine are often used as biomarkers of exposure, but in reality the amount of a compound or metabolite excreted in the urine is only a proportion of the absorbed compound. In the absence of a thorough understanding of the processes involved, this approach needs to be carefully evaluated.

## F. Intra- and Interindividual Variability and Interaction

Among others, individual susceptibility may be affected by host factors such as age, gender, disease state, and social influences. Together with dietary factors, related to the food matrix, these are the major cause of the observed overall high inter- and intraindividual variabilities.

### 1. Genetic Variability

Even if oral bioavailability were to be consistent within an individual, many compounds show considerable interindividual variations in their bioavailability. Genetic factors—polymorphisms in transporters or metabolic enzymes—are one of the major reasons for such interindividual differences. It has been estimated that approximately 40% of human phase 1 enzymes (CYP2D6, CYP2C19, CYP2A6, CYP2C9, CYP3A4) are polymorphic. Important phase 2 enzymes, notably *N*-acetyltransferase and transporters such as the efflux pump P-gp, are also subject to genetic polymorphism. Stable duplication, multiduplication, or amplification of active genes have been proposed to explain degrees of polymorphism high enough to cause quantitative or qualitative alterations in xenobiotic metabolism. This may also be a response to dietary components that have resulted in a selection of alleles with multiple noninducible genes [25]. Some genetic polymorphisms are prevalent in certain ethnic groups; thus, 5–10% of the Caucasian population are poor metabolizers of CYP2D6 substrates, but only 1–2% of Asians show this polymorphism. The opposite has been observed for CYP2C19: while poor metabolizers comprise 18–22% in Asian populations, this figure is only 2–6% among Caucasians [25,26].

## 2. Host Factors

It is widely agreed that the factors that affect the structure and function of the gastrointestinal tract are manifold, especially those parts of the gut involved in absorption and first-pass xenobiotic metabolism. The most significant of these are described below.

*Age.* Absorption and first-pass metabolism of phytochemicals may be age-dependent, thus neonates exhibit reduced hepatic metabolism and renal excretion as a direct consequence of the immaturity of liver and kidney function. The susceptibility of the elderly to xenobiotics may be affected by age-related alterations in absorption, intestinal and hepatic metabolism, renal clearance, or volume of distribution [27].

*Gender.* Studies of certain xenobiotics have shown inconsistencies in pharmacokinetic parameters between male and female subjects that are suggestive of gender-dependent differences in bioavailability, volume of distribution, activity of phase 1/2 metabolizing enzymes, renal clearance, or physiological characteristics. Normalization of pharmacokinetic parameters for body weight has been proposed as a means of assessing apparent differences between male and female patients involved in studies of drug interactions [28].

*Disease States.* Chronic diseases of the small and large intestine (a well-known example is Crohn's disease) can cause structural changes of the epithelium, with consequent changes in the expression and activity of enzymes and transporters. Furthermore, patients with renal impairment may be at increased risk of metabolic interactions; due to a diminished contribution of the excretory component to the overall elimination process, highly reactive metabolites could accumulate and cause damage at different sites of the organism.

*Social Influences.* Tobacco smoke is an inducer of CYP1A1, CYP1A2, and possibly CYP2E1 [29], while chronic alcohol consumption results in the induction of CYP2E1 [30]. These are two examples of numerous social parameters, including exercise, that are known to affect LADME.

In summarizing the current knowledge on host factors, it must be stressed that only limited and insufficient data are available on how host factors influence LADME and thus bioavailability of phytochemicals. It is unclear to what extent and how such parameters should be considered. From the perspective of the efficacy and risk assessment, more knowledge is certainly needed to allow identification of sub-populations at particular risk, who would benefit from appropriate dietary advice or treatment.

## 3. Effect of the Diet

Despite the fact that a plethora of dietary factors could, and will, affect the absorption characteristics of phytochemicals, this area has not been systematically explored. One reason might be the complexity of dietary factors and their interactions that could affect absorption. A nonexhaustive list would include the volume and composition of the food consumed, pH, caloric density, viscosity, nutrients (carbohydrates, protein, fat, fibers), alcohol, caffeine, and the presence of other phytochemicals. Such dietary factors affect the functional status, motility, and acidity of the gastrointestinal tract in a complex manner and modify the physico-chemical properties, formulation, and dissolution characteristics of the compound of interest. Calcium in dairy products, for example, has the potential to chelate tetracyclines and fluoroquinolones and, thereby, reduce their bioavailability and biological activity [31].

As mentioned above, specific food components may also interact with the enterocytes in the intestinal cell lining to modulate gene expression of transporter and metabolizing enzymes [19,32,33]. Flavonoids in grapefruit juice have been shown to be potent inhibitors of CYP3A4 in enterocytes [32,34]; bergapten, quercetin, naringenin, and naringin inhibited CYP3A4 by 67%, 55%, 39%, and 6%, respectively, at 100 $\mu$M concentration. These compounds have been suggested to be modulators of efflux systems and responsible for marked drug–food and food–food interactions [35].

Children (especially young) and the elderly may be considered special groups in terms of their absorptive and metabolic capacities because of adaptive processes that can differ significantly from a ''standard'' adult diet. There is little information currently available about adaptive processes to nutritional conditions such as vitamin or mineral status.

## III. METHODS TO STUDY LADME

Numerous methods have been proposed to study the individual processes of LADME as well as the final systemic bioavailability of phytochemicals as a result of LADME. Most have originated from the pharmaceutical research/new drug development sector, but may be applied and optimized for phytochemical studies. The absorption and metabolism of phytochemicals share similar principles and metabolic pathways (common transport systems) as drugs. The major points that should be borne in mind when extending such studies to phytochemicals are:

The food matrix is much more complex than that associated with drug formulations.

The prevalence of synergistic effects.

Competition with other food constituents for common metabolic enzymes and transporters.

The much lower concentrations of compounds in food compared with those of active components in drugs.

The longer-term application of low doses of compounds in food compared to higher doses from drugs; the long-term exposure to food constituents will require consideration of possible adaptation to exposure, which is applicable only to some drugs.

Four major methodologies have emerged for investigating the principal mechanisms of LADME, namely in silico, in vitro, in situ, and in vivo procedures. A selection of the most appropriate will depend on the specific question to be addressed, the chemical and biological nature of the compound being studied, and the matrix within which the compound is incorporated.

Since intestinal absorption and intestinal and hepatic first-pass metabolism are the major processes limiting bioavailability, the majority of the methods described below will focus on appropriate models for these processes. Nevertheless, it should be emphasized that other barriers, notably the blood-brain barrier (BBB) and virtually all organ-related membrane barriers, also need to be considered when investigating the distribution of a compound throughout the body and ultimately its tissue bioavailability. A recent study by the European Centre for the Validation of Alternative Methods (ECVAM) compared in vitro BBB methods and characterized several models in relation to in vivo studies: immortalized BBB-derived endothelial cell lines (SV-ARBEC, MBEC4), non–BBB-derived cell lines (MDCK, Caco-2, ECV-C6), and primary cells derived from BBB. In all cases, the correlations between in vitro and in vivo assays were low [36], and, moreover, most in vitro methods lacked at least some of the features of the in vivo barrier [37,38]. Therefore, animal studies will continue to play a major role in studies of the BBB.

The following listing of methodologies is not intended to be comprehensive; rather it represents an approach to summarize, compare, and analyze the different methods and their ability to predict LADME and, finally, the bioavailability of phytochemicals. Particular attention will be drawn to novel and alternative methods, even if these have not yet been extensively applied and validated.

## A. In Silico Methods

In silico methods have by far the highest throughput but are, currently, largely peripheral to the study of phytochemical bioavailability. The great challenge for in silico methods remains the development of models that correlate closely with in vivo systems, and their subsequent validation.

Quantitative in silico predictions for several pharmacokinetic parameters, particularly absorption and distribution, are now available; while in many cases no worse than in vitro tests, they are far less expensive, time-consuming, and labor intensive. Descriptors such as numbers of rotatable bonds and aromatic rings, branching behavior, and polar surface area are commonly used to predict the permeability of a given compound [39].

The bioavailability of a phytochemical often depends upon its interactions with specific proteins, such as metabolizing enzymes or transport and binding proteins. A detailed knowledge of these proteins is necessary for the development and validation of tools for predicting their pharmacokinetics. Several databases describing specific classes of ADME-associated proteins are available; one such, ADME-AP [40], provides comprehensive information on all classes of ADME-associated proteins described in the literature, including physiological function, pharmacokinetic effects, ADME classification, direction and driving force of disposition, location and tissue distribution, substrates, synonyms, gene name, and protein availability in other species.

Valuable information can also be gleaned from chemical structure: thus, within a group of compounds, highly lipophilic compounds would be expected to be most actively metabolized. If the metabolic potential and pathway of a compound containing a specific chemical moiety is known, the metabolism of structurally related compounds may be predicted with reasonable accuracy using appropriate computational tools. As more information is gained on the specific conformational constraints imposed by the catalytic sites of metabolic enzymes, the ability to predict in vivo metabolic events based solely on chemical structure will be increased.

Given that a large number of individual phytochemicals belong to distinct chemical classes and, therefore, have similar physicochemical properties, in silico methods could, and need to be developed for screening phytochemical libraries, with the ultimate aim of predicting bioactivities based on the knowledge of ADME and activity criteria. The priority at the present time should be to use mechanistic in vitro and in vivo data on phytochemicals to further develop and optimize in silico methods by identifying transport mechanisms, metabolic pathways, and binding behaviors to endogenous macromolecules (proteins, especially albumin).

The promise of predicting membrane permeability and metabolism from chemical structure alone is enticing; however, in silico methods are not yet at a level of sophistication required to supplant experimental methods. Their limitations become obvious when the system under investigation is a whole food or diet, rather than the individual, isolated compound. Food–food interactions, synergies, and canceling-outs of biological activities among food constituents remain very difficult to model. Consequently, once an initial screening has yielded a list of individual phytochemicals or the basic ADME properties of a given compound have been calculated, a second step must involve the consideration of the food matrix. The importance of such a complex food approach becomes obvious whenever intervention trials with isolated phytochemicals fail to show the predicted effects in humans.

## B.  In Vitro Methods

Artificial membranes, cells, tissues, and organ culture enable researchers to study a broad range of endogenous processes under carefully controlled conditions outside the body. In cell culture models, individual cell types are grown in special media that allow the cells to be maintained artificially. The compound to be studied can then be added directly to this medium. Tissue cultures are fragments of tissue, commonly lung, heart, and liver, that are taken at autopsy or during an operation and maintained in a medium; frequently organ cultures contain a full range of cell and tissue types of an individual organ, but they suffer from the disadvantage that they can only be kept for a short time in a special medium.

In vitro methods enable studies to be conducted on the function and response of individual cell types, tissue types, or whole organs separate from the potentially confounding influences of other body systems. Since the compound of interest is not distributed throughout the body, the amount of test substance required is correspondingly reduced. If human cells are used, results are directly relevant to the conditions in humans. The most important methods are summarized below.

## 1.  Artificial Membranes

Immobilized artificial membranes (IAM) are solid membrane mimetics that are covalently bound to the surface of a silica chromatographic support to generate a phospholipid monolayer. IAM chromatography can be applied to measure membrane partitioning of a given compound or to predict bile salt-membrane interactions; they may also be used in studies on transcellular absorption. Artificial membranes appear to be well correlated with 1-octanol/water partition coefficients; however, since the latter can be well predicted in silico, the real value of AIMs is the ability to study complex molecules or extracts, where in silico prediction is poor. IAM can also be applied to generate large data sets, that in turn can be applied to train and, therefore, improve in silico models by an extensive data input [41].

The parallel artificial membrane permeability assay (PAMPA) is a recent development in the area of artificial membranes that appears to offer considerable potential. Measuring the flux values (membrane permeation levels) of a range of test compounds by PAMPA and relating these values to the flux curves obtained in Caco-2 studies have shown good correlations, indicating that the PAMPA assay could be a good alternative to Caco-2 cells for the measurement of passively diffusing compounds.

## 2.  Microsomes

Hepatic microsomes are among the most popular and widely employed systems in current use. These preparations retain activity of enzymes that reside in the smooth endoplasmic reticulum, such as cytochrome P450s, flavin monooxygenases, and glucuronosyltransferases. However, isolated hepatocytes and cell models in general retain a broader spectrum of enzymatic activities, including cytosolic and mitochondrial enzymes.

## 3.  Cell Models

As a response to diverse pressures to find alternatives to animal studies in pharmacokinetic studies, a number of cell culture–based methods have been developed, extensively validated, and applied. Considerable attention has been paid to the use of epithelial cell cultures for studying transport mechanisms, but culturing enterocytes has proved difficult because of their limited viability and the loss of important in vivo anatomical and biochemical features. Therefore, human adenocarcinoma cell lines (HT-29 and Caco-2) have been employed: these reproducibly display a number of properties characteristic of differentiated intestinal cells [42]. Although the limitations of cell models must always be borne in mind, they offer the advantage of relative simplicity and are suitable for automated procedures in high throughput screening (HTS). As these cell lines are tumor-derived and removed from their in vivo physiological environment, care must be taken when extrapolating data to the in vivo situation.

*Small Intestinal Cell Lines.*    The most notable, and certainly best characterized, in vitro tool for studying absorption and, to a less extent, intestinal metabolism utilizes Caco-2 cells grown in a confluent monolayer on porous membrane filters and, for the experiments, mounted in diffusion chambers. Under these growing conditions they differentiate spontaneously into polarized enterocyte-like cells possessing an apical brush border and tight junctions between adjacent cells, thereby retaining many characteristics of the intestinal brush border. The permeability of a compound is determined by the rate of its appearance in the basolateral compartment.

The cells express functional transport proteins and metabolic enzymes, but the degree of expression depends on factors such as the stage of differentiation, the age of the cells, and passage number. Accordingly, Caco-2 cells express membrane-bound peptidases and disaccharidases of the small intestine, enable the active transport of amino acids, sugars (GLUT1, GLUT3, GLUT5, GLUT2, SGLT1), vitamins, and hormones as well as containing ion channels ($Na^+/K^+$

ATPase, $H^+/K^+$ ATPase, $Na^+/H^+$ exchange, $NaI/K^+/Cl^-$ co-transport, apical $Cl^-$ channels), and nonionic membrane transporters (P-gp, multidrug-resistant protein, etc.). They also express a number of receptors (vitamin $B_{12}$, vitamin $D_3$, epidermal growth factor) and phase 1 and 2 enzymes. The application of different inhibitors can clarify which transporter is involved in the active transport or which enzyme is required for metabolism of a given compound [43].

The use of Caco-2 cells is not without its limitations. They do not express LPH, the enzyme catalyzing the deglycosylation of many flavonoid glycosides [3,44]; CYP3A4, present in almost all intestinal cells, is only very weakly expressed in Caco-2 cells. While different treatments have shown to increase CYP3A4 expression levels, in vivo levels were not attained [45]. It should also be noted that Caco-2 cells express higher levels of P-pg than are found in the human intestine [46,47]. Caco-2–derived TC7 cells express CYP3A4 at higher and P-gp at lower levels than the parental line and are a particularly useful alternative to the Caco-2 cells for studying first-pass metabolism in human enterocytes. However, they do not express either CYP2B6 or CYP2D6.

When using Caco-2 or TC7 cells, parameters such as transepithelial electrical resistance (TEER), differentiation marker, morphology, P-gp expression, and monolayer integrity need to be tested; failure to do so is often the reason for wide interlaboratory variations. Caco-2 cell lines contain only a single cell type and do not form a mucus layer. As this can limit the absorption of lipophilic compounds in particular, the co-culturing of Caco-2 cells and mucus-secreting lines such as HT29-MTX has been proposed to overcome this limitation [48].

Significant differences were revealed when comparing overall gene expression profiles and permeability data obtained from Caco-2 cells with data derived from human enterocytes (GeneChip analysis); these were consistent with observed differences in carrier-mediated transport. It has been concluded that in vivo/in vitro (Caco-2) permeability measurements correlated well for passively absorbed compounds ($R^2 = 85\%$), but carrier-mediated compounds showed on average a 3- to 35-fold higher absorption in humans [49].

A comparison of the permeability coefficients of a series of xenobiotics using Caco-2 cells and an in situ human jejunal perfusion showed that the permeability of compounds with high or complete absorption differed 2- to 4-fold between the in vitro and in situ models, but poorly absorbed compounds differed as much as 30- to 80-fold [18]. Measurement of permeability using Caco-2 cells, therefore, affords only a qualitative comparison.

According to information made available by the INCELL Corp., cell lines from human small intestine and colon have been isolated, characterized for organotypic and cell-specific markers, and grown under conditions that induce growth and/or differentiation. However, only limited information is currently available, and these cell lines have not yet been extensively applied; thus it will be some time before they can be confidently applied to study intestinal absorption or metabolism.

*Nonintestinal Cell Lines.* Madin Darby canine kidney cells (MDCK [50]) differentiate into columnar epithelial cells forming tight junctions when cultured on semi-permeable membranes; they are commonly applied to study cell growth regulation, metabolism, toxicity, and transport at the level of the distal renal tubule epithelia [51]. MDCK cells, like Caco-2 cells, are suitable for molecular permeability screening studies; MDCK cells have an advantage over Caco-2 cells in that they do not need 3 weeks in culture before differentiation, but a disadvantage is that they do not express P-gp.

*Human Liver Cell System.* As has been mentioned, the liver is the major site of metabolism for most xenobiotics. There is, therefore, a need for a highly characterized model for human liver metabolism. While preliminary data may be gained from studies using animal cell lines and liver microsomal fractions, these data need to be confirmed in human liver preparations or cells. Human hepatocytes are a commonly used and valuable tool for studying short-term hepatic metabolism and for cytotoxicity evaluation. However, their preparation requires both skill and experience, so that the yield of successful isolations following cryopreservation may be low. It is important that the cryopreservants are added slowly, that freezing is performed at a controlled rate with adjustment for the heat of crystallization if the cells are stored at $-150°C$, and that the cells are rapidly thawed. Cryopreserved human hepatocytes have cytochrome P450 isoform activities and phase 2 conjugation enzyme activities similar to those of freshly isolated hepatocytes [52]. The use of pooled hepatocytes from multiple donors (five at least) is a means to minimize the effects of interindividual variations in metabolic activity.

The disadvantages of primary hepatocytes are mainly related to their inability to multiply in culture. Human liver cell lines (such as the ACTIVTox® system) that retain all of the major liver metabolic pathways have been developed; ACTIVTox® cells are reported to be a highly selected subclone of HepG2 that has retained many of the properties of normal adult hepatocytes, including metabolic activity. If cultured in small hollow fiber devices, they offer potential for long-term interaction and secondary metabolite studies [53].

Another novel technology employs three-dimensional co-cultures of primary human liver stroma and parenchymal cells. These are claimed by the manufacturers to function similarly to human liver tissue in vivo (CuDoS Ltd., UK, company information). The utilization of a stem cell–like liver reserve cell and a specific liver-regenerating growth factor has been reported to enable long-term proliferation of hepatocytes and retention of liver-specific function.

These are examples of recent developments that have yet to be validated and generally cited in the scientific literature. While important in demonstrating the nature and diversity of current attempts, for the near future, primary human hepatocytes will remain the standard model for studying liver function.

## 4. Organotypic Models

Organotypic models, available for all organs involved in ADME, are used to study food matrix effects, intestinal metabolism/stability, and regional differences in permeability. The latter is of particular importance since it has been shown that permeability to various marker molecules varies along the intestinal tract; in general, permeability decreases in the order jejunum > ileum > colon [54]. The half-lives of organotypic models limits the duration of possible studies to 1–3 hours.

In vitro cross-comparison of metabolic turnover rates in various tissues from different species and those from animals treated with enzyme inducers or inhibitors offer considerable potential for improving nonhuman models. Commonly applied organotypic methods are summarized below.

*Everted Gut Sacs and BBMV.* Everted intestinal sacs and brush-border membrane vesicles (BBMV) are used widely for assessing membrane permeability and to determine kinetic parameters with high reliability and reproducibility [55,56]. BBMV are prepared by removing the brush-border surface from the intestine and molding it into vesicles by homogenization and differential centrifugation.

Everted gut sacs are based on one of the earliest in vitro absorption systems in which an intestinal segment, everted and suspended in buffer, was used to measure mucosal-to-serosal transfer. Barthe et al. [57] developed and improved this system by incubating an everted gut sac from rat small intestine in tissue culture medium. Under these conditions, metabolic activity was preserved for up to 2 h at 37°C, but permeability measurements were only valid up to 15–30 minutes [22].

Everted gut sacs are used mainly to quantify paracellular transport of hydrophilic molecules and to estimate the effects, including toxicity (release of intracellular enzymes, histological characterization), of potent absorption enhancers. They have also been exploited for studying transport of macromolecules and liposomes [22,56,58]. The transport of paracellular markers (e.g., mannitol) shows the same apparent permeability as has been reported for low molecular weight hydrophilic compounds in human perfusion studies. This similarity also applies to highly permeable molecules that cross the epithelium by a transcellular route.

Everted gut sacs are suitable for comparative studies of absorption at different sites in the gastrointestinal tract [54,55], for evaluating first-pass metabolism in enterocytes, and for studying the role of certain transporters in xenobiotic transport through the intestinal barrier or efflux back into the lumen.

Since the muscularis mucosa is usually not removed from everted sac preparations, this model does not reflect the actual intestinal barrier. The compound to be studied must pass from the lumen into the lamina propria (where blood and lymph vessels are found) and across the muscularis mucosa; as a result the transport of compounds capable of binding strongly to muscle cells may be underestimated.

*Isolated and Perfused Organs.*   Isolated perfused organs have the advantage that physiological cell-cell contacts and normal intracellular matrices are preserved. Using an isolated, vascularly perfused rat small intestine, Levet-Trafit et al. [59] were able to simultaneously study the appearance in the artificial bloodstream and the intestinal metabolism of an orally administered drug. Unlike in vivo studies, the use of isolated organs permits internal and external parameters, such as type and composition of the perfusate (nutrient supply), to be controlled before and during the experiment. The concentration of the compound applied to isolated organs can be well controlled; samples of the perfusate (venous effluate) or the organ itself may be readily and frequently obtained. The experimental setup facilitates continuous and simultaneous monitoring of a number of (patho)-physiological characteristics under identical genetic and experimental conditions. Such control is difficult, if not impossible, to achieve with laboratory animals. This model offers the possibility of investigating the administration of multiple agents by different routes and, by adding a compound to the perfusate, allows the formation of metabolites to be monitored in real time enabling essential pharmacokinetic information to be determined. The major limitation is the short duration of the experiments that are possible, since irreversible changes occur rapidly.

At present a wide range of isolated organ systems is available. The use of isolated and perfused organs, such as the intestine, to study ADME in vitro has been reported to be highly predictive of the in vivo situation, including absorption at the organ level [55,60].

*Ussing Chamber.*   The Ussing chamber technique, originally developed by Ussing and Zehran in 1951, has been used to study transepithelial transport via para- and transcellular routes through human or animal intestinal mucosa. This technique allows measurement of the intestinal barrier function in physiological or pathological tissues and under the influence of several factors [55]. The compounds can be applied at either the mucosal or serosal side. Ussing chambers also offer the potential for studying the effects of a compound on electrophysiological characteristics of the intestinal barrier. However, decreasing viability causes changes in the transepithelial potential; therefore TEER values must be monitored constantly and the duration of an experiment is limited [61]. In common with the everted gut sac model, the Ussing chamber has the disadvantage that the compound must traverse the complete intestinal wall; this may be rate limiting in vitro, but not so in vivo, and transport values may be underestimated. Far less material is needed for the Ussing chamber (microsystems are available) compared to the everted gut sac and the degree of automation, and, hence, throughput is higher.

All in vitro techniques are rather simplistic models to study highly complex systems; a proper understanding of the underlying physiological principles, their impact on the measurement and their limitations are necessary if the correct conclusions are to be drawn from the experimental data.

## C.  In Situ Models

In situ models, such as continuous, single-pass perfusion in animals and humans, provide a powerful research tool for the investigation of intestinal transport and metabolism and have shown good predictive performance for passively absorbed and actively transported compounds. Compound permeability is derived from the rate of disappearance from the perfusate. In situ methods have significant advantages over in vivo models in whole organisms: first, bypassing the stomach means that acidic compounds are less likely to precipitate, so that dissolution rates do not confound intestinal compound concentrations and plasma levels; second, in situ methods allow the assessment of organ-specific processes under isolated and controlled conditions.

The volume of the luminal solution, which may change as a result of absorption or secretion of water, is an important parameter when applying in situ techniques. A non- or low-absorbable volume marker, such as radiolabeled polyethyleneglycol 4000 or a fluorescent marker (lucifer yellow), and a marker of paracellular absorption (mannitol) need to be added and monitored as internal standards.

### 1.  Perfused Animal Preparations

Single-pass intestinal perfusion involves isolating a small segment of the intestine in a living animal and perfusing the intestinal lumen with a known quantity of the compound of interest. The concentrations of the compound and its metabolites are then measured in the perfusate as it leaves the segment. Although the mesenteric blood flow is still intact after the animal has been anaesthetized and surgically manipulated, caution must be taken with the choice of anaesthetic, which can have a significant effect on intestinal absorption [62]. Since disappearance of a compound from the perfusate does not necessarily equal the amount absorbed, following analysis needs to include the monitoring of concentration changes of the parent compound and/or metabolites in the perfusate as well as in the blood. By comparing blood samples obtained from the portal and hepatic veins, additional information about the first-pass effect in the liver can be obtained.

Although labor-intensive, in situ intestinal perfusions are widely used because of the perceived clinical relevance of permeability data that directly reflect the absorption potential of a compound in combination with intestinal metabolism [63].

## 2. Single-pass Human Perfusion

An in vivo human jejunal perfusion remains the method of choice for studying small intestinal absorption, first-pass metabolism, and efflux of metabolites in the human intestine. This technique has been validated by several means and one of the most important findings to emerge is that, even for carrier-mediated transport, a good correlation exists between measured human effective permeability values ($P_{eff}$, measured by single-pass, regional perfusion) and the extent of absorption of a number of compounds (mostly drugs) in humans, determined by pharmacokinetic studies [64,65]. The method is regarded as the "gold standard" for permeability studies [66], being also atraumatic, safe, reproducible, and accurately reflecting biochemical processes taking place in the small intestine. The drawback is that it is both expensive and difficult; it cannot be used routinely, and, at present, results are limited to relatively few (around 20) compounds.

Loc-I-Gut®, a perfusion instrument developed for human jejunal perfusions, comprises a six-channel tube with two occluding latex balloons placed 10 cm apart from one another. Once the instrument is inserted (this being monitored by fluoroscopy), the balloons are inflated to form a 10 cm long, isolated segment in the jejunum; a separate tube allows gastric suction. $^{14}$C-PEG 4000 is used both as a volume marker and to ensure that there is no leakage to or from the segment (Fig. 2).

The majority of shed enterocytes collected during the perfusion are still functionally active and show no signs of apoptosis [67,68]. On that basis, Petri et al. [19] were able to study the cellular response to phytochemicals in human enterocytes in vivo, by measuring differences in mRNA levels of phase 2 metabolizing enzymes. Indeed, the authors observed an immediate cellular response of the enterocytes to phytochemicals present in a crude vegetable extract [19].

## D.  In Vivo Methods

It is difficult to study complex interactions in simplistic models. Only in vivo models provide the opportunity to integrate the dynamic components of the mesenteric blood circulation, the mucous layer, and all other factors that can influence LADME. The disadvantage of these methods, especially of human studies, is that it is impossible to separate the variables and individual processes leading to the systemic bioavailability of a compound; in other words, it is impossible to identify the individual rate-limiting processes and factors. Because the individual processes of LADME (especially those involving metabolism) are subject to high interindividual variation, a large number of subjects or animals need to be included

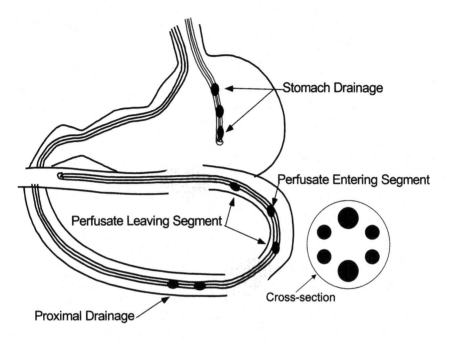

**Figure 2** The Loc-I-Gut® system, a multichannel tube system with double balloons, enables segmental jejunal perfusion in humans. (From Ref. 64.)

into a study to ensure statistical significance. In vivo methods are, therefore, expensive and time-consuming.

## 1. In Vivo Animal Studies

Although new in vitro techniques have reduced the need for animal trials, the latter still contribute significantly to studies on the bioavailability of xenobiotics. Numerous data demonstrate interspecies differences and similarities in gastrointestinal anatomy and function between different animal species and humans. Far less is known about how these translate into interspecies differences in LADME and bioavailability. It is assumed that for passively transported compounds the gastrointestinal tract and membrane barriers from different species behave and function in similar ways leading to similar permeabilities. However, a very different situation pertains when considering paracellular passive diffusion via the tight junctions; it is known, for example, that the dog has wider pores than the rat or human and, therefore, exhibits higher permeability. For ethical and other pragmatic reasons, the most frequently used animal model is the rat; for many processes it is also the closest model to the human situation [69]. However, oral studies in rats have important limitations and tend to provide false-positive results [43]. Therefore, Wilding et al. have concluded that "the only real model for man is man" [70].

Species differences become inevitable when active processes (metabolism or active transport) have to be taken into consideration. Differences in anatomy and physiology (gut length and motility), amount and composition of gut microflora, coprophagy, enterohepatic recycling, as well as intestinal transit time may all profoundly affect absorption in different species. Additionally, one of the major questions remains whether the dose of a compound used in test animals can be scaled up (or down) to the human situation and, if so, on what basis? Detailed comparative knowledge of anatomical, physiological, and pharmacological parameters affecting absorption in an animal species and in humans is required to answer this question, and this is not always available.

The development of transgenic animal models for heterologous expression of human drug-metabolizing enzymes or for gene inactivation by homologous recombination may be seen as a promising in vivo approach to investigate the role of specific enzymes involved in metabolism. Once again, however, interspecies differences mean that the endpoints of such studies may not reflect the human situation.

## 2. Human Studies

At dietary levels, food constituents—including phytochemicals—are recognized as safe. As a result, human studies of dietary phytochemicals at dietary doses should always be considered as the method of choice, which provides a great advantage over drug studies. The disadvantage and challenge is (1) the complexity of the food matrix that needs to be studied to simulate dietary conditions, (2) the low concentrations of most phytochemicals in food, and (3) the complexity of the human organism. The gaps in knowledge on the individual processes of LADME have been mentioned and urgently need to be addressed. In the meantime it is necessary to complement human in vivo data with appropriate in vitro data obtained by studies on human tissues, cells, or subcellular fractions.

*Mass Balance Studies.* Pharmacokinetic mass balance studies apply unlabeled, stable isotopes or radiolabeled compounds to study the extent of absorption and first-pass metabolism, distribution, and excretion of a given compound. In the "microdosing approach," a $^{14}C$-labeled compound is administered to human volunteers at doses from as low as one microgram; blood, urine, and fecal samples are collected over time and analyzed for $^{14}C$ content by accelerator mass spectroscopy to determine half-life, plasma AUC, and maximal concentration ($C_{max}$). However, these methods are not very popular even when very low doses of radioactivity are involved. Highly sensitive, and more readily available, techniques for separation and analysis (e.g., LC-MS, LC-MS/MS) are frequently used alternatives that enable pharmacokinetic investigations and metabolite profiling of nonradiolabeled compounds.

The interindividual variability of the parameter under investigation needs to be determined in a smaller initial study prior to the main experiment. Based on these findings, a sufficient number of subjects need to be enrolled to provide reliable data. Mass balance studies tend to provide highly variable data for many phytochemicals and necessitate large subject numbers, with associated labor, time, and financial requirements.

Human data on phytochemicals concentrate largely on measurement of urinary metabolites; in combination with stable isotope labeling this is a valid method for studying mineral bioavailability, since the difference between the amount absorbed and excreted is retained in the body and, excluding the amount stored in body pools, assumed to be bioavailable.

Because most phytochemicals are extensively metabolized and labeled substrates are generally unavailable, these methods yield only low recovery figures. A thorough understanding of the individual processes involved (LADME) is once again vital if the correct interpretation of the results is to be made.

*Absolute Bioavailability Studies.* Oral bioavailability determinations using intravenous administration as a reference are intensively used in drug development. As is the case for mass balance studies, high numbers of subjects must be enrolled in a study to provide a reliable estimate of the extent of absorption. Since phytochemicals undergo extensive first-pass metabolism, absolute bioavailability studies using intravenous dosing need to be carefully planned. First-pass metabolism is a very efficient endogenous detoxification mechanism of the organism. Circumventing this mechanism may raise toxicological issues that need to be clarified in advanced.

*Ileostomy Studies.* Ileostomy patients have proved to be an extremely valuable and generous subject group; they have participated willingly in many studies designed to determine small intestinal absorption and metabolism independently from processes in the colon, e.g., colonic fermentation. As a human model system, they do have some limitations. A bacterial population associated specifically with the unusual physiology resulting from ileostomy reproduces and can cause bacterial overgrowth in the small intestine. Although this has not yet been well characterized, it may include bacteria from the external environment, including soil, that would normally not be viable in the small intestine. Data obtained from ileostomy studies may thus reflect bacterial degradation and metabolism. Results from such studies should be carefully interpreted and confirmed by other models; in particular, breakdown and metabolism of the bacterial population should be tested in parallel with the main study.

## E.  Compound Analysis

A major advantage of in vitro models is their capability to be highly automated and to allow high-throughput screening. Once an in vitro system has been established and optimized, the analytical method employed to identify and quantify the parent compound and its metabolites becomes the rate-determining step. In situ and in vivo studies also depend on highly sensitive analytical methods that enable a sufficient throughput. LC-MS and LC-MS/MS are fast becoming the analytical technique of choice for determining ADME of phytochemicals. Due to their inherent sensitivity and selectivity, hybrid quadrupole time-of-flight (TOF) mass spectrometers have gained considerable popularity. The integration of data collected from ion trap, triple quadrupole, and quadrupole/TOF instruments facilitates a comprehensive evaluation of biotransformation products. Sophisticated and expensive instruments, such as MALDI Qq-TOF or the API QSTAR(TM) Pulsar Hybrid LC-MS/MS, can operate in two modes; by using the technique initially in MS mode, a TOF spectrum is obtained, while operation in MS/MS mode allows specific mass (or more specifically, mass-to-charge, m/e) ranges to be selected for further analysis.

In order to increase throughput in pharmacokinetic screening studies, simultaneous dosing of numerous compounds followed by multiple component analysis using LC-MS/MS (the so-called $N$-in-1 approach) has been developed and proved to be effective [71].

To speed up postanalytical data mining, dedicated software, such as MetaboLynx™, has been developed: this software extracts meaningful information from complex data sets and converts large LC-MS raw data files into simple, manageable reports. By applying sample comparison algorithms and combining chromatographic and spectral data, the software manufacturer claims that new components can be identified and highlighted as potential metabolites or impurities. Such metabolite identification software packages greatly simplifies the process of searching for expected metabolites or characteristic isotopic patterns; it provides lists of potential metabolites and assists with setting up further experiments to confirm their identity.

## IV.  CONCLUSION

The close correlation between diet and health has been well recognized and is the object of intensive and expanding research. Bioactive compounds, especially those of plant origin (phytochemicals), together with vitamins and minerals are responsible for many of the observed effects.

It is important to recognize that any protective or toxic effect is highly dependent on the concentration and the structure of a compound at a target site.

Such variables obviously relate to, but do not equate to the concentration and structure of the compound *in the food ingested*. Appropriate biomarkers of exposure need to be developed from a thorough understanding of individual processes leading to the bioavailability of a compound/metabolite and validated. Individual susceptibility must also be taken into account. Application of these biomarkers will allow intervention trials to be conducted that will provide conclusive data on the effects of a dietary phytochemical and makes it possible to develop food products to maintain and improve consumer health by dietary intervention. Furthermore, studies directed towards the effects of host factors would make it possible to identify subject groups that could specifically benefit from a particular dietary intervention, or who may be at risk from a particular food or constituent.

Among the reasons for the gaps in knowledge on LADME for many phytochemicals are the large number and complexity of interactions of individual parameters and processes leading to the bioavailability of a phytochemical. To address this situation there is a pressing need for appropriate methods to study individual processes of LADME and also for determining the complexity of interactions. A wide range of methods is currently available for studying LADME and bioavailability (mostly adapted from pharmaceutical research), but detailed knowledge is necessary to select the right combination of methods to gain a complete picture of parameters affecting the bioavailability of phytochemicals. Failure to understand and appreciate these will inevitably compromise the conclusions reached. The choice of model depends on the nature of the compound and its availability, the complexity of the matrix, the knowledge already available for LADME, the questions to be answered, and, last but not least, the resources available.

Because of time and cost constraints, current trends in method development are directed towards in silico and highly automated in vitro methods. Although the number of individual phytochemicals in foods is very high, most belong to distinct chemicals classes (e.g., glucosinolates, flavonoids, allicins). Experiences gained in pharmacology should be applied to phytochemical research in order to predict LADME properties of structurally related compounds and their homologues. The Biopharmaceutical Classification System (BCS), a systematic approach for classifying drugs according to their absorption and solubility behavior, could provide guidance for a systematic investigation of the bioavailability of phytochemicals.

Since dietary phytochemicals at dietary doses are recognized as safe, the opportunity is afforded to perform human studies; this should be taken as a challenge to gain mechanistic information on LADME as well as the effects and dose-effect correlations in humans. A combination of genomics and proteomics approaches could supply additional mechanistic insight, e.g., by identifying genes regulated by food components and by linking of common regulatory mechanisms (such as the coordinated regulation of metabolic enzymes and transporters).

Against this background it is to be expected that much detailed information on phytochemical metabolism and absorption will emerge over the next decade and will provide a firm basis for industrial exploitation of dietary phytochemicals and societal health benefit.

## REFERENCES

1. Kong AN, Yu R, Chen C, Mandlekar S, Primiano T. Signal transduction events elicited by natural products: role of MAPK and caspase pathways in homeostatic response and induction of apoptosis. Arch Pharm Res 2000; 23:1–16.
2. Stahl W, van den Berg H, Arthur J, Bast A, Dainty J, Faulks RM, Gartner C, Haenen G, Hollman P, Holst B, Kelly FJ, Polidori MC, Rice-Evans C, Southon S, van Vliet T, Vina-Ribes J, Williamson G, Astley SB. Bioavailability and metabolism. Mol Aspects Med 2002; 23:39–100.
3. Day AJ, Canada FJ, Diaz JC, Kroon PA, Mclauchlan R, Faulds CB, Plumb GW, Morgan MRA, Williamson G. Dietary flavonoid and isoflavone glycosides are hydrolysed by the lactase site of lactase phlorizin hydrolase. FEBS Lett 2000; 468: 166–170.
4. Hollman PCH, Bijsman MNCP, van Gameren Y, Cnossen EPJ, de Vries JHM, Katan MB. The sugar moiety is a major determinant of the absorption of dietary flavonoid glycosides in man. Free Rad Res 2000; 31:569–573.
5. Scalbert A, Williamson G. Dietary intake and bioavailability of polyphenols. J Nutr 2000; 130:2073S–2085S.
6. Welling PG. Interactions affecting drug absorption. Clin Pharmacokinet 1984; 9: 404–434.
7. Lombardo F, Shalaeva M, Tupper KA, Gao F, Bissett BD. The anxieties of HT lipophilicity determination: plate formats, detection, purity level and data handling. Annual Symposium on Chemical and Pharmaceutical Structure Analysis (CPSA), 10th Proceedings, 2001.
8. Michaelsen S, Otte J, Simonsen LO, Sørensen H. Absorption and degradation of individual intact glucosinolates in the digestive-tract of rodents. Acta Agric Scand Sec A Animal Sci 1994; 44:25–37.
9. Slominski BA, Campbell LD, Stanger NE. Extent of hydrolysis in the intestinal tract and potential absorption of intact glucosinolates in laying hens. J Sci Food Agric 1988; 42:305–314.
10. Nugon-Baudon L, Rabot S, Flinois JP, Lory S, Beaune P. Effects of the bacterial status of rats on the changes in some liver cytochrome P450 (EC 1.14.14.1) apoproteins consequent to a glucosinolate-rich diet. Br J Nutr 1998; 80:231–234.
11. Pacha J. Development of intestinal transport function in mammals. Physiol Rev 2000; 80:1633–1667.
12. Lipinski CA, Lombardo F, Dominy BW, Feeney PJ. Experimental and computational approaches to estimate solubility and permeability in drug discovery and development settings. Adv Drug Del Rev 1997; 23:3–25.

13. Levin RJ. Assessing small intestinal function in health and disease in vivo and in vitro. Scand J Gastroenterol Suppl 1982; 74:31–51.
14. Pade V, Stavchansky S. Link between drug absorption solubility and permeability measurements in Caco-2 cells. J Pharm Sci 1998; 87:1604–1607.
15. Lennernas H. Human intestinal permeability. J Pharm Sci 1998; 87:403–410.
16. Lennernas H. Human jejunal effective permeability and its correlation with preclinical drug absorption models. J Pharm Pharmacol 1997; 49:627–638.
17. Lennernas H, Ahrenstedt O, Hallgren R, Knutson L, Ryde M, Paalzow L. Regional jejunal perfusion, a new in vivo approach to study oral drug absorption in man. Pharmaceut Res 1992; 9:1243–1251.
18. Lindahl A, Sandstrom R, Ungell AL, Abrahamsson B, Knutson TW, Knutson L, Lennernas H. Jejunal permeability and hepatic extraction of fluvastatin in humans. Clin Pharmacol Ther 1996; 60:493–503.
19. Petri N, Tannergren C, Holst B, Mellon FA, Bao Y, Plump GW, Bacon J, O'Leary KA, Kroon PA, Knutson L, Forsell P, Eriksson T, Lennernas H, Williamson G. Absorption/metabolism of sulforaphane and quercetin, and regulation of phase II enzymes, in human jejunum in vivo. Drug Metab Dispos 2003; 31:1–9.
20. Cooper DA, Webb DR, Peters JC. Evaluation of the potential for olestra to affect the availability of dietary phytochemicals. J Nutr 1997; 127:S1699–S1709.
21. Zhang Y, Callaway EC. High cellular accumulation of sulphoraphane, a dietary anticarcinogen, is followed by rapid transporter-mediated export as a glutathione conjugate. Biochem J 2002; 364:301–307.
22. Gee JM, DuPont MS, Day AJ, Plumb GW, Williamson G, Johnson IT. Intestinal transport of quercetin glycosides in rats involves both deglycosylation and interaction with the hexose transport pathway. J Nutr 2000; 130:2765–2771.
23. Day AJ, DuPont MS, Ridley S, Rhodes M, Rhodes MJC, Morgan MRA, Williamson G. Deglycosylation of flavonoid and isoflavonoid glycosides by human small intestine and liver β-glucosidase activity. FEBS Lett 1998; 436:71–75.
24. Bock KW, Eckle T, Ouzzine M, Fournel-Gigleux S. Coordinate induction by antioxidants of UDP-glucuronosyltransferase UGT1A6 and the apical conjugate export pump MRP2 (multidrug resistance protein 2) in Caco-2 cells. Biochem Pharmacol 2000; 59:467–470.
25. Ingelman-Sundberg M, Oscarson M, McLellan RA. Polymorphic human cytochrome P450 enzymes: an opportunity for individualized drug treatment. Trends Pharmacol Sci 1999; 20:342–349.
26. Lin JH, Lu AY. Inhibition and induction of cytochrome P450 and the clinical implications. Clin Pharmacokinet 1998; 35:361–390.
27. Hammerlein A, Derendorf H, Lowenthal DT. Pharmacokinetic and pharmacodynamic changes in the elderly. Clinical implications. Clin Pharmacokinet 1998; 35:49–64.
28. Beierle I, Meibohm B, Derendorf H. Gender differences in pharmacokinetics and pharmacodynamics. Int J Clin Pharmacol Ther 1999; 37:529–547.
29. Zevin S, Benowitz NL. Drug interactions with tobacco smoking. An update. Clin Pharmacokinet 1999; 36:425–438.
30. Oneta CM, Lieber CS, Li J, Ruttimann S, Schmid B, Lattmann J, Rosman AS, Seitz HK. Dynamics of cytochrome P4502E1 activity in man: induction by ethanol and disappearance during withdrawal phase. J Hepatol 2002; 36:47–52.

31. Fleisher D, Li C, Zhou Y, Pao LH, Karim A. Drug, meal and formulation interactions influencing drug absorption after oral administration. Clinical implications. Clin Pharmacokinet 1999; 36:233–254.

32. Ameer B, Weintraub RA. Drug interactions with grapefruit juice. Clin Pharmacokinet 1997; 33:103–121.

33. Zhang H, Wong CW, Coville PF, Wanwimolruk S. Effect of the grapefruit flavonoid naringin on pharmacokinetics of quinine in rats. Drug Metabol Drug Interact 2000; 17:351–363.

34. Ho PC, Saville DJ, Wanwimolruk S. Inhibition of human CYP3A4 activity by grapefruit flavonoids, furanocoumarins and related compounds. J Pharm Pharm Sci 2001; 4:217–227.

35. Wagner D, Spahn-Langguth H, Hanafy A, Koggel A, Langguth P. Intestinal drug efflux: formulation and food effects. Adv Drug Deliv Rev 2001; 50(suppl 1): S13–S31.

36. Worth AP, Balls M. Alternative (non-animal) methods for chemicals testing: current status and future prospects. A report prepared by ECVAM and the ECVAM Working Group on Chemicals. ATLA 2002; 30(suppl 1):55–70.

37. Foster KA, Roberts MS. Experimental methods for studying drug uptake in the head and brain. Curr Drug Metab 2000; 1:333–356.

38. Foster KA, Mellick GD, Weiss M, Roberts MS. An isolated in situ rat head perfusion model for pharmacokinetic studies. Pharm Res 2000; 17:127–134.

39. Darvas F, Keseru G, Papp A, Dorman G, Urge L, Krajcsi P. In silico and ex silico ADME approaches for drug discovery. Curr Top Med Chem 2002; 2:1287–1304.

40. Sun LZ, Ji ZL, Chen X, Wang JF, Chen YZ. ADME-AP: a database of ADME associated proteins. Bioinformatics 2002; 18:1699–1700.

41. van der Waterbeemd H. Intestinal permeability: prediction from theory. In:. Dressman JB, Lennernas H, Eds. Oral Drug Absorption: Prediction and Assessment. New York: Marcel Dekker, Inc., 2000:31–49.

42. Kedinger M, Haffen K, Simon-Assmann P. Intestinal tissue and cell cultures. Differentiation 1987; 36:71–85.

43. Ferrec EL, Chesne C, Artusson P, Fabre G, Gires P, Guillou F, Rousset M, Rubas W, Scarino ML. In vitro models of the intestinal barrier. ATLA 2002; 29:649–668.

44. Day AJ, Gee JM, DuPont MS, Johnson IT, Williamson G. Absorption of quercetin-3-glucoside and quercetin-4'-glucoside in the rat small intestine: the role of lactase phlorizin hydrolase and the sodium-dependent glucose transporter. Biochem Pharmacol 2003; 65:1199–1206.

45. Hu M, Li Y, Davitt CM, Huang SM, Thummel K, Penman BW, Crespi CL. Transport and metabolic characterization of Caco-2 cells expressing CYP3A4 and CYP3A4 plus oxidoreductase. Pharm Res 1999; 16:1352–1359.

46. Hunter J, Hirst BH, Simmons NL. Drug absorption limited by P-glycoprotein-mediated secretory drug transport in human intestinal epithelial Caco-2 cell layers. Pharm Res 1993; 10:743–749.

47. Burton PS, Conradi RA, Hilgers AR, Ho NF. Evidence for a polarized efflux system for peptides in the apical membrane of Caco-2 cells. Biochem Biophys Res Commun 1993; 190:760–766.

48. Hilgendorf C, Spahn-Langguth H, Regardh CG, Lipka E, Amidon GL, Langguth P. Caco-2 versus Caco-2/HT29-MTX co-cultured cell lines: Permeabilities via diffusion, inside- and outside-directed carrier-mediated transport. J Pharm Sci 2000; 89: 63–75.

49. Sun D, Lennernas H, Welage LS, Barnett JL, Landowski CP, Foster D, Fleisher D, Lee KD, Amidon GL. Comparison of human duodenum and Caco-2 gene expression profiles for 12,000 gene sequences tags and correlation with permeability of 26 drugs. Pharm Res 2002; 19:1400–1416.

50. Gaush CR, Hard WL, Smith TF. Characterization of an established line of canine kidney cells (MDCK). Proc Soc Exp Biol Med 1966; 122:931–935.

51. Cho MJ, Thompson DP, Cramer CT, Vidmar TJ, Scieszka JF. The Madin Darby canine kidney (MDCK) epithelial cell monolayer as a model cellular transport barrier. Pharm Res 1989; 6:71–77.

52. Li AP, Gorycki PD, Hengstler JG, Kedderis GL, Koebe HG, Rahmani R, de Sousas G, Silva JM, Skett P. Present status of the application of cryopreserved hepatocytes in the evaluation of xenobiotics: consensus of an international expert panel. Chem Biol Interact 1999; 121:117–123.

53. Kelly JH, Spierling AL, Sussman NL. The use of a human liver cell system in ADME/PK studies. In Vitro Cell Dev Biol Animal 1998; 34(3 part II):T11.

54. Davis GR, Santa Ana CA, Morawski SG, Fordtran JS. Permeability characteristics of human jejunum, ileum, proximal colon and distal colon: results of potential difference measurements and unidirectional fluxes. Gastroenterology 1982; 83:844–850.

55. Barthe L, Woodley J, Houin G. Gastrointestinal absorption of drugs: methods and studies. Fundam Clin Pharmacol 1999; 13:154–168.

56. Leppert PS, Fix JA. Use of everted intestinal rings for in vitro examination of oral absorption potential. J Pharm Sci 1994; 83(7):976–981.

57. Barthe L, Woodley JF, Kenworthy S, Houin G. An improved everted gut sac as a simple and accurate technique to measure paracellular transport across the small intestine. Eur J Drug Metab Pharmacokinet 1998; 23:313–323.

58. Andlauer W, Kolb J, Furst P. Absorption and metabolism of genistin in the isolated rat small intestine. FEBS Lett 2000; 475:127–130.

59. Levet-Trafit B, Gruyer MS, Marjanovic M, Chou RC. Estimation of oral drug absorption in man based on intestine permeability in rats. Life Sci 1996; 58:L359–L363.

60. Acra SA, Ghishan FK. Methods of investigating intestinal transport. J Parenter Enteral Nutr 1991; 15:93S–98S.

61. Soderholm JD, Hedman L, Artursson P, Franzen L, Larsson J, Pantzar N, Permert J, Olaison G. Integrity and metabolism of human ileal mucosa in vitro in the Ussing chamber. Acta Physiol Scand 1998; 162:47–56.

62. Yuasa H, Matsuda K, Watanabe J. Influence of anaesthetic regimens on intestinal absorption in rats. Pharm Res 1993; 10:884–888.

63. Bohets H, Annaert P, Mannens G, Van Beijsterveldt L, Anciaux K, Verboven P, Meuldermans W, Lavrijsen K. Strategies for absorption screening in drug discovery and development. Curr Top Med Chem 2001; 1:367–383.

64. Lennernas H. Human jejunal effective permeability and its correlation with preclinical drug absorption models. J Pharm Pharmacol 1997; 49:627–638.

65. Lennernas H. Human intestinal permeability. J Pharm Sci 1998; 87:403–410.
66. Lipka E, Amidon GL. Setting bioequivalence requirements for drug development based on preclinical data: optimizing oral drug delivery systems. J Control Release 1999; 62:41–49.
67. Ahrenstedt O, Knutson F, Knutson L, Krog M, Sjoberg O, Hallgren R. Cell recovery during segmental intestinal perfusion in healthy subjects and patients with Crohn's disease. Gut 1991; 32:170–173.
68. Glaeser H, Drescher S, Van Der KH, Behrens C, Geick A, Burk O, Dent J, Somogyi A, von Richter O, Griese EU, Eichelbaum M, Fromm MF. Shed human enterocytes as a tool for the study of expression and function of intestinal drug-metabolizing enzymes and transporters. Clin Pharmacol Ther 2002; 71:131–140.
69. Kararli TT. Comparison of the gastrointestinal anatomy, physiology, and biochemistry of humans and commonly used laboratory animals. Biopharm Drug Dispos 1995; 16:351–380.
70. Wilding IR, Kenyon CJ, Hooper G. Gastrointestinal spread of oral prolonged-release mesalazine microgranules (Pentasa) dosed as either tablets or sachet. Aliment Pharmacol Ther 2000; 14:163–169.
71. Wu JT, Zeng H, Qian M, Brogdon BL, Unger SE. Direct plasma sample injection in multiple-component LC-MS-MS assays for high-throughput pharmacokinetic screening. Anal Chem 2000; 72:61–67.
72. Wang K, Shindoh H, Inoue T, Horii I. Advantages of in vitro cytotoxicity testing by using primary rat hepatocytes in comparison with established cell lines. J Toxicol Sci 2002; 27:229–237.

# 3
# Characterization of Polyphenol Metabolites

**Andrea J. Day, Joseph A. Rothwell, and R. A. Morgan**
*University of Leeds, Leeds, United Kingdom*

## I. INTRODUCTION

Dietary phytochemicals, food plant secondary metabolites, are a diverse range of low molecular weight compounds that include the flavonoids, hydroxycinnamates, glucosinolates, and allylsulfides. Consumption of plant foods is associated with a reduced risk of chronic diseases such as cardiovascular disease and cancer, as demonstrated by the results of epidemiological cohort and case-control studies [1,2]. It is likely that at least part of the protection from plant foods results from the action of the phytochemical component [3]; this may arise through mechanisms involving (direct or indirect) antioxidant activity, modulation of enzyme activity (including inhibition of phase I enzymes and induction of phase II enzymes), and modulation of gene expression [4,5].

In addition to this epidemiological data, the putative evidence for the biological activity of phytochemicals arises from animal studies, cell and tissue culture studies, and in vitro experiments. Frequently these studies are carried out with the phytochemicals of interest not in the form present in the diet, as metabolites (e.g., nonglycosylated or nonconjugated forms), and/or at concentrations considerably higher than those obtainable through a normal diet. Not only could such factors have a profound effect on the interpretation of the results from in vitro studies, but experimentation with nonphysiologically relevant compounds may also influence the rate and extent of absorption and the range of metabolites produced in the animal and cell culture studies. Furthermore, model systems may display qualitatively and quantitatively different metabolic pathways for phytochemicals when compared to their behavior in humans. It is also worth noting that there is considerable inter-individual variation in metabolism of phyto-chemicals due to enzyme polymorphisms and variations in microbial populations; metabolism may also be influenced by other dietary components when the phyto-chemicals are included in complex mixtures (as is the case with whole food experiments). All of these factors should be taken into account when assessing the potential bioactivity of phytochemicals.

Metabolism of phytochemicals can occur throughout the entire gastrointes-tinal tract, for example: (1) in the mouth, by action of resident microflora or secreted salivary enzymes, (2) in the stomach, due to its acidic environment, (3) within the small or large intestinal lumen, by action of pancreatic, brush-border, or microbial enzymes, (4) during passage across the enterocytes, by endogenous phase I and phase II enzymes, (5) in the liver, by hepatic phase I and phase II enzymes, or (6) by phase I and phase II enzymes at various tissues within the body (Fig. 1). The most important sites for metabolism of phytochemicals are expected to be the gut (via endogenous and microbial enzymes) and liver. Metabo-lism can involve changes in the functional groups present, partial or complete breakdown of the compound, or conjugation with another molecule.

Metabolism will modify or completely alter the physico-chemical properties of the parent compound: this, in turn, may affect biological activity, the ability to interact with cell transporters, receptors, enzymes and plasma proteins, tissue distribution and half-life, and excretion rate and pathway. However, metabolism is frequently neglected by researchers, both in in vitro experiments and when analyzing phytochemicals in biological fluids and tissues. This neglect is often a result of necessity, for example, to increase the concentration to more favorable levels required for accurate analysis, or because the compounds of interest (either in the form found in the food or as metabolites) are not available in the quantities demanded for such studies. However, the question of metabolism cannot be ig-nored if the potential mechanisms by which phytochemicals generate bioactivity are to be understood and most effectively exploited. This chapter will summarize the range of metabolites that have been characterized from an important class of dietary phytochemicals, the polyphenols. Information generated from their metabolism may have potential for our understanding of the behavior of other phytochemicals in similar systems.

## II. POLYPHENOL METABOLISM

Polyphenols include flavonoids, proanthocyanidins, stilbenes, microbial metabo-lites of lignan, and hydroxycinnamates (Fig. 2). Flavonoid metabolism, while still far from being fully understood, has been the most widely studied and will therefore form the basis of this chapter. Six main subclasses of flavonoids are widely consumed by humans: flavonols, flavones, flavanones, isoflavonoids, fla-

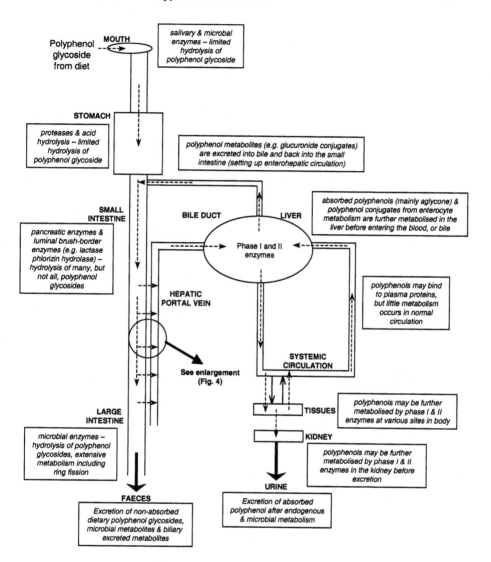

**Figure 1** Scheme showing sites of metabolism of dietary polyphenol glycosides in humans. Dotted arrows represent pathways followed by the polyphenols.

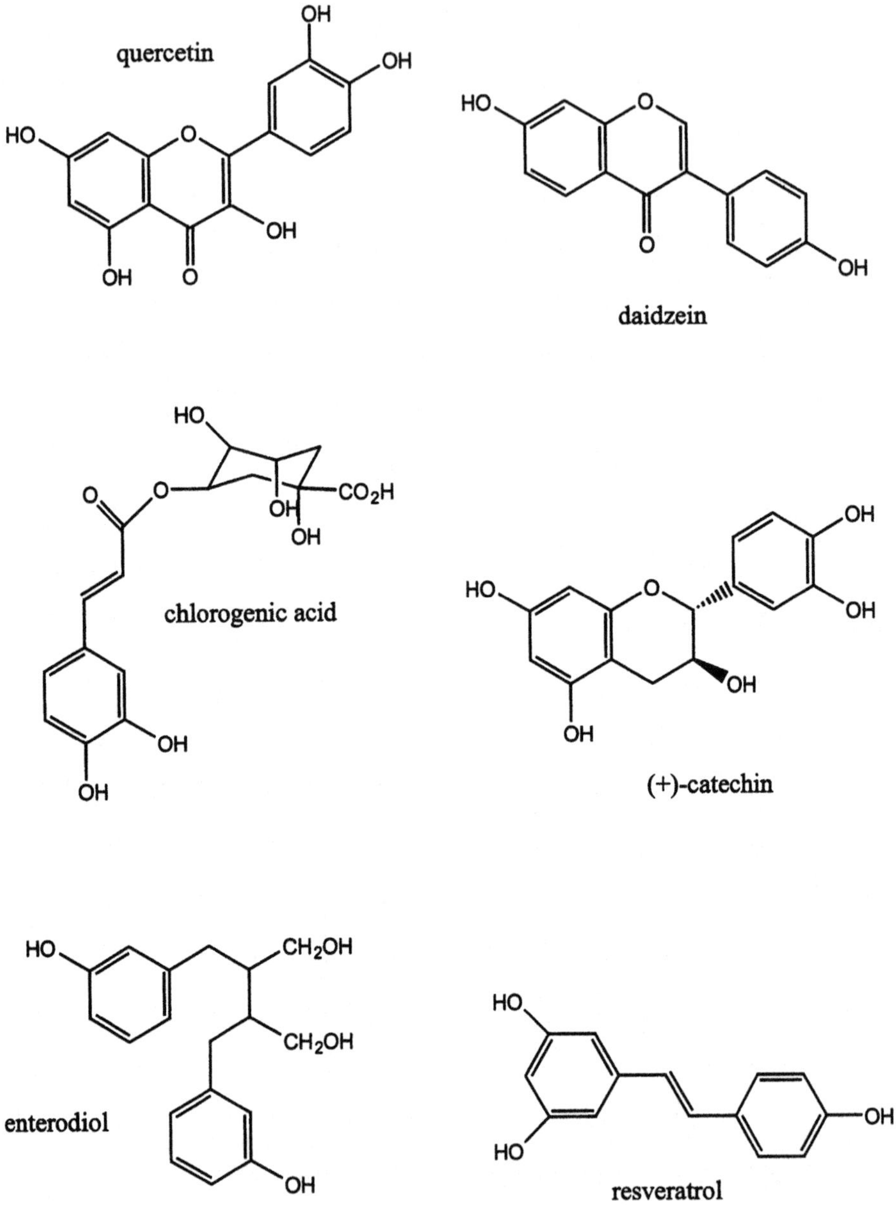

**Figure 2** Chemical structures of selected plant polyphenols. Structures include a flavo-nol (quercetin), isoflavone (daidzein), cinnamic acid (chlorogenic acid), flavan-3-ol (cate-chin), a lignan microbial metabolite (enterodiol), and a stilbene (resveratrol).

vanols (catechins), and anthocyanins; these posses the generic structure shown in Fig. 3. These classes differ in the degree of saturation and the nature and position of reactive groups on their three rings: examples of substitution patterns for selected flavonoids are given in Table 1.

With the exception of catechins, flavonoids in nature are almost always found in a glycosylated form, i.e., conjugated to a sugar. Aglycones are not usually found in plant food, but processing the plant food (e.g., fermentation) may increase the level of aglycone (as in the case of wine and miso) [6,7]. Flavonols are found in nearly all fruits and vegetables, with quercetin (glycosides) the most abundant in the diet. Flavones are found in fruit, vegetables, and herbs; flavanones in citrus fruits; isoflavonoids in leguminous plants such as soybean; catechins are notable for their presence in tea and are also widely found in fruits; and anthocyanins are natural colorants in berry fruits and red wine.

## A. Absorption

In order to elucidate polyphenol metabolic pathways, the primary site for metabolism needs to be addressed. The intestinal microflora will extensively metabolize (and even completely degrade) most compounds arriving in the colon; however, early absorption from the gastrointestinal tract, in the stomach or small intestine, will result in a differing metabolite profile. Therefore, the mechanism of absorption of dietary polyphenols is inextricably linked to characterization of the resulting metabolites within the systemic circulation or individual tissues.

The pharmacokinetic profiles of plasma quercetin following administration of quercetin-3- or -4′-glucosides and quercetin-3-rhamnoglucoside (rutin) are very different; the monoglucosides are rapidly absorbed while the diglycoside is absorbed at a time corresponding to its presence in the large intestine [8]. The

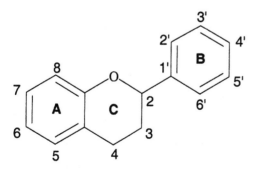

**Figure 3** The generic structure of flavonoids.

**Table 1**   Substitution Patterns of Selected Flavonoids

| Subclass | Flavonoid | Substitution pattern |
|---|---|---|
| Flavonol | Kaempferol | OH: 3, 5, 7, 4' |
|  | Quercetin | OH: 3, 5, 7, 3', 4' |
|  | Myricetin | OH: 3, 5, 7, 3', 4', 5' |
| Flavone | Chrysin | OH: 5, 7 |
|  | Apigenin | OH: 5, 7, 4' |
|  | Luteolin | OH: 5, 7, 3', 4' |
|  | Tangeretin | $OCH_3$: 5, 6, 7, 8, 4' |
| Flavanone | Naringenin | OH: 5, 7, 4' |
|  | Eriodyctiol | OH: 5, 7, 3', 4' |
| Isoflavone | Daidzein | OH: 7, 4' |
|  | Genistein | OH: 5, 7, 4' |
|  | Biochanin A | OH: 5,7; $OCH_3$: 4' |
| Flavan-3-ol | Catechin | OH: 3, 5, 7, 3', 4' |
|  | Epicatechin | OH: 3, 5, 7, 3', 4' |
|  | Epigallocatechin | OH: 3, 5, 7, 3', 4', 5' |
|  | Epicatechin gallate | OH: 3, 5, 7, 3', 4'; O-galloyl: 3 |
|  | Epigallocatechin gallate | OH: 3, 5, 7, 3', 4', 5'; O-galloyl: 3 |

Differences between flavonoid subclasses: flavonols have a 2–3 double bond, 3-hydroxyl and 4-ketone group; flavones have a 2–3 double bond and 4-ketone group; flavanones have a 4-ketone group; isoflavones have a 4-ketone group and have the B-ring attached at the 3-position; flavan-3-ols are hydroxylated or galloylated at the 3-position.

mechanism of absorption for the monoglucosides is also substantially different to that of rutin, and is reflected by the plasma concentrations of quercetin, which are higher for the monoglucosides, and the amount excreted, which is significantly lower for rutin [9]. This can be explained by the inability of rutin to be absorbed in the small intestine due to lack of a specific transporter and because it is not a substrate for brush-border enzymes [10]. The colon microflora will, therefore, substantially metabolize rutin, thus reducing the amount of quercetin available for absorption. In contrast, quercetin-3- and -4'-glucosides are absorbed in the small intestine, either via the sodium-dependent glucose transporter, SGLT1, or after glucoside hydrolysis by lactase phlorizin hydrolase, LPH [11], and plasma concentrations of quercetin metabolites will, therefore, be relatively high.

Whether flavonoid glycosides are absorbed intact has been a matter of controversy [12]. However, several researchers have now shown that quercetin glucosides cannot be found in human plasma after subjects consumed meals rich in flavonol glucosides [13–17] or supplements containing quercetin-4'-glucoside or rutin [14], quercetin-3- or -4'-glucoside [18], or rutin [19]; such results provide evidence for pre- or post-absorption deglycosylation. Furthermore, both Graefe et al. [14] and Sesink et al. [18] observed similar qualitative metabolic profiles in subjects after consumption of quercetin glycoside derived from different sources. Likewise, Morand et al. [20] found a similar quercetin metabolic profile after rats were fed quercetin, rutin, or quercetin-3-glucoside. These data suggest that postdeglycosylation quercetin follows the same metabolic pathway regardless of administered form of quercetin glycosides, although different isoforms of the phase II conjugating enzyme UDP-glucuronosyltransferase may have been involved. However, quantitative differences do exist as the amount absorbed depends on the site of absorption from the gastrointestinal tract.

The isoflavone genistin, genistein-7-glucoside, was not found to cross the enterocyte into the blood when investigations were conducted using the bile duct-cannulated rat model [21]; genistin was infused into the femoral and portal veins and its metabolites characterized in the bile. Approximately 20% of the intravenously infused genistin was transported into bile without modification, the remainder being hydrolyzed to genistein and subsequently conjugated with glucuronic acid. Similar results were found following administration to either the femoral or portal veins, indicating that biliary genistin was a marker of genistin entering the blood stream. When the same dose was infused into the small intestine in the rat model, no genistin was detected in bile. Although indicating that genistin does not cross the enterocyte, this does not exclude the possibility that the compound is taken up by the enterocyte and hydrolyzed and glucuronidated before being transferred basolaterally into the blood.

Deglycosylation of flavonoid glycosides may result from the action of (1) cytosolic β-glucosidase within the enterocyte [22], (2) LPH, present on the luminal side of the brush-border and shown to be active on a range of flavonoid glycosides [10,11], or (3) microbial β-glycosidases (Fig. 4). The human gut microflora are concentrated (approximately $10^{12}$ microorganisms/ g) mainly within the colon, with relatively few microorganisms being found in the small intestine. Many colonic microorganisms exhibit some β-glycosidase activity, contributing to the hydrolysis of the O-glycosidic (but not the C-glycosidic) bond, although some microorganisms will also metabolise the flavonoids further (see below).

## B. Enterohepatic Circulation

Absorbed flavonoids are metabolized by the enterocytes during passage across the small intestine and then further metabolized by the liver before entering the systemic circulation or being excreted into the bile. Bile is secreted directly into the upper small intestine through the gallbladder and compounds in the bile may, therefore, be reabsorbed. This continuous cycle is referred to as enterohepatic circulation (see Fig. 1). Evidence from animal studies suggests that flavonoid

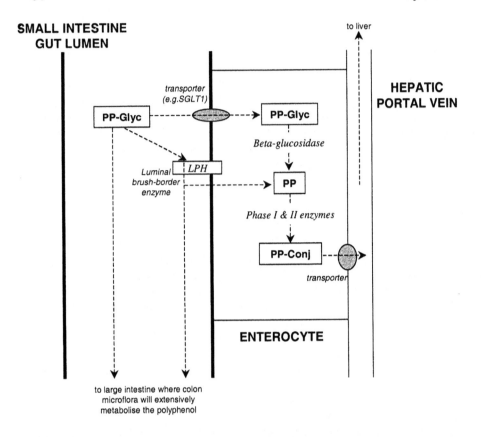

**Figure 4**   Enlargement of intestinal brush border, showing that polyphenol glycosides are not absorbed, transported into the enterocyte by a transporter, or hydrolyzed by a brush-border enzyme. Dotted arrows represent pathways followed by the polyphenols. PP-Glyc are polyphenol glycosides, PP-Conj are polyphenol conjugates, SGLT1 is the sodium-dependent glucose transporter, and LPH is lactase phlorizin hydrolase.

metabolites are likely to be excreted through the bile; thus Sfakianos et al. [23] showed that 75% of genistein-7-glucuronide was excreted in the bile within 4 hours after infusion of genistein into the intestine. As the conjugated metabolites derived from the bile are hydrophilic, they are unlikely to be absorbed back across the intestinal epithelium. Only hydrophobic compounds can readily diffuse across biological membranes, so hydrophilic compounds generally require an active transport mechanism. Specific transporters have not been identified for typical phase II metabolites (e.g., glucuronide conjugates); hence, flavonoid metabolites will continue along the gastrointestinal tract until they reach the microflora in the distal parts of the small intestine or the colon. The vast array of microorganisms in the colon constitutes enormous catalytic power, readily hydrolyzing the conjugated forms. The released aglycone may then either be reabsorbed or undergo further metabolism (including ring fission).

## C.  Phase I Metabolism

Once absorbed from the intestine, flavonoids will be subjected to metabolism involving phase I or phase II enzymes. Hydroxylation and demethylation are examples of cytochrome P450 mono-oxygenase–dependent (phase I) activities that may be involved in the metabolism of flavonoids. Nielsen et al. [24] showed that microsomes prepared from normal and Aroclor-treated rats were capable of hydroxylating certain flavonols and flavones. The requirement for metabolic activity was that one or no hydroxyl groups were present in the B-ring; two or more hydroxyl groups in the B-ring prevented further hydroxylation. Demethylation in the same system was observed when a methyl group was present in the 4'-, but not in the 3'-, position. Human breast cancer cell lines [25] exhibited cytochrome P450 activity, leading to the conversion of biochanin A (4'-methoxygenistein) to genistein. When phase I metabolism was investigated in vivo in rats, phase I metabolites were only observed for highly methylated flavonoids such as tangeretin [26]. There was no evidence for conversion of quercetin to either myricetin (introduction of a 5'-hydroxyl group) or kaempferol (removal of the 3'-hydroxyl), which was expected given that quercetin already has two hydroxyl groups in the B-ring (Table 1). However, 4'-monohydroxylated flavonoids, such as apigenin and naringenin, were also not found as hydroxylated metabolites in urine [27], which was surprising based on the results from the rat microsome study described above. Galijatovic et al. [28] found no evidence of phase I metabolism of chrysin and apigenin (flavones with zero or one hydroxyl in the B-ring, respectively) in either 3-methylcholanthrene–induced HepG2 cells or primary rat hepatocytes, although phase II metabolism occurred rapidly.

No evidence for P450-mediated hydroxylation of apigenin was found in a human intervention study, where only the parent compound and not luteolin (possessing two hydroxyl groups in the B-ring) was excreted after consumption of parsley [29]. Human studies involving naringenin provided no evidence of eriodictyol (with two hydroxyl groups in the B-ring) in the urine after consumption of either pure naringin (the rhamnoglucoside present in grapefruit juice) or grapefruit juice [30]. Chrysin was conjugated to sulfate and glucuronide in human plasma and urine after a supplement was consumed but, in agreement with the results from cell culture model experiments, no hydroxylation to apigenin occurred [31]. These results suggest that most dietary flavonoids do not undergo phase I metabolism in vivo to a significant degree, possibly because the hydroxyl groups already exposed on the flavonoid provide sites are readily susceptible to phase II conjugation.

## D. Phase II Metabolism

Metabolism of absorbed flavonoids involves conjugation, mainly with glucuronide and sulfate, although methylation of the catechol group occurred to a limited extent. Conjugation is a common detoxification reaction, which reduces the number of reactive hydroxyl groups, leading to the increased solubility and molecular weight that is necessary for biliary or urinary excretion. Glucuronidation requires the enzyme UDP-glucuronosyltransferase (UGT) and the cofactor UDP-glucuronic acid; the latter is abundant within cells and ensures that the conjugation is unlikely to become saturated at high concentrations. Sulfotransferase requires the cofactor 3'-phosphoadenosine-5'-phosphosulfate (PAPS), which can be in limited supply in cells; sulfation may thus tend to predominate only at low polyphenol concentrations. However, as flavonoids from the diet are present at relatively low concentrations, the level of expression of conjugating enzymes in the small intestine (varying between individuals) may be the determining factor for which metabolites are formed, rather than any saturation of individual conjugating enzymes.

Sulfation of flavonoids may occur, although the level of sulfotransferase activity may be dependent on the model system being used. For example, phenol sulfotransferases show sexual dimorphism in rodents, with higher activities being found in male animals [32]; Moreover, these enzymes are absent in pigs [33]. Sulfation of isoflavones is prominent in some, but not all, human breast cancer cell lines [25,34], and HepG2 have higher levels of sulfotransferase activity than do normal human hepatocytes [35,36]. Methylation by catechol-O-methyltransferase also occurs for flavonoids such as quercetin, which possess a catechol functional group. There is no evidence for methylation at any other hydroxyl group on the flavonoid structure.

Boutin et al. [37] found that the 5-hydroxyl group of flavonoids was generally unavailable for conjugation unless it was the only possible site available. Conjugation to glucuronic acid occurred at one or more of four other positions on quercetin [38,39], with UDP-glucuronossyltransferase showing the highest affinity for the 4'-position using liver cell-free extracts [38]. For kaempferol only two monoglucuronides were detected after incubation with UDP-glucuronosyltransferase [39]. Quercetin-3-glucuronide, 3'-methylquercetin-3-glucuronide, and quercetin-3'-sulfate were found to be the major metabolites in human plasma after quercetin-rich onions were consumed [13]. Other quercetin- and methylquercetin metabolites were not fully characterized, although it is likely that both diglucuronides and mixed glucuronides-sulfate conjugates were present, as shown earlier by Manach et al. [40]. Quercetin-3-glucuronide was confirmed in rat plasma after quercetin was fed to rats [41].

Conjugation of flavonoids can occur at any single hydroxyl group on the polyphenolic structure or at multiple sites. The position of conjugation can greatly affect the bioactivity of the resulting conjugate, reducing the antioxidant capacity, for example, by conjugating with the catechol group on the B-ring of quercetin (the catechol group having been shown to greatly increase the antioxidant capacity in structure-function studies [42]), or reducing the ability of the compound to interact with receptors, for example, by conjugating with the 7-hydroxyl group of isoflavones that is necessary for binding to estrogen receptors $\alpha$ and $\beta$. The bioactivity of flavonoid metabolites is altered compared to the aglycone, but in different assays individual metabolites exhibited significant activity. For example, 3′-conjugated quercetin inhibited the action of xanthine oxidase to a similar extent as quercetin [38], whereas quercetin-3-glucuronide was almost as good as quercetin at delaying oxidation of copper-induced LDL [40]. It is also possible that other metabolites may have enhanced activity at certain sites; thus equol, a microbial metabolite of daidzein, is better at binding to estrogen receptor than is the parent isoflavone [42].

O-Methoxylation appears to play a more prominent role in the metabolism of quercetin in rats compared to humans. Both 3′- and 4′-methylquercetin have been found in rat urine, plasma, and bile; the 3′-position appearing to be the preferential site of catechol O-methyltransferase activity with a ratio of >2:1 [44,45]. 4′-Methylquercetin appears to decline more rapidly than the 3′-methylquercetin with little detectable in plasma after 12 hours; only 3′-methylquercetin was identified in the urine or bile of rats after prolonged feeding [46]. Nielsen et al. [27] suggested that active demethylation of 4′-methylquercetin by P450 enzymes (phase I activity) may be responsible for the accumulation of 3′-methylquercetin conjugates observed over time. Methylation appears only to be a minor metabolic pathway of quercetin in humans compared to rats, accounting for only about 20–30% of its total metabolites [38,40]. Most human studies measuring quercetin have only detected the 3′-methylated derivative [47], although DuPont et al. [48] recently reported the presence of 4′-methylquercetin after low levels of quercetin from cider were consumed. 3′-Methylquercetin and 4′-methylquercetin have similar hydrophobicity and hence chromatographic properties, thus without an adequate solvent gradient the two compounds may not be resolved by HPLC, especially at the low levels often encountered. However, the presence of 4′-methylquercetin in plasma after consumption of cider may be a result of the different dietary form given to subjects or a low amount of quercetin consumed (compared to other human intervention studies) altering the reactivity to catechol-O-methyltransferase.

The four main tea catechins (flavan-3-ols) are epicatechin, epigallocatechin, epicatechin-3-gallate, and epigallocatechin-3-gallate; all were found conjugated to glucuronides or sulfate in the urine and plasma after green tea consumption in a human study [49]. Methylation of the catechol group was found to occur rapidly (within 0.25 h), with epicatechin metabolized to 3'- and 4'-methylepicatechin, epigallocatechin methylated in the 4'-position, and a methylated epigallocatechin gallate identified. All these methylated compounds were also conjugated to glucuronides or sulfate. 4'-Methyl-epigallocatechin was the major metabolite, with a longer half-life than that of the nonmethylated parent compound (4 h vs. 1 h) and a 4–6 times higher maximum plasma concentration, again demonstrating the importance of taking the metabolites into consideration when conducting human intervention studies. In rat plasma the main metabolite of epicatechin was found to be 3'-methyl-epicatechin-5-glucuronide [50].

Hepatocytes possess β-glucuronidase activity [51], and O'Leary et al. [36] have carried out experiments with HepG2 cells that show quercetin glucuronides to be hydrolyzed with a transient release of aglycone and then reconjugated (to form other metabolites) within the liver. Inflammatory cells (e.g., neutrophils) also produce β-glucuronidase; these and neutrophil-like HL-60 cells were both capable of hydrolyzing luteolin glucuronides when the cells were stimulated to have a respiratory burst [52]. It is thus possible that flavonoid glucuronides from the circulation could be hydrolyzed at sites of inflammation, releasing the possibly more potent aglycone in localized vicinity.

Another type of polyphenol metabolism has been described by Boersma et al. [53]. Inflammatory cells produce high concentrations of the hypohalogenous acids HOCl and HOBr from $H_2O_2$ that may subsequently chlorinate or brominate tyrosine residues in proteins of surrounding tissues. The isoflavones daidzein, genistein, and biochanin A all reacted with HOCl in vitro to form a mixture of mono- and dichlorinated isomers. Binsack et al. [54] found that quercetin was similarly converted to 6-chloro- and 6,8-dichloroquercetin. The formation of these halogenated flavonoids can occur when neutrophils or HL-60 cells [55] are stimulated to have a respiratory burst. Such halogenated flavonoid metabolites have not be measured in vivo, but they may be formed, and have significant biological activity, at sites of inflammation [56].

## E.  Microbial Metabolism

It has been estimated that $10^7–10^9$ bacteria/g dry mass are able to utilize quercetin-3-glucoside as an energy source either by releasing quercetin or by producing lower molecular weight phenols such as 3,4-dihydroxyphenylacetic acid through ring fission [57]. Figure 5 shows an example of quercetin glucoside microbial metabolites. Isoflavones, somewhat surprisingly, appear to be more resistant to ring fission, but are capable of further metabolism. The majority of daidzein (60%) is excreted in urine without ring modification [58], although it may be reduced to equol (which may have a similar, or increased, phytoestrogenic activity

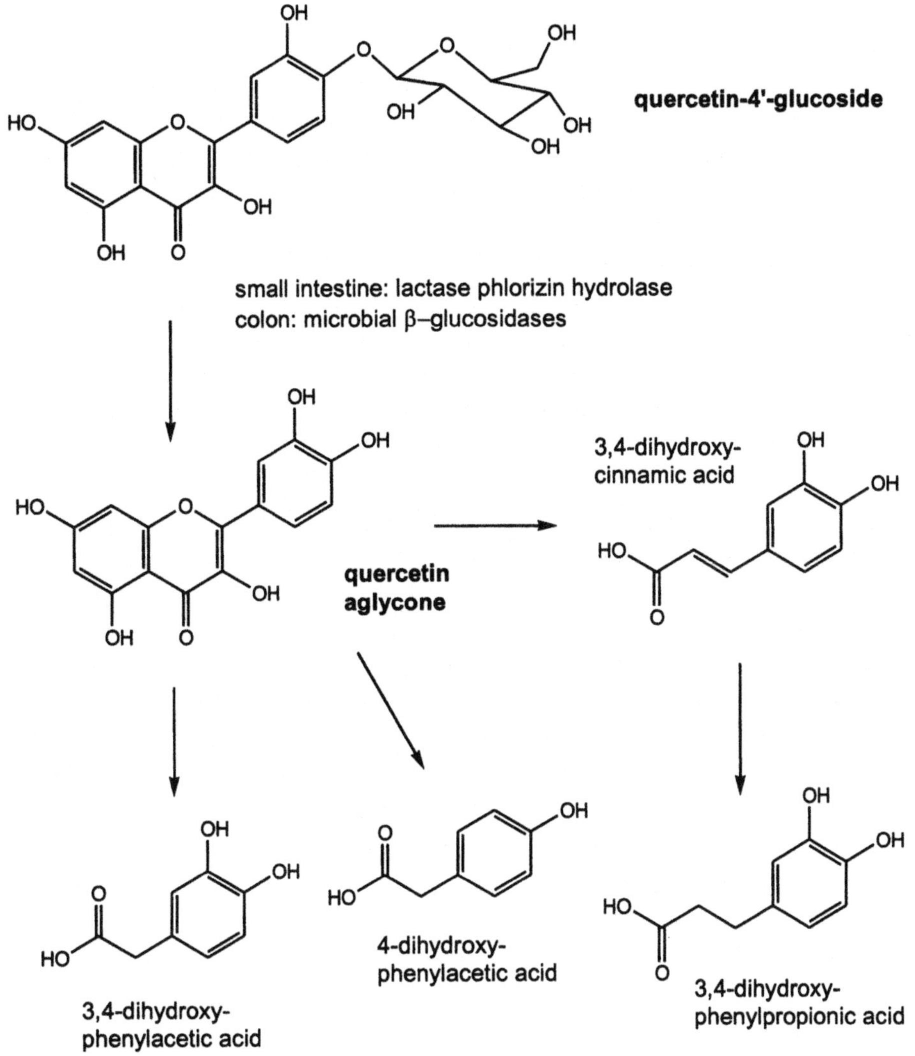

**Figure 5** Microbial metabolism of quercetin-4′-glucoside.

compared to the parent compound, depending on the analytical method employed), or dihydrodaidzein. Alternatively, daidzein could be metabolized to O-desmethylangolensin through ring opening or 4-ethylphenol by ring cleavage (Fig. 6). Genistein tends to undergo a greater degree of ring modification, with only 20% excreted in the urine with an unmodified ring structure. The major metabolite of genistein in rats was found to be 2-(4-hydroxyphenyl)propionic acid, a ring-cleaved product [59]. As these products all require microbial metabolism for formation, uptake of the flavonoids may be strongly influenced by the composition of the gut bacteria, which is known to vary significantly between individuals. Flavonoids, or other compounds in the food matrix from which they are derived, may themselves regulate the composition of the microflora. Furthermore, habitual diet will play a role in maintenance of the microflora composition [60], influencing both the rate of deglycosylation and the extent of further flavonoid metabolism and degradation.

Biliary excretion and enterohepatic circulation ensures that flavonoids are continually subjected to further metabolism and degradation by colonic microflora. Only 50% of a dose of rutin was excreted in the urine after 48 h [61]; less than 1% was in the form of its aglycone (quercetin), with the remainder being characterized as phenolic acids. In a recent study involving administration of $^{14}$C-quercetin to human volunteers, Walle et al. [62] showed that up to 81% of the oral dose was recovered as carbon dioxide in a 72-h period; only 2–5% was excreted in the faeces, with 3–6% excreted in the urine. These studies demonstrate that quercetin undergoes extensive intestinal, bacterial, and systemic metabolism, resulting in a complex mixture of metabolites with altered bioactivity.

## III.  CHARACTERIZATION OF POLYPHENOL METABOLITES

Metabolites of polyphenols can be produced and characterized using different model systems, each having advantages and disadvantages, but all generating information that may help elucidate the true metabolites in vivo—information that is vital for determining putative biological activity. In the most simple form of model, tissue homogenate extracts can be used to identify enzyme activity towards specific polyphenols. Kinetics, rates of reactions, and competition can be measured to determine likely metabolites. Absence of activity in controlled systems would suggest (but no more) that the reaction is unlikely to occur in vivo. Conjugation pathways are both competitive and complementary, and, given current levels of understanding, monitoring specific enzyme activities of tissue extracts cannot provide an overall picture of metabolism. Furthermore, metabolism is dependent on the ability of a compound to cross the cell membrane and reach the appropriate conjugating enzymes within cells. Cell culture can provide

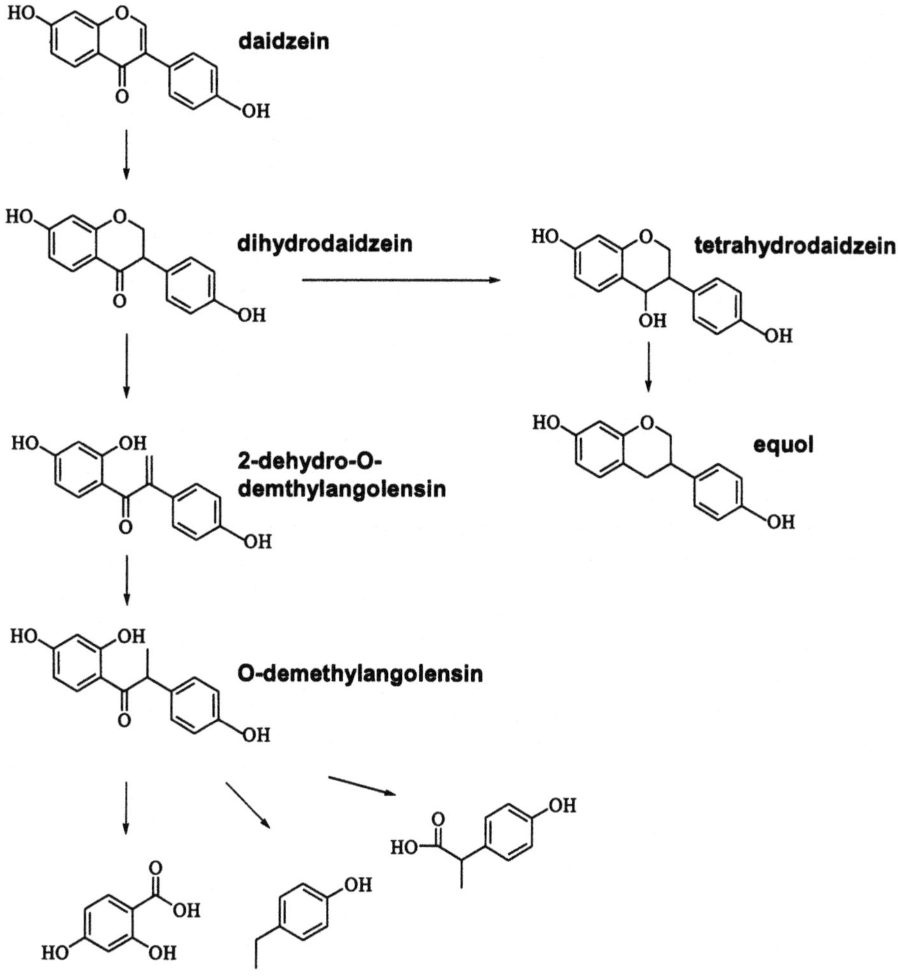

**Figure 6** Microbial metabolism of daidzein-7-glucoside.

additional evidence for the types of polyphenol metabolites that could be expected in vivo. However, depending on the cell type and passage number, enzyme expression may be significantly different to normal cells within living tissues. For example, Caco-2 cells are often used to measure the ability of a compound to be absorbed across the small intestine, but these cells have a very low expression of LPH [63], and so deglycosylation of polyphenol glucosides (a route for absorption of some compounds [9]) will not occur.

Primary hepatocytes or liver slices can be used to measure metabolism, but only for a short period of time after the liver sample has been removed from the body; however, both models have problems associated with their use. In liver slices, cell-to-cell contact and three-dimensional structure are maintained with a full compliment of cell types (including Kupffer cells); primary hepatocytes have lost the orientation, organization, and nonhepatocyte cells which may contribute to the metabolic activity of the whole liver. Liver slices may suffer from the presence of damaged or dead cells, restricted access of culture media to internal cells, thereby reducing oxygen and nutrient supplies, and from a build-up of toxic products that may result in impaired metabolism. Perfusion of tissue in situ can ameliorate these problems, but of necessity such experiments can normally only be carried out in animals, and these will have different metabolic profiles due to differences in enzyme and transporter expression.

Human intervention studies can provide plasma and urinary metabolites, but these can be difficult to characterise due to lack of sample, or low (often meaning ''normal'') levels of compound administered. Furthermore, information on metabolism (and mechanisms) within cells or inaccessible tissues will restrict the information generated. Metabolism is a dynamic process; in order to fully characterize polyphenol metabolites in particular and phytochemical metabolites in general, there is no substitute at present for detailed studies utilizing advanced and validated analytical techniques, such as LC-MS/MS and LC-NMR-MS, that take account of dietary intake in a normal range and operate within the limitations of the experimental system under study.

## IV. CONCLUSIONS

The history of the characterization of plant secondary metabolites has been intimately associated with developments in chemical analysis: extraction, acid hydrolysis, derivatization (as appropriate), and quantification. Such an approach—essential at the time—has had the unfortunate consequences that (1) biologically important information has been lost in the hydrolysis and (2) compounds that may have similar chemical structures and reactivities but different bioactivities have been grouped together without differentiation. Now that interest has shifted from plant composition to diet–health relationships, it is clear that new approaches to analytical methodology are required, that the behavior of *individual compounds* needs to be considered, and that a full understanding of the processes of absorption and metabolism is necessary. In this way, the contributions of dietary polyphenols to health will become clearer, notwithstanding the significant complications of genetic and environmental variability in and between individuals. While significant progress has been made, it is also clear that yet greater effort lies ahead.

## REFERENCES

1. Block G, Patterson B, Subar A. Fruit, vegetables and cancer prevention: a review of the epidemiological evidence. Nutr Cancer 1992; 18:1–29.
2. Steinmetz KA, Potter JD. Vegetables, fruit, and cancer prevention: a review. J Am Diet Assoc 1996; 96:1027–1039.
3. Knekt P, Kumpulainen J, Jarvinen R, Rissanen H, Heliovaara M, Reunanen A, Hakulinen T, Aromaa A. Flavonoid intake and risk of chronic diseases. Am J Clin Nutr 2002; 76:560–568.
4. Nijveldt RJ, van Nood E, Boelens PG, Norren K. van Leeuwen PAM. Flavonoids: a review of probable mechanisms of action and potential applications. Am J Clin Nutr 2001; 74:418–425.
5. Yang CS, Landau JM, Huang MT, Newmark HL. Inhibition of carcinogenesis by dietary polyphenolic compounds. Annu Review Nutr 2001; 21:381–406.
6. Crozier A, Burns J, Aziz AA, Stewart AJ, Rabiasz HS, Jenkins GI, Edwards CA, Lean MEJ. Antioxidant flavonols from fruits, vegetables and beverages: measurements and bioavilability. Biol Res 2000; 33:79–88.
7. Wang HJ, Murphy PA. Isoflavone content in commercial soybean foods. J Agr Food Chem 1994; 42:1666–1673.
8. Olthof MR, Hollman PCH, Vree TB, Katan MB. Bioavailabilities of quercetin-3-glucoside and quercetin-4'-glucoside do not differ in humans. J Nutr 2000; 130:1200–1203.
9. Hollman PCH, Buysman MNCP, van Gameren Y, Cnossen EPJ, deVries JHM, Katan MB. The sugar moiety is a major determinant of the absorption of dietary flavonoid glycosides in man. Free Rad Res 1999; 31:569–573.
10. Day AJ, Cañada FJ, Diaz FC, Kroon PA, Mclauchlan R, Faulds C, Plumb GW, Morgan MRA, Williamson G. Role of lactase phlorizin hydrolase in the deglycosylation of dietary flavonol and isoflavone glycosides. FEBS Lett 2000; 478:71–75.
11. Day AJ, Gee JM, DuPont MS, Johnson IT, Williamson G. Absorption of quercetin-3-glucoside and quercetin-4'-glucoside in the rat small intestine: the role of lactase phlorizin hydrolase and the sodium-dependent glucose transporter. Biochem Pharmacol 2003; 65:1199–1206.
12. Day AJ, Williamson G. Biomarkers for exposure to dietary flavonoids: a review of the current evidence for identification of quercetin glycosides in plasma. Br J Nutr 2001; 86:S105–S110.
13. Day AJ, Mellon F, Barron D, Sarrazin G, Morgan MRA, Williamson G. Human metabolism of dietary flavonoids: identification of plasma metabolites of quercetin. Free Rad Res 2001; 35:941–952.
14. Graefe EU, Wittig J, Mueller S, Riethling AK, Uehleke B, Drewelow B, Pforte H, Jacobasch G, Derendorf H, Veit M. Pharmacokinetics and bioavailability of quercetin glycosides in humans. J Clin Pharmacol 2001; 41:492–499.

15.  Moon J-H, Nakata R, Oshima S, Inakuma T, Terao J. Accumulation of quercetin conjugates in blood plasma after short-term ingestion of onion by women. Am J Physiol 2000; 279:R461–R467.

16.  Walle T, Otake Y, Walle UK, Wilson FA. Quercetin glucosides are completely hydrolyzed in ileostomy patients before absorption. J Nutr 2000; 130:2658–2661.

17.  Wittig J, Herderich M, Graefe EU, Veit M. Identification of quercetin glucuronides in human plasma by high-performance liquid chromatography-tandem mass spectrometry. J Chrom B 2001; 753:237–243.

18.  Sesink ALA, O'Leary KA, Hollman PCH. Quercetin glucuronides but not glucosides are present in human plasma after consumption of quercetin-3-glucoside or quercetin-4'-glucoside. J Nutr 2001; 131:1938–1941.

19.  Erlund I, Kosonen T, Alfthan G, Maenpaa J, Perttunen K, Kenraali J, Parantainen J, Aro A. Pharmacokinetics of quercetin from quercetin aglycone and rutin in healthy volunteers. Eur J Clin Pharmacol 2000; 56:545–553.

20.  Morand C, Manach C, Crespy V, Remesy C. Quercetin 3-O-beta-glucoside is better absorbed than other quercetin forms and is not present in rat plasma. Free Rad Res 2000; 33:667–672.

21.  Barnes S, Xu J, Smith M, Kirk M, Lack of evidence for the intestinal absorption of isoflavone β-glucosides in the intact rat. 221st American Chemical Society National Meeting, April 1–5, 2001, San Diego, CA.

22.  Day AJ, DuPont MS, Ridley S, Rhodes M, Rhodes MJC, Morgan MRA, Williamson G. Deglycosylation of flavonoid and isoflavonoid glycosides by human small intestine and liver β-glucosidase activity. FEBS Lett 1998; 436:71–75.

23.  Sfakianos J, Coward L, Kirk M, Barnes S. Intestinal uptake and biliary excretion of the isoflavone genistein in rats. J Nutr 1997; 127:1260–1268.

24.  Nielsen SE, Breinholt V, Justesen U, Cornett C, Dragsted LO. In vitro biotransformation of flavonoids by rat liver microsomes. Xenobiotic 1998; 28:389–401.

25.  Peterson TG, Coward L, Kirk M, Falany CN, Barnes S. Isoflavones and breast epithelial cell growth: the importance of genistein and biochanin A metabolism in the breast. Carcinogenesis 1996; 17:1861–1869.

26.  Nielsen SE, Breinholt V, Cornett C, Dragsted LO. Biotransformation of the citrus flavone tangeretin in rats. Identification of metabolites with intact flavane nucleus. Food Chem Toxicol 2000; 38:739–746.

27.  Nielsen SE. Metabolism and biomarker studies of dietary flavonoids, Ph.D. dissertation. Danish Veterinary and Food Administration, Denmark, 1998.

28.  Galijatovic A, Otake Y, Walle UK, Walle T. Extensive metabolism of the flavonoid chrysin by human Caco-2 and Hep G2 cells. Xenobiotica 1999; 29:1241–1256.

29.  Nielsen SE, Young JF, Daneshvar B, Lauridsen ST, Knuthsen P, Sandstrom B, Dragsted LO. Effect of parsley (*Petroselinum crispum*) intake on urinary apigenin excretion, blood antioxidant enzymes and biomarkers for oxidative stress in human subjects. Br J Nutr 1999; 81:447–455.

30.  Ameer B, Weintraub RA, Johnson JV, Yost RA, Rouseff RL. Flavanone absorption after naringin, hesperidin and citrus administration. Clin Pharm Ther 1996; 60:34–40.

31.  Walle T, Otake Y, Brubaker JA, Walle UK, Halushka PV. Disposition and metabolism of the flavonoid chrysin in normal volunteers. Br J Clin Pharmacol 2001; 51: 143–146.

32. Klaassen CD, Liu L, Dunn RT. Regulation of sulfotransferase mRNA expression in male and female rats of various ages. Chem Biol Interactions 1998; 109:299–313.

33. Gibson GG, Skett P. Factors affecting drug metabolism: internal factors In:. Introduction to Drug Metabolism. London: Blackie Academic and Professional, 1994: 107–132.

34. Peterson TG, Ji G-P, Kirk M, Coward L, Falany CN, Barnes S. Metabolism of the isoflavones genistein and biochanin A in human breast cancer cell lines. Am J Clin Nutr 1998; 68:1505–1511.

35. Day AJ. Human Absorption and Metabolism of Flavonoid Glycosides, Ph.D. dissertation University of East Anglia, UK, 2000.

36. O'Leary KA, Day AJ, Needs PW, Mellon FA, O'Brien NM, Williamson G. Metabolism of quercetin-7- and quercetin-3-glucuronides by an in vitro hepatic model: the role of human β-glucuronidase, sulfotransferase, catechol-O-methyltransferase and multi-resistant protein 2 (MRP2) in flavonoid metabolism. Biochem Pharmacol 2003; 65:479–491.

37. Boutin JA, Meunier F, Lambert P-H, Hennig P, Bertin D, Serkiz B, Volland J-P. In vivo and in vitro glucuronidation of the flavonoid diosmetin in rats. Drug Met Dispos 1993; 21:1157–1161.

38. Day AJ, Bao Y-P, Morgan MRA, Williamson G. Conjugation position of quercetin glucuronides and effect on biological activity. Free Rad Biol Med 2000; 29:1234–43.

39. Oliveira EJ, Watson DG. In vitro glucuronidation of kaempferol and quercetin by human UGT-1A9 microsomes. FEBS Lett 2000; 471:1–6.

40. Manach C, Morand C, Crespy V, Démigné C, Texier O, Régérat F, Rémésy C. Quercetin is recovered in human plasma as conjugated derivatives which retain antioxidant properties. FEBS Lett 1998; 426:331–336.

41. Moon JH, Tsushida T, Nakahara K, Terao J. Identification of quercetin 3-O-beta-D-glucuronide as an antioxidative metabolite in rat plasma after oral administration of quercetin. Free Rad Biol Med 2001; 30:1274–1285.

42. Rice-Evans C. Flavonoid antioxidants. Current Med Chem 2001; 8:797–807.

43. Morito K, Hirose T, Kinjo J, Hirakawa T, Okawa M, Nohara T, Ogawa S, Inoue S, Muramatsu M, Masamune Y. Interaction of phytoestrogens with estrogen receptors α and β. Biol Pharm Bull 2001; 24:351–356.

44. Manach C, Morand C, Texier O, Favier ML, Agullo G, Demigune C, Regerat F, Remesy C. Quercetin metabolism in plasma of rats fed diets containing rutin or quercetin. J Nutr 1995; 125:1911–1922.

45. Manach C, Texier O, Regerat F, Agullo G, Demigne C, Remesy C. Dietary quercetin is recovered in rat plasma as conjugated derivates of isorhamnetin and quercetin. J Nutr Biochem 1996; 7:375–380.

46. Manach C, Morand C, Démigné C, Texier O, Régérat F, Rémésy C. Bioavailability of rutin and quercetin in rats. FEBS Lett 1997; 409:12–16.

47. Hollman PCH, Katan MB. Absorption, metabolism and bioavailability of flavonoids. In: Rice-Evans CRE, Packer L, Eds. Flavonoids in Health and Disease. New York: Marcel Dekker, 1998:483–522.

48. DuPont MS, Bennett RN, Mellon FA, Williamson G. Polyphenols from alcoholic apple cider are absorbed, metabolized and excreted by humans. J Nutr 2002; 132: 172–175.

49. Meng X, Lee M-J, Li C, Sheng S, Zhu N, Ho C-T, Yang CS. Formation and identification of 4'-O-methyl-( − )-epigallocatechin in humans. Drug Met Dispos 2001; 29: 789–793.

50. Harada M, Kan Y, Naoki H, Fukui Y, Kageyama N, Nakai M, Miki W, Kiso Y. Identification of the major antioxidative metabolites in biological fluids of the rat with ingested ( + )-catechin and ( − )-epicatechin. Biosci Biotechnol Biochem 2000; 63:973–977.

51. O'Leary KA, Day AJ, Needs PW, Sly WS, O'Brien NM, Williamson G. Flavonoid glucuronides are substrates for human liver beta-glucuronidase. FEBS Lett 2001; 503:103–106.

52. Shimoi K, Saka N, Kaji K, Nozawa R, Kinae N. Metabolic fate of luteolin and its functional activity at focal site. Biofactors 2000; 12:181–186.

53. Boersma BJ, Patel RP, Kirk M, Darley-Usmar VM, Barnes S. Chlorination and nitration of soy isoflavones. Arch Biochem Biophys 1999; 368:265–275.

54. Binsack R, Boersma BJ, Patel RP, Kirk MC, White CR, Darley-Usmar VM, Barnes S, Zhou F, Parks DA. Enhanced antioxidant activity following chlorination of quercetin by hypochlorous acid. Alcoholism Clin Exp Res 2001; 25:434–443.

55. Boersma B, Barnes S, Kirk M, Wang C-C, Smith M, Kim H, Xu J, Patel R, Darley-Usmar VM. Soy isoflavonoids and cancer—metabolism at the target site. Mutat Res 2001; 480:121–127.

56. Boersma BJ, Patel RP, Kirk MC, Botting N, Oldfield M, Darley-Usmar VM, Barnes S, Chlorination of soy polyphenols enhances antioxidant properties compared to parent compounds 221$^{st}$ American Chemical Society National Meeting, April 1–5, 2001, San Diego.

57. Schneider H, Simmering R, Hartmann L, Pforte H, Blaut M. Degradation of quercetin-3-glucoside in gnotobiotic rats associated with human intestinal bacteria. J Appl Microbiol 2000; 89:1027–1037.

58. King RA, Bursill DB. Plasma and urinary kinetics of the isoflavones daidzein and genistein after a single soy meal in humans. Am J Clin Nutr 1998; 67:867–872.

59. Coldham NG, Sauer MJ. Pharmacokinetics of $^{14}$C-Genistein in the rat: gender-related differences, potential mechanisms of biological action, and implications for human health. Toxicol App Pharmacol 2000; 164:206–215.

60. Rowland IR, Wiseman H, Sanders TAB, Adlercreutz H, Bowey EA. Interindividual variation in metabolism of soy isoflavones and lignans: influence of habitual diet on equol production by the gut microflora. Nutr Cancer 2000; 36:27–32.

61. Olthof MR, Siebelink E, Hollman PCH, Katan MB. Metabolism of chlorogenic acid, quercetin-3-rutinoside and black tea polyphenols in healthy volunteers. In: Johnson IT, Fenwick GR, Eds. Dietary Antioxidants and Antimutagens. Cambridge: Royal Society of Chemistry, 2000:73–75.

62. Walle T, Walle UK, Halushka PV. Carbon dioxide is the major metabolite of quercetin in humans. J Nutr 2001; 131:2648–2652.

63. Chantret I, Rodolosse A, Barbat A, Dussaulx E, Brotlaroche E, Zweibaum A, Rousset M. Differential expression of sucrase-isomaltase in clones isolated from early and late passages of the cell-line Caco-2—evidence for glucose-dependent negative regulation. J Cell Sci 1994; 107:213–225.

# 4

# Microarray Profiling of Gene Expression Patterns of Genistein in Tumor Cells

**Ching Li and Biehuoy Shieh**
*Chung Shan Medical University, Taichung, Taiwan*

## I. INTRODUCTION

### A. Genistein

The widespread infertility in captive wild animals [1] and grazing livestock [2] led to the discovery of phytoestrogens, chemicals synthesized by vegetables, fruits, and legumes that elicit weak estrogenic effects after dietary consumption [2,3]. Some phytoestrogens, such as the isoflavones daidzein and genistein (Fig. 1), are considered to possess a structural similarity to the animal estrogen, estradiol [4]. Phytoestrogens have been shown to compete with estradiol for its receptor binding sites (ER) in animal models and are considered antiestrogenic because they cannot elicit a full estrogenic response upon forming a complex with the estrogen receptor in vivo [2,5,6]; such understanding led to phytoestrogens being evaluated as preventive or therapeutic agents against hormone-dependent diseases, including certain cancers and heart disease [7–9].

Daidzein and genistein have been most intensively investigated because of their high levels of occurrence in soybean, a widely used animal feedstuff and a component of traditional Asian diets consumed daily in high quantities by Japanese [10] and Chinese populations [8,11]. The consumption of products derived from soy has, therefore, been considered as the major dietary factor contributing to the lower rates of hormone-dependent diseases, including breast, prostate, and colon cancers and coronary heart disease, in many Asian countries [12–14]. Almost 10 years after Akiyama et al. [15] first identified genistein as a tyrosine protein kinase inhibitor [15], it was entered into clinical chemoprevention trials in 1996 [16,17], followed by formal human trials as a treatment for acute lymphoblastic leukemia and breast cancer [18–20].

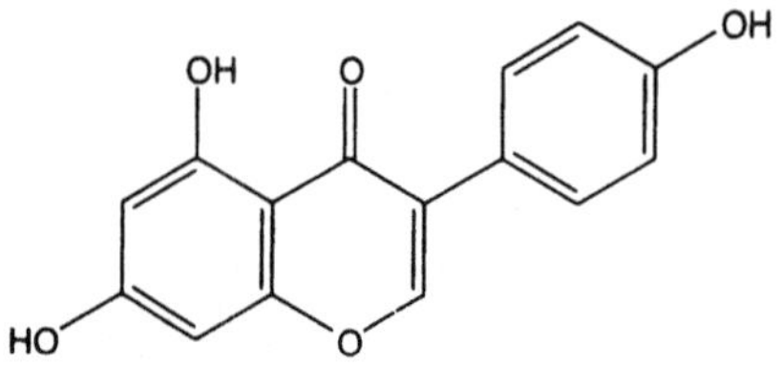

Genistein

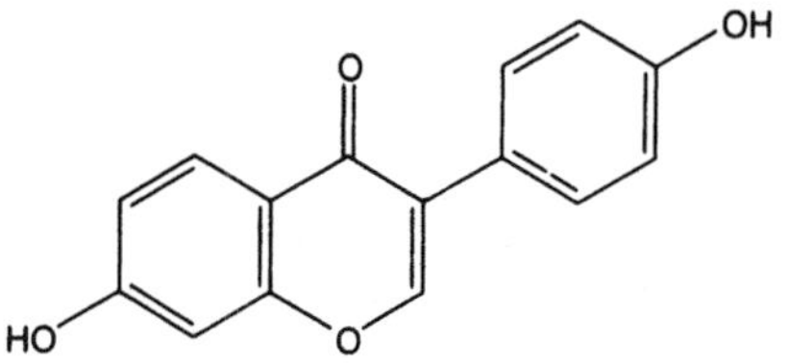

Daidzein

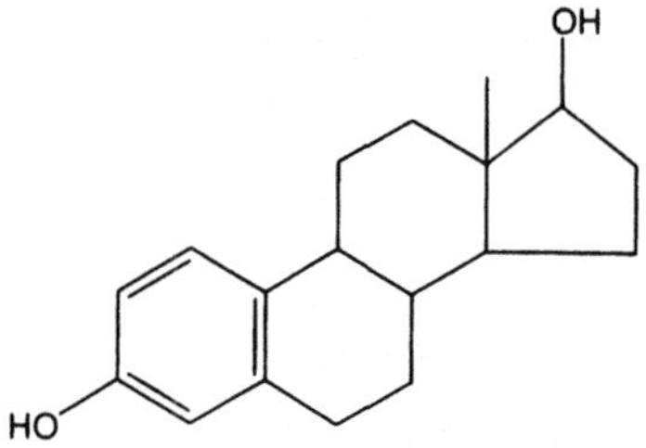

Estradiol

**Figure 1** Chemical structures of two major isoflavones and an estrogen. The structures of genistein (4H-benzopyran-4-one-5, 7-dihydroxy-3-(4-hydroxyphenyl) or 4′,5,7-trihydroxyisoflavone), daidzein (4H-benzopyran-4-one-7-hydroxy-3-(4-hydroxyphenyl) or 4′, 7-dihydroxyisoflavone), and estradiol are shown. The purification schemes of the isoflavones and their physical properties are reported in Ref. 98.

## B.  Action of Genistein

Despite the structural similarity between daidzein and genistein, only the latter is able to regulate cellular activities, and for this reason it has received attention as a potential cancer chemopreventive agent and treatment [21]. The actions of genistein and its role in the possible mechanisms involved in inhibiting carcinogenesis have been comprehensively reviewed [9,22]; in summary, it has inhibitory activities against the estrogen receptor [23], growth factor–associated tyrosine protein kinases [15], DNA topoisomerase II function [24], and the angiogenesis process (including endothelial cell proliferation) in vitro [25]. It is an antioxidant [26], stimulates the synthesis of sex hormone–binding globulin [23], promotes differentiation of certain leukemia cell lines [27], and induces other types of tumor cells to undergo apoptosis [28]. The interested reader will find many more reports describing its biochemistry, nutritional value, and other functions in the recent scientific literature.

Today, in the era of functional genomics, scientists are developing and exploiting many effective and sophisticated tools to gain increasingly deeper insights into the molecular mechanisms of carcinogenesis and tumor suppression. This chapter is concerned with current knowledge about genistein-induced gene expressions and their possible roles in regulating the proliferation of different types of cancer. Using the bladder tumor as an example, results are presented from ongoing cDNA microarray studies on gene expression profiling in tumor cells.

## C. Genistein-Induced Gene Expression in Tumors

The literature details at least 10 different types of cancer cell line that have been employed to investigate the mechanism(s) involved in suppression of tumor cell growth by genistein, including cell lines from the bladder, blood, breast, intestine, liver, lung, pancreas, prostate, skin, and stomach. This section will summarize current knowledge of genistein-induced gene expression and its effects on inhibiting the growth of certain cancers.

### 1. Breast Cancer

Cancer of the breast is, after lung and bronchus, the second most lethal tumor in the U.S. female population [29], and the most common in women worldwide [30] in recent years. Breast cancer–derived cell lines frequently require estrogen to grow; there are two estrogen receptors inside these cells, ER$\alpha$ [31] and ER$\beta$ [32], and studies have shown genistein to be more potent against the latter [33]. Irrespective of whether the tumor cells are ER-positive/estrogen-dependent (e.g., MCF-7, T47D) or ER-negative/estrogen-independent (e.g., MDA-MB-231, BT20), their treatment with high doses of genistein ($>20$ $\mu$M) activates apoptosis [34–37]. In contrast to its effect on these tumor cells, genistein did not induce apoptosis in nonneoplastic human mammary epithelial cells [38]. Because MCF-7 and MDA-MB-231 human breast cancer cell lines express wild-type and mutant p53 gene products, respectively, it is generally considered that genistein-induced apoptosis is both ER− and p53-independent. Recently, however, Xu and Loo [37] have suggested that genistein elicits differential apoptotic effects in ER+ and ER− cells, and other researchers have proposed that its antiproliferative capability is estrogen-dependent [39]. At genistein concentrations in the nanomolar range, genistein appears to function as a true estrogen by stimulating growth of estrogen-dependent human tumor cells both in vitro [36] and in tumor cell–implanted mice [40,41]. Genistein has also been found to induce the expression of four mammalian estrogen-regulated markers (pS2, transforming growth factor $\beta$3, monoamine oxidase A, and $\alpha_1$-antichymotrypsin in MCF-7 cells) but with a lowered potency than endogenous estrogen [35,40,42]; it also increased levels of cathepsin-D and nuclear ER [34]. These differential, concentration-dependent effects of genistein on the physiology and pathology of breast tissue provide a justifiable cause for concern over the elevated risk of potential tumor growth in women with preexisting breast cancer who regularly consume soy-containing products [41,42].

The molecular mechanism of the genistein-induced apoptotic pathway has been intensively investigated using both ER+ and ER− cell lines [30,37,38]. Although apoptosis of breast cancer cells is mediated through the p53-independent pathway, levels of tumor suppressor p53 may be affected by genistein [30,37]. It is already known that p53 modulates the apoptosis process physiologically by trans-activating numerous apoptosis-related genes including p21$^{WAF1/CIP1}$ [43], a component of cyclin:CDK (cyclin-dependent kinase) complexes [44], and Bax [45]. However, with or without the presence of wild-type p53, genistein induces the expression of p21$^{WAF1/CIP1}$, elevated levels of which are believed to lead to inhibition of CDK activity, cessation of damaged (tumor) cells from entering the cell cycle progression and restriction of these cells to the relevant apoptotic pathway [43,46]. Furthermore, induction of Bax, a pro-apoptotic factor, and activation of the caspase-3 pathway in genistein-treated cells has also been reported [35,36]. In its discrimination of normal breast epithelial cells from cancer cells, genistein has been observed to arrest cell growth at the G2/M phase, presumably a result of the combined effects of inhibiting Cdc2 (p34) kinase activity, downregulating cell cycle–associated phosphatase Cdc25 expression and suppressing cell cycle progression by inducing p21$^{WAF1/CIP1}$ [34–36,38,46]. The exact nature of the genistein-regulated molecular switch associated with the apoptotic or cell growth arrest pathway in tumor cells, however, remains to be determined.

## 2.  Prostate Cancer

The incidence of prostate cancer, one of the most common cancers and causes of death in males globally, in the male populations of the United States and Europe is higher than that in Asia. Studies have shown that dietary genistein caused expression of both the androgen receptor (AR) and ER to be downregulated in the rat prostate, suggesting that suppressed sex steroid receptor expression may be responsible for the reduced rate of prostate cancer among Asian men [47]. In androgen-dependent and -independent human prostate cancer lines (e.g.,

LNCaP and PC-3, respectively), low doses ($\leq$20 µM, equivalent to physiological concentrations) of genistein induced growth arrest, but higher concentrations (>20 µM) induced cell apoptosis, and even necrosis, in vitro [48,49]. However, regardless of concentration, genistein-inhibited cell growth has been demonstrated to be associated with upregulation of p21$^{WAF1/CIP1}$, downregulation of cyclin B1 and Bcl-XL, and induction of caspase-3 protein syntheses in an AR- and p53-independent manner, resulting in cell growth arrest at the G2/M phase (or at the G1 phase at lower concentrations) or apoptosis [48–55]. In both LNCaP and PC-3 cells, genistein also abrogates the activation of a potent transcription factor NF-κB by reducing IκB phosporylation, thereby blocking NF-κB from nuclear translocation. Under such circumstances, NF-κB will not respond to signals from DNA damaging agents $H_2O_2$ or tumor necrosis factor (TNF)-α and thus leads to cell destruction [56]. Genistein exhibits an additional very important biological function in prostate tumor cells, notably the inhibition of ERK1/2 activation and blocking of the signal transduction pathway [51]. This phytochemical may, therefore, exploit protein kinase inhibition to counteract prostate tumor progression; this process is frequently the result of loss of the functional androgen receptor and enhancement in expression of the *erb B* family of cellular growth factor receptors.

## 3.  Gastrointestinal and Colonic Cancers

Piontek et al. [57] have shown human gastric cancer cell line AGS to be able to respond to growth stimulation by the epidermal growth factor (EGF) in a dose-dependent manner and that cell growth could be inhibited by genistein. The growth of another gastric cancer cell line, HGC-27, was also reported to be arrested by genistein at the cell cycle G2/M phase [58]. It is reasonable to surmise that genistein may also upregulate expression of p21$^{WAF1/CIP1}$ in gastric cancer cells to arrest cell growth, as found in breast and prostate cancer cells.

The effect of dietary genistein on human intestinal Caco-2 cells was first reported by Kuo and coworkers in 1999 [59,60]; these researchers found that genistein, in the concentration range of 10–100 µM, significantly upregulated expression of metallothionein (MT, a protein possessing metal-binding and anti-oxidant properties) in a time- and dose-dependent manner. MT is responsible for neutralizing cellular free radicals and is, hence, involved in protecting against carcinogenesis. While this would be consistent with the observation that genistein treatment reduced levels of oxidative DNA damage in vitro and in vivo [61,62], a later study in inbred male Wistar rats reported an exactly opposite effect [63]. Although the role of sex hormones in preventing colon or colorectal cancer is still controversial, Arai et al. [64] have shown that the growth of five human colon cancer lines selectively expressed mRNA for the ERβ isoform rather than the ERα isoform was unaffected by the presence of exogenous estrogen. When these five cell lines were treated with genistein (10 $\mu$M, the concentration that usually stimulates proliferation of estrogen-dependent MCF-7 breast cancer cells), growth rates for HT29, Colo320, and Lovo cells were found to decrease [64]. In another study, Salti et al. [65] observed genistein at concentration $\geq$10 $\mu$M to induce DNA topoisomerase II–mediated DNA damage in HT29 cells, but at 60 $\mu$M or greater, to arrest cell cycle progression at the G2/M phase in HT29, SW-620, and SW1116 colon cancer cell lines or induce cell apoptosis in a DNA topoisomeras II–independent manner. The assumption that this effect was a result of induction of p21$^{WAF1/CIP1}$ and downstream effectors regulated by this protein has now been confirmed in subsequent studies using Colo320-HSR colorectal carcinoma cells [66].

## 4. Lung and Liver Cancers

Liver cancer, manifested primarily as non–small-cell lung cancer (NSCLC), is a serious chronic disease in many Asian countries; as with breast and prostate cancer cells, genistein has been found to upregulate the expression of p21$^{WAF1/CIP1}$ and Bax, leading to arrested cell growth at the G2/M phase and apoptosis in human NSCLC cell lines H460 (wild-type p53) and H322 (mutant p53) [67,68]. Although both these processes were p53-independent, genistein only induced endogenous wild-type p53 expression in H460 cells, being without effect on the mutant protein in H322 cells [68].

Genistein is known to induce cell growth arrest and apoptosis in the human hepatocellular carcinoma cell line HepG2 [66] and not only inhibited expression of *c-fos* proto-oncogene induced by the tumor promoter phorbol 12-myristate 13-acetate [69], but significantly increased production of SHBG (sex hormone–binding globulin) [70]. SHBG is yet another factor that has been linked to lowered incidence of hormone-dependent breast and prostate cancers, being associated with an increased duration of sex hormones circulating in the blood and with higher SHBG plasma levels in vegetarians who possess lower incidence rates for these cancers [71].

## 5. Bladder and B-Cell Lymphoma

Two important oncogenic factors, EGFR and p21$^{ras}$, are associated with advanced muscle invasive bladder cancer that frequently leads to death [72–74]. In clinical cases, a large number of patients were found to overexpress these factors during the phenotypic transition of non–muscle invasive superficial bladder tumors into the life-threatening advanced tumor form. Theodorescu et al. [75] demonstrated that by blocking EGFR activity, genistein could effectively inhibit the motility and growth of a variety of bladder cancer cell lines harboring diverse EGFR and p21$^{ras}$ expression patterns that correspond to the invasive phenotype; in many cases tumor cells were induced to undergo apoptosis, and genistein was suggested to exert its effect through the EGFR signaling pathway that is known to involve expression of $p21^{WAF1/CIP1}$ [72]. It was further shown that, over the concentration range 0–50 μM, genistein inhibited the growth of murine (MB49 and MBT-2) and human (HT-1376, UM-UC-3, RT-4, J82, and TCCSUP) bladder cancer cell lines in a dose-dependent manner; these cells were induced into growth arrest at the G2/M phase followed by apoptosis [76]. Mice inoculated subcutaneously with MB49 cells and treated with genistein showed reduced tumor size, increased apoptosis, and decreased angiogenesis [76]. Genistein would also appear to have potential as a chemopreventive agent for patients with urinary tract cancer as it is able to suppress growth of pediatric renal tumor cells in vitro [77].

Chen et al. [78] have presented encouraging results from clinical tests of murine monoclonal anti-CD19 antibody B43-genistein conjugate (B43-Gen) in patients with B-lineage lymphoid malignancies; this approach resulted from an earlier finding that CD19 is highly expressed on the surface of B-lineage lymphoid cells and is physically associated with the tyrosine protein kinase Lyn, a member of proto-oncogene *src* family. B43-Gen specifically targeted CD19-positive tumor cells and induced rapid cell apoptosis in radiation-resistant p53-Bax-Ramos-BT B-lineage lymphoma cells overexpressed with anti-apoptosis factor Bcl-2. A similar result was observed in SCID mice xenografted with this cell line [79]. At present, however, no molecular mechanism has been proposed to explain the effects of genistein in this cancer system.

## D. Genistein and Biomedical Research

There are few publications addressing genistein-induced gene expression in other cancer types, and rather than discuss these here, mention will be made of the role and properties of genistein in other biological systems. The most prominent area of research focuses on the ability of this isoflavone to inhibit tyrosine protein kinases (often associated with receptors), providing a means to dissect and study the signal transduction pathway. As examples, this inhibitory property of genistein assists in demonstrating the role of the endogenous *c-src* protein with 60 kDa mass in modulating voltage-operated calcium channels in vascular smooth muscle cells [80]; the protein is involved in activating both wild-type and mutant cystic fibrosis trans-membrane conductance regulator (CFTR). This is significant since cystic fibrosis is known to be caused by mutation in CFTR [81]. Genistein has been exploited in clarifying the involvement of the phospholipase Cβ subfamily in mediating signals transduced from hormone receptors to induce cardiac hypertrophy, a major cause of heart failure [82]. It has also been used in studies with the α-adrenergic agonist phenylephrine in cultured ventricular myocytes from neonatal rats that demonstrated the involvement of mitogen-activated protein kinase activity (e.g., Erk1 and Erk2) in the induction of hypertrophy [83].

The focus of this chapter is the effect of genistein on gene expression and tumoric disease; however, thus far the literature contains only limited detail on the genistein-induced genes whose expressions are associated with tumor cell growth inhibition and/or apoptosis. However, over the last decade, the initiation and progression of the Human Genome Project (HGP) has led to the emergence of new tools of previously unimaginable power. Functional genomics provides an opportunity to investigate many genes simultaneously in a highly parallel and rapidly serialized manner. The development of microarray technology has been a crucial aspect of unlocking the potential of this new area of specialization; this technology facilitates the examination of global cellular gene expression patterns in the growth inhibition of tumor cells. The remainder of this chapter addresses globally monitored gene expression alternations in bladder transitional cell carcinoma cells induced by genistein; these studies hold out promise for an understanding of the mechanisms underlying chemical-mediated tumor cell growth inhibition.

## II.  MICROARRAY PROFILING OF GENE EXPRESSION PATTERNS

In the past few years, research has been focused on gene expression alterations in human bladder tumor cells induced by genistein in order to gain insight into the molecular mechanism(s) of growth inhibition mediated by this compound [84]; in this way, potential genes may be identified for further evaluation of their gene therapeutic values. To determine gene expression alterations, cDNA microarrays (cDNA-chip), which consist of miniaturized arrays containing tens and thousands of DNA fragments (cDNA) or synthesized oligonucleotides for human genes, were employed. The invention of cDNA-chip is a direct outcome of HGP and gene-related cloning studies performed in the current era of molecular biological technology. These studies had identified and sequenced thousands of functionally "known" gene fragments and millions of Expression Sequence Tags, which were then properly collected in frozen cultures of *E. coli* [85]; the IMAGE Consortium [86] is an example of such gene library. Many research laboratories and commercial organizations are now using PCR to obtain human cDNA from the gene libraries for fabricating cDNA microarrays containing more than 10,000 human known genes and ESTs on glass slides or nylon membranes with the size of a postage stamp [87]. Until now, our laboratories have produced two small human cDNA-chips [84,w3.csmu.edu.tw/~chingli-Biochip/] and a 9600-dot gene-chip is currently under construction. All will contribute to the overall armory of techniques available to study gene expression in bladder TCC cells following treatment with genistein.

## A.   Materials and Methods

Three bladder transitional cell carcinoma cell lines and normal primary bladder epithelial cells (BDEC) were involved in profiling the gene expression patterns induced by genistein. Tumor cell lines TCCSUP [88], T24 [89], and TSGH-8301 [90] and BDEC were all maintained as recommended by the previous reports or the supplier. The growth characteristics of the cells in the presence of genistein were monitored using a CellTiter 96® AQ$_{ueous}$ One Solution Cell Proliferation Kit. While T24 and BDEC cells were unaffected by genistein, the other cell lines had their growth rates reduced to approximately one third of the normal culturing condition (Fig. 2). This finding indicates, first, that genistein inhibits the growth of tumor, but not normal, cells, an essential property for an anticancer drug and that, second, different tumors may arise from the actions of different carcinogenic pathways and thereby lead to differential growth inhibition responses following treatment with genistein. The mRNA samples were isolated from these tumor and normal cells after treatment with genistein (50 μM final concentration) for 0, 0.5, 1, 2, 4, 12, and 24 hours; 1 μg of mRNA was taken to one-step cDNA conversion and biotin labeling, followed by hybridizing to cDNA-chips containing 884 human gene spots (approximately 10 ng of PCR DNA in a round area of 100 μm in diameter and 150 μm apart) [84,87]. The computer images of a representative set of cDNA-chips (scanned by PowerLook 3000 scanner, UMAX; Taipei, Taiwan) hybridized to biotin-labeled cDNA samples derived from TSSCUP cells and treated with genistein for various periods of time are shown in Fig. 3. The colorimetric detection procedure produces a blue image when biotin-labeled cDNA targets hybridize to human gene probes spotted on cDNA-chips; the darker the color, the higher the transcriptional levels of genes in the cells. In this study, the cDNA-chip hybridization experiments have been performed at least three times with every mRNA sample.

Analysis of digital hybridization signals generated by ScanAlyze program (www.microarrays.org/) [91] was facilitated by exploiting computer software that normalized signal variations originating from inconsistencies in the measurement and/or the quality of the experimental materials employed. In contrast to programs that cluster a large quantity of genes possessing similar gene expression patterns, this software assigns "expression scores" to all of the genes on a cDNA-chip, based upon their levels and patterns of induction/reduction, in folds (Table 1); higher scores indicate more interesting or greater gene expression alterations. The software thus rapidly eliminates numerous genes with unchanged expression levels and lists the gene names with the top 10% expression scores; users may, therefore, select the genes of greatest interest or those possessing the most potential for further characterization and/or clustering of expression patterns. Together with colorimetric detection, this cDNA chip technology is capable of quantifying the molar ratio changes in any mRNA species, before or after treatment.

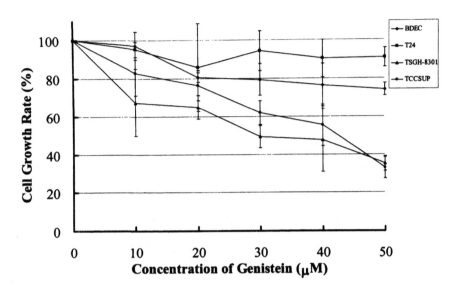

**Figure 2** Cell proliferation assay of bladder tumor cells treated with genistein. Approximately 5000 bladder primary epithelial cells (BDEC) and tumor cell lines of TCCSUP, TSGH-8301, and T24 were incubated separately in 24-well culture dishes containing media with an increased amount of genistein (in DMSO), as indicated. The cells were treated for 72 hours and then subjected to cell proliferation assay with CellTiter 96® AQ$_{ueous}$ One Solution Cell Proliferation Kit. The figure shows growth of BDEC (♦) and T24 (■) to be slightly inhibited and unchanged, respectively, with respect to increasing genistein concentration, whereas growth of TSGH-8301 cells (▲) and TCCSUP (●) was strongly inhibited by this compound in a dose-dependent manner. Experiments were performed twice with duplicated samples. The standard deviations for these experiments are also shown.

## B. Alterations in Gene Expression Induced by Genistein

From the work reported above, it was clear that there are some genes whose expression levels in TCCSUP cells were altered upon genistein stimulation [84]. The growth pattern of TSGH-8301 is similar to TCCSUP, but since that of T24 and BDEC show an opposite response to genistein treatment (Fig. 2), a study was conducted to identify similar and dissimilar gene expression patterns among these cells to elucidate the specific gene(s) mediating the genistein growth inhibitory effect. Table 1 shows the gene list derived from cDNA-chip hybridizations with mRNA samples isolated from TCCSUP treated with 50 μM genistein for 0, 0.5, 1, 2, 4, 12, and 24 hours and analysed as described above. The numbers in the time point columns indicate either fold induction (positive values) or reduction

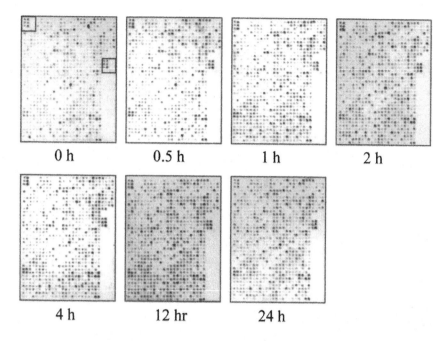

0 h      0.5 h      1 h      2 h

4 h      12 hr      24 h

**Figure 3** cDNA microarray profiling gene expression patterns in TCCSUP cells treated with genistein. One μg of mRNA derived from genistein-treated bladder tumor TCCSUP cells (treatment times indicated) was used each time for hybridization with a cDNA-chip (6.5 × 8.0 mm) containing 884 human genes (for gene list, see w3.csmc.edu.tw/~chingli-Biochip/ [87]) and 11 control genes from plant or *E. coli* bacteriophage lambda (located in the boxed areas of all cDNA-chips). Images of hybridized cDNA-chips were scanned into a computer and digitized using ScanAlyze [91], as the darker the color (actually blue), the higher the level of gene transcription, followed by comparison and analysis with the software developed by the authors. Three identical experiments were performed, producing consistent gene expression patterns listed in Table 1.

(negative figures) when compared with time 0. The Expression Score column represents the degree of kinetic change, with higher scores indicative of greater gene expression alterations with period of genistein stimulation.

Microarray hybridizations were carried out with mRNA samples derived from TSGH-8301, T24, and BDEC treated with the same amount of genistein; comparisons with similar gene expression patterns in these bladder tumor cells and BDEC are presented in Table 2. The similar behavior of TCCSUP and TSGH-8301 cells during genistein treatment suggested that they would probably exhibit somewhat similar gene expression patterns; however, only 11 such genes were

**Table 1**   Alterations in Gene Expression Levels in TCCSUP Cells Induced by Genistein

| Gene product name | Hs. number | Fold increase/decrease | | | | | | | Exp. score |
|---|---|---|---|---|---|---|---|---|---|
| | | 0 h | 0.5 h | 1 h | 2 h | 4 h | 12 h | 24 h | |
| IAP-2 | 127799 | 1.0 | −12.1 | −12.1 | −12.1 | −12.1 | −12.1 | −12.1 | 744.0 |
| PLK-1 | 77597 | 1.0 | 5.3 | 6.8 | 2.4 | 4.4 | 6.4 | 7.0 | 130.4 |
| Highly similar to P13Kγ | 32942 | 1.0 | 3.2 | 3.1 | 6.5 | 5.1 | 8.1 | 3.6 | 113.8 |
| DNA TOPO IIα | 3378 | 1.0 | 4.5 | 5.0 | 3.0 | 4.1 | 6.0 | 6.5 | 97.6 |
| PAI2 | 75716 | 1.0 | 5.0 | 6.4 | 4.0 | 3.6 | 4.1 | 4.3 | 81.9 |
| Highly similar to IPL | 154036 | 1.0 | 4.5 | 4.6 | 3.0 | 4.0 | 5.6 | 5.0 | 74.7 |
| p55$^{cdc}$ | 82906 | 1.0 | 4.2 | 3.6 | 2.3 | 4.5 | 6.6 | 4.3 | 73.6 |
| CKS2 | 83758 | 1.0 | 3.2 | 4.0 | 3.2 | 3.8 | 4.8 | 4.3 | 52.1 |
| CCNA2 | 85137 | 1.0 | 2.3 | 2.8 | 1.6 | 4.2 | 6.1 | 4.1 | 51.4 |
| MSS1 | 61153 | 1.0 | 2.8 | 2.9 | 3.0 | 4.2 | 4.8 | 3.0 | 39.4 |
| NDR | 8724 | 1.0 | −4.8 | −4.3 | −2.5 | −3.2 | −1.4 | −3.4 | 38.4 |
| N-Methylpurine-DNA glycosylase | 79396 | 1.0 | 3.5 | 3.1 | 3.4 | 3.5 | 3.9 | 3.7 | 38.2 |
| CCNB1 | 23960 | 1.0 | 2.5 | 3.6 | 1.9 | 2.4 | 3.9 | 4.4 | 31.6 |
| VDUP1 | 179526 | 1.0 | 1.7 | 2.1 | 3.5 | 4.6 | 3.0 | 3.5 | 31.3 |
| UbcH7 | 108104 | 1.0 | 2.4 | 3.4 | 2.7 | 3.0 | 4.5 | 2.8 | 30.0 |
| c-Fos | 25647 | 1.0 | 6.1 | 2.5 | 1.0 | 1.0 | 1.5 | 1.0 | 28.8 |
| GADD153 | 129913 | 1.0 | −3.0 | 1.2 | −1.6 | −5.3 | −1.3 | −2.9 | 26.8 |
| Highly similar to E2-17 | 4890 | 1.0 | 2.3 | 2.1 | 2.0 | 2.5 | 5.2 | 2.5 | 26.4 |
| CDKN2C | 4854 | 1.0 | 2.6 | 2.4 | 1.3 | 2.4 | 3.8 | 4.4 | 25.6 |
| AKAP1 | 78921 | 1.0 | 1.9 | 2.8 | 3.4 | 3.6 | 3.2 | 2.6 | 24.5 |
| Proteasome component C13 precursor | 1550 | 1.0 | 1.9 | 3.0 | 2.6 | 2.3 | 4.5 | 2.7 | 24.4 |
| Forkhead-like protein | 93468 | 1.0 | 1.9 | 2.9 | 2.0 | 4.0 | 3.5 | 2.3 | 22.2 |
| CCNG2 | 79069 | 1.0 | 1.5 | 2.8 | 1.6 | 1.9 | 4.1 | 3.8 | 21.6 |
| Egr-1 | 738 | 1.0 | 4.1 | 4.2 | 1.0 | 1.0 | 1.0 | 1.0 | 19.8 |
| Proteasome subunit p112 | 3887 | 1.0 | 2.2 | 2.9 | 2.3 | 2.5 | 3.9 | 2.3 | 19.2 |
| Creatine kinase B | 147553 | 1.0 | 3.0 | 2.5 | 2.6 | 3.0 | 2.8 | 2.5 | 18.4 |
| UbcH6 | 7766 | 1.0 | −1.1 | 2.1 | 2.3 | 1.7 | 4.8 | 1.7 | 18.4 |
| Highly similar to G2/M-specific cyclin B2 | 20483 | 1.0 | 2.2 | 2.5 | 1.9 | 2.7 | 4.0 | 2.3 | 17.8 |
| RPA3 | 1608 | 1.0 | 1.6 | 2.5 | 1.9 | 2.6 | 3.9 | 2.8 | 17.8 |
| DEK | 110713 | 1.0 | 1.7 | 3.3 | 1.8 | 2.0 | 3.2 | 3.2 | 17.1 |
| PFKP | 99910 | 1.0 | 2.0 | 2.2 | 3.0 | 2.9 | 3.2 | 2.3 | 16.8 |
| Dihydrolipoamide dehydrogenase precursor | 74635 | 1.0 | 2.1 | 3.1 | 2.4 | 2.3 | 2.8 | 2.9 | 16.4 |
| Highly similar to ATP synthase β chain | 114912 | 1.0 | 1.9 | 3.3 | 2.1 | 2.6 | 2.8 | 2.8 | 16.0 |
| CDC7-like protein 1 | 28853 | 1.0 | 2.4 | 2.3 | 2.3 | 2.2 | 3.0 | 3.3 | 16.0 |
| PYK2B | 20313 | 1.0 | −1.2 | 2.2 | 3.4 | 2.2 | 3.6 | 1.4 | 15.8 |
| DNMT1 | 77462 | 1.0 | 2.1 | 2.5 | 2.0 | 2.9 | 2.9 | 2.6 | 14.9 |
| Contactin | 143434 | 1.0 | 1.8 | 1.9 | 1.9 | 3.0 | 3.2 | 2.9 | 14.8 |
| PKC inhibitor I | 43721 | 1.0 | 1.9 | 2.4 | 3.0 | 2.8 | 2.7 | 2.4 | 14.8 |
| Highly similar to VAV | 37331 | 1.0 | 1.5 | 2.1 | 1.2 | 1.8 | 3.6 | 3.4 | 14.6 |
| Diacylglycerol kinase, α subuni | 64084 | 1.0 | 1.7 | 2.2 | 1.7 | 2.4 | 3.1 | 3.4 | 14.5 |

(*Continued*)

**Table 1**  Continued

| Gene product name | Hs. number | Fold increase/decrease | | | | | | | Exp. score |
|---|---|---|---|---|---|---|---|---|---|
| | | 0 h | 0.5 h | 1 h | 2 h | 4 h | 12 h | 24 h | |
| Highly similar to CDK inhibitor 1B | 3561 | 1.0 | 2.2 | 2.3 | 1.7 | 2.5 | 3.2 | 2.9 | 14.2 |
| CKS1 | 348669 | 1.0 | 2.2 | 2.4 | 2.8 | 2.2 | 2.8 | 2.6 | 14.1 |
| E2F transcription factor 1 | 96055 | 1.0 | 2.6 | 2.3 | 2.1 | 2.7 | 2.8 | 2.6 | 14.0 |
| hSOX20 | 95582 | 1.0 | 1.7 | 2.6 | 1.8 | 3.2 | 2.7 | 2.5 | 13.4 |
| RAD6 homolog | 80612 | 1.0 | 1.7 | 2.4 | 2.4 | 2.6 | 3.1 | 2.4 | 13.4 |
| Calgranulin A | 100000 | 1.0 | 1.0 | 1.0 | 1.4 | 3.8 | 3.0 | 2.2 | 13.3 |
| MARCK substrate | 75607 | 1.0 | 1.6 | 1.6 | 1.3 | 2.3 | 3.4 | 3.2 | 13.0 |
| PK C substrate 80K-H | 1432 | 1.0 | 2.4 | 2.2 | 3.2 | 2.1 | 2.7 | 1.7 | 12.8 |
| UFDIL | 10298 | 1.0 | 1.8 | 2.8 | 1.9 | 2.2 | 3.0 | 2.5 | 12.6 |
| Cystatin A | 2621 | 1.0 | −1.4 | 1.9 | 3.3 | 1.7 | 3.0 | 2.1 | 12.4 |
| Glutaredoxin | 28988 | 1.0 | 2.2 | 2.1 | 1.4 | 2.7 | 2.5 | 3.0 | 12.1 |
| CDK 2 | 19192 | 1.0 | 1.3 | 1.6 | 1.5 | 2.4 | 3.9 | 1.7 | 11.6 |
| PCNA | 78996 | 1.0 | 1.9 | 2.3 | 2.0 | 2.0 | 2.4 | 3.2 | 11.5 |
| PDGF receptor β | 76144 | 1.0 | −2.4 | −2.2 | −2.4 | −2.4 | −2.4 | −2.4 | 11.2 |
| LIM domain kinase 1 | 36566 | 1.0 | 2.3 | 1.9 | 1.5 | 2.8 | 2.7 | 2.5 | 11.0 |
| TEL | 171262 | 1.0 | 2.1 | 1.8 | 1.5 | 3.1 | 2.8 | 2.0 | 10.6 |
| Caspase 7 | 9216 | 1.0 | 1.0 | 1.8 | 2.3 | 2.2 | 3.4 | 2.1 | 10.6 |
| SLK | 320955 | 1.0 | 1.8 | 1.7 | 1.9 | 2.3 | 3.0 | 2.7 | 10.5 |
| PIC1 | 81424 | 1.0 | 1.4 | 2.2 | 2.5 | 1.7 | 2.9 | 2.5 | 10.5 |
| CDC25C | 656 | 1.0 | 2.1 | 2.2 | 1.0 | 1.7 | 3.2 | 2.6 | 10.4 |
| ELK1 | 119850 | 1.0 | 2.2 | 1.9 | 2.2 | 2.6 | 2.7 | 2.1 | 10.3 |
| RYK receptor-like tyrosine kinase | 79350 | 1.0 | 1.1 | 2.0 | 2.4 | 2.1 | 3.0 | 2.4 | 10.1 |
| CD31/PECAM1 | 115474 | 1.0 | 1.2 | 2.2 | 1.4 | 2.0 | 3.7 | 1.4 | 10.0 |
| RAD21 homolog | 81848 | 1.0 | 1.7 | 2.3 | 1.9 | 1.6 | 2.7 | 2.8 | 9.6 |
| Sp3 | 44450 | 1.0 | 1.5 | 1.9 | 2.1 | 2.4 | 2.6 | 2.5 | 9.2 |
| TTF-I | 54780 | 1.0 | 1.7 | 2.5 | 1.7 | 2.8 | 2.2 | 2.2 | 9.2 |
| Caspase 10 | 5353 | 1.0 | 2.0 | 1.9 | 1.5 | 2.0 | 2.1 | 3.2 | 9.2 |
| Highly similar to PGK1 | 78771 | 1.0 | 2.1 | 2.3 | 2.2 | 2.1 | 2.3 | 2.3 | 9.1 |
| STE24p | 25846 | 1.0 | 1.2 | 2.2 | 1.9 | 2.2 | 3.2 | 1.3 | 8.8 |
| CHEK1 | 20295 | 1.0 | 2.0 | 2.0 | 2.1 | 1.7 | 1.8 | 3.1 | 8.7 |
| eIF4E | 79306 | 1.0 | 1.4 | 2.4 | 2.1 | 1.8 | 2.7 | 2.4 | 8.7 |
| BAK | 93213 | 1.0 | 1.3 | 2.0 | 1.6 | 2.6 | 2.9 | 2.0 | 8.7 |
| ICAM-1 precursor | 51061 | 1.0 | −1.1 | 2.2 | 2.7 | 1.6 | 2.5 | 2.3 | 8.6 |
| IL-10 receptor | 327 | 1.0 | −2.3 | −2.0 | −2.3 | −2.1 | −2.1 | −2.3 | 8.6 |
| MAPKAPK2 | 75074 | 1.0 | −1.0 | 1.6 | 2.5 | 1.9 | 3.2 | 1.6 | 8.6 |

The digitalized signals for all genes on cDNA-chips detected at 0–24 hours (7 time points) were compared to those at time 0; the fold changes obtained and listed under different time point columns. Negative numbers represent decreased gene expression levels; all others indicate the induced transcription levels. Every gene has a group of seven numbers (fold changes), were further used to calculate Expression Score, that were obtained by inserting seven grouped numbers in the formula $\Sigma(N_i-N_0)^2$, where $N_i$ represents the absolute values of the six numbers at time points 0.5 to 24 for the particular genes and $N_0$ represents absolute values of the last numbers at time 0. The table lists 75 gene product names (approximately one-tenth the total number of genes on the cDNA-chip) in descending order of Expression Score. The Hs. numbers identify the genes sorted in the NCBI UniGene database (National Institutes of Health).

**Table 2** Comparison of Gene Expression Patterns in Cell Lines Treated with Genistein

| TCCSUP & TSGH-8301 | | TCCSUP & BDEC | | TCCSUP & T24 | | Hs. number |
|---|---|---|---|---|---|---|
| c-Fos | ↑ | c-Fos | ↑ | c-Fos | ↑ | 25647 |
| Egr-1 | ↑ | Egr-1 | ↑ | Egr-1 | ↑ | 738 |
| RPA3 | ↑ | RPA3 | ↑ | | | 1608 |
| IPL-like kinase | ↑ | IPL-like kinase | ↑ | | δ | 154036 |
| GADD153 | ↓ | | | GADD153* | | 129913 |
| CHEK1 | ↑ | | | | | 20295 |
| CKS2 | ↑ | | | | | 83758 |
| CCNB1 | ↑ | | | | | 23960 |
| CCNA2 | ↑ | | | | | 85137 |
| VDUP1 | ↑ | | | | | 179526 |
| PYK2B | ↑ | | | | | 20313 |
| | | DEK | ↑ | | | 110713 |
| | | CDKN2C | ↑ | | | 4854 |
| | | | | CLK1 | ↑ | 2083 |

Four different bladder cell lines, TCCSUP, TSGH-8301, T24, and BDEC, responded differently to genistein-mediated growth inhibition (Fig. 2) and were subjected to microarray profiling of gene expression patterns after treated with this isoflavone. A total of 88 genes (10% of total number of genes on the cDNA-chip) possessing the highest Expression Scores were selected from each cell line for comparing between cell line pairs. Gene product names with arrowheads ↑ or ↓ represent upregulated or downregulated gene expression patterns, respectively. Designation * on GADD153 indicates differential gene expression regulation since upregulation and downregulation of gene expression were detected in T24 and TCCSUP, respectively.

found. This number should be compared with the situation found when comparing TCCSUP/BDEC (6 genes) and TCCSUP/T24 (4 genes) cell pairs.

## C. Confirmation and Characterization of Gene Expression Patterns

Analysis of the mRNA levels of *egr-1*, *c-fos*, and VDUP1 genes by PCR or Northern blotting was used to confirm the expression patterns derived from microarray experiments. The proto-oncogenes *c-fos* and *egr-1* encode the immediate early transcription factors that regulate cell growth, differentiation, and/or other biological evens [92,93]; they exhibit a rapid (usually within 30 minutes) and transient (maximal activity at 1–2 hours, followed by a dramatic decrease) inducible transcriptional kinetic pattern. VUDP1 (vitamin $D_3$ upregulated protein 1) was first identified in human leukemic cells HL-60 following stimulation with

1,25-dihydroxyvitamin $D_3$ and later found associated with and suppressed thioredoxin protein (TRX) in cellular redox regulatory machinery [94–96].

Figure 4A shows *egr-1* mRNA to be dramatically synthesized 0.5–1 hour after TCCSUP cells were treated with 50 μM genistein; a rapid decrease follows, with undetectable levels after 2 hours. Synthesis of Egr-1 protein was consistent with its transcription levels (Fig. 4B). PCR results (Fig. 4C) reveal a similar transient induction of *c-fos* mRNA synthesis within 0.5–1 hour, and a gradual increase of VUDP1 mRNA levels from 0.5 to 12 hours posttreatment. Transcriptional patterns for all three genes are, therefore, consistent with results from cDNA microarray experiments (Table 1). A small cDNA-chip has been fabricated as a means of investigating the relationship between transcriptional levels and genistein dose. Levels of *egr-1* mRNA were found to be increased linearly up to approximately 7.5-fold following treatment of SGH-8301 cells with increasing amounts of genistein (Fig. 5), implying induction of *egr-1* gene expression to be a dose-dependent process, and the gene has an important role in mediating the growth inhibitory effect of genistein.

Anti-sense technology was used to reduce synthesis of Egr-1, c-Fos, and VUDP-1 proteins and to provide an experimental system for examining the differential response, if any, of bladder tumor cells to genistein growth inhibition. Phosphorothioate oligonucleotides (ODNs) possessing nucleotide sequences of the sense or anti-sense strands from the above genes (Fig. 6) were synthesized, loaded into liposomes, and added to culture medium that included either TCCSUP or TSGH-8301 bladder cells, both susceptible to growth inhibition by genistein. Cell growth rates, in the presence of 50 μM genistein, were determined; the data showed both to behave similarly to ODNs and genistein treatments. Abrogating expression of *c-fos*, *egr-1*, or VUDP1 genes had no effect on growth rates (Fig. 6), suggesting a lack of involvement in the growth inhibition mechanism. In contrast, anti-*H-ras*-ODN reduced growth of T24 bladder TCC cells, known to overexpress oncogenic *H-ras* [73,89], to relative the genistein-sensitive phenotype (unpublished data), indicating that this gene has a central role in the susceptibility of this cell line to drug treatment. Before selecting genistein as a potential chemotherapeutic drug, it will be important to determine whether bladder tumors carried in patients harbor the *H-ras* mutation; this is because the mutation accompanying onco-protein overexpression has been frequently detected in bladder TCC, but these are resistant to genistein [73].

Using cDNA microarrays, many gene expression alterations induced by genistein in bladder TCC cell lines have been simultaneously identified, and these genes may be involved in the growth inhibition molecular pathway governed by genistein. There is reason to be optimistic about the exploitation of microarray technology within the overall armory of technologies to be applied to gene discovery for cancer biology and gene therapy. Although the results presented here

A

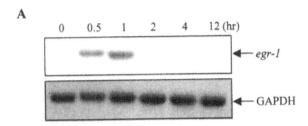

**Figure 4** Gene expression level confirmation for the *egr-1, c-fos*, and the VDUP1 genes. Our cDNA microarray experiments showed that expression levels for *egr-1, c-fos*, and the VDUP1 genes exhibited high degrees of alternations. Northern blotting or reverse transcription (RT)-PCR was, therefore, used to confirm their gene expression patterns. (A) Northern blotting of mRNA samples (2 μg each) derived from bladder tumor TCCSUP cells treated for various periods of time, as indicated, hybridized with $^{32}$P-labeled *egr-1* or glyceraldehydes-phosphate dehydrogenase (GAPDH) gene probe. Only mRNA samples isolated from cells treated with the drug for 0.5 and 1 hours contained the *egr-1* transcript, consistent with the results of microarray studies. The control GAPDH mRNA remained unchanged. (B) The Egr-1 protein synthesis in genistein-treated TCCSUP cells was examined using SDS-polyacrylamide gel electrophoresis. The Egr-1 protein reached the maximal level at 1.5 hours poststimulation, followed by a decrease to the background level. However, β-actin protein synthesis was not affected by drug stimulation. Therefore, the protein synthesis pattern was parallel, but delayed 30–60 minutes, to the time course of the mRNA production, suggesting that the expression level changes detected by microarray hybridization were faithful for this gene. RT-PCR determinations of the expression levels of *c-fos* and the VDUP-1 genes are shown (C). These experiments were performed by adding one-tenth the amount of cDNA converted (by RT) from 1 μg each of cellular RNA to a PCR mixture containing primers for *c-fos* (forward: AAGGAGAATCCGAAGG-GAAA; reverse: GCTGCTGATGCTCTTGACAG), VDUP-1 (forward: CCCTGATT-TAATGGCACCTG; reverse: TTGGATCCAGGAACGCTAAC), or the β-actin (forward: ATCATGTTTGAGACCTTCAA; reverse: CATCTCTTGCTCGAAGTCCA) gene. The PCR products were then loaded on agarose gel electrophoresis for separation and visualization. The results show that the *c-fos* mRNA was induced at 30 minutes followed by a decrease at 1 hour, whereas the VDUP-1 mRNA synthesis showed a linear increase. The results also agree with cDNA-chip hybridization detection data.

focus on research with genistein, future opportunities obviously encompass a broader span of phytochemical research.

## III.  DISCUSSION AND CONCLUSION

### A.  Problem of Genistein

Although genistein possesses promising cancer cell growth–inhibiting effects, many uncertainties remain about its potential for clinical use. Given its broad

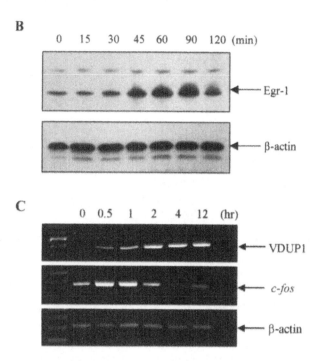

**Figure 4** Continued.

range of biological and/or biochemical functions, it is difficult to determine precisely which function(s) actually play the key role in tumor suppression. Evidence presented here points to genistein being able to induce the expression of many genes, but it is unclear whether these tumor-suppressing effects result from direct biochemical action, are the indirect consequences of mediation by other gene products, or reflect a combination of both. At low concentrations, genistein can stimulate tumor cell growth in vivo and in vitro so that taking an insufficient amount of genistein as a dietary supplement to prevent cancer may actually have deleterious consequences associated with growth stimulation of preexisting cancer cells.

In reviewing its biochemical targets in tumor cells, Peterson [97] examined the efficacy of genistein in vivo by comparing dietary achievable serum level and $IC_{50}$ values in vitro as if the drug acts an inhibitor for topoisomerase II or protein kinase activities in a range of signal transduction pathways (including *c-src*, *v-abl*, epidermal growth factor receptor, insulin receptor, p94 protein tyrosine kinase, Grb2/mSOS-*ras* pathway kinases, phospolipase $C_\gamma$, S6-K, and serine/threonine kinases); the conclusion was that the target of genistein action in tumor cells remained to be elucidated. Almost a decade later, this conclusion still stands.

A

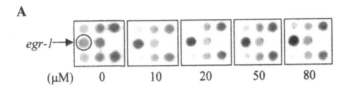

B

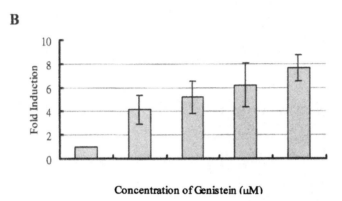

Concentration of Genistein (uM)

**Figure 5** Dose-dependent induction of *egr-1* in TSGH-8301 cells treated with genistein. (A) *egr-1* and 7 other nonrelevant genes were fabricated onto a small nylon membrane for hybridizing with cDNA derived from bladder tumor TSGH-8301 cells treated with increasing amounts of genistein, as indicated. The results show that the *egr-1* signals linearly increased as additional genistein was added to the cell cultures. The signal levels for the nonrelevant genes remained unchanged. (B) Fold-induction quantification of the *egr-1* gene expressions in four independent experiments yielded a linear increase to approximately 7.5-fold at the maximal concentration of 80 μM. It is concluded that this cDNA microarray can be used in a kinetic study of any gene expression of interest.

## B.  Gene Expression and Genistein Function

A number of features of cDNA microarray technology make it particularly valuable in gene expression profiling research; it is:

   Relatively cheap and considerably less laborious than other techniques for
      screening such large numbers of genes
   Flexible and universal—one chip can be adapted for a variety of biological
      systems providing that they derive from the same species
   Fast and effective in screening a large number of genes simultaneously
   Requires only a limited number of simple techniques

# Egr-1

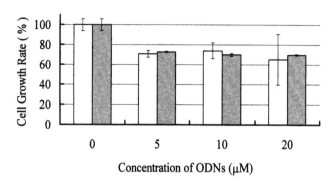

# c-Fos

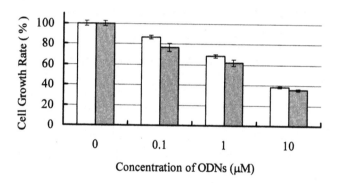

**Figure 6** Analyses of the roles of the *egr-1*, *c-fos*, and VDUP1 genes in tumor cell growth suppression induced by genistein. Expression patterns for *egr-1*, *c-fos*, and the VDUP1 genes were greatly induced when the cells were treated with genistein; the study was conducted to determine whether these genes were involved in mediating the growth inhibition effect exerted by genistein. Phosphorothioate oligonucleotides (ODNs) for *egr-1* (antisense: GsCsGsGsGGTGCAGGGCsAsCsAsCsT; sense: AsGsTsGsTGCCCCTG-CACsCsCsGsC), *c-fos* (antisense: GsAsAsGsCsCsCsGAGAsAsCsAsTsCsAsT; sense: AsTsGsAsTsGsTsTCTCsGsGsGsCsTsTsC), and the VDUP-1 gene (antisense: TsTsCsTsTGAACATCAsCsCsAsT; sense: AsTsGsGsTGATGTTCAsAsGsAsA) were synthesized for suppressing target gene product synthesis in vivo, followed by assaying growth rates of TCCSUP cells in the presence of 50 μM genistien. The figure shows the results from triplicate experiments as cell growth averages (in percentages) and their standard deviations.

# VDUP1

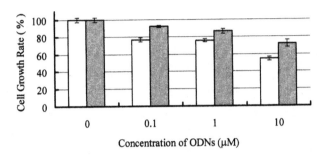

**Figure 6**  Continued.

In summary, microarray analysis with cDNA-chips containing a large number of known genes and EST cDNA (usually >10,000-dot) is a valuable means of investigating alterations in gene expression.

Difficulties in pinpointing the individual biochemical targets associated with the antitumor activity of genistein action demanded the development of alternative approaches. Microarray profiling of gene expression patterns in tumor cells has been found to be both feasible and fast. Significant results have been obtained using a small cDNA microarray, and the introduction of larger chips, containing hundreds or thousands of sequence-verified human cDNA, would considerably broaden our search power and allow simultaneous monitoring of global gene expression patterns in different tumor cells. The major goal of our research still remains the identification of those genes involved in inhibiting tumor cell growth mediated by genistein; results eagerly anticipated in this particular area will, it is hoped, contribute to an even bigger goal, the discovery of the target genes for gene therapy.

## ACKNOWLEDGMENTS

The authors wish to express our gratitude to Chao-Chin Hu, Chin-Chuan Chen, and Shang-Kai Hung for their expert technical assistance. The BDEC, normal primary bladder epithelial, cells described in this chapter were a gift from Dr. Hsing-Jien Kung, University of Davis, CA. This research was support partly by funds from National Health Research Institutes (DD01-86IX-MG601P) and National Science Council (NSC88-2318-B-006-006-M51, NSC89-2318-B-006-006-M51, and NSC90-2318-B-040-004-M51) of R.O.C. to CL. CSMU also contributed a travel fund for CL to present a part of this chapter at the 6th International Symposium on Predictive Oncology and Intervention Strategies held in Paris, France (2/9–12/2002).

## REFERENCES

1. Setchell KDR, Gosselin SJ, Welsh MB, Johnston JO, Balistreri WF, Kramer LW, Dressner BL, Tarr MJ. Dietary estrogens—a probable cause of infertility and liver disease in captive cheetah. Gastroenterol 1987; 93:225–233.
2. Shutt DA, Cox RI. Steroid and phytoestrogen binding to sheep uterine receptors in vivo. J Endocrinology 1972; 52:299–310.
3. Shutt DA. The effects of plant oestrogens on animal reproduction. Endeavour 1976; 35:110–113.
4. Setchell KDR, Adlercreutz H. Mammalian lignans and phytoestrogens. In: Rowland I, Ed. The Role of the Gut Flora in Toxicity and Cancer. London: Academic Press, 1988:315–346.
5. Tang BY, Adams NR. Effects of equol on oestrogen receptors and on synthesis on DNA and protein in the immature rat uterus. J Endocrinol Metabol 1979; 49:152–154.
6. Thomson MA, Lasley BL, Rideout BA, Kasman WH. Characterization of the estrogenic properties of a non-steroidal estrogen, equol, extracted from the urine of pregnant macaques. Biol Reprod 1984; 31:705–713.
7. Steele VE, Pereira MA, Sigman CC, Kelloff GJ. Cancer chemoprevention agent development strategies for genistein. J Nutr 1995; 125(suppl):713S–716S.
8. Barnes S. Effect of genistein on in vitro and in vivo models of cancer. J Nutr 1995; 125(suppl):777S–783S.
9. Barnes S, Peterson TG, Coward L. Rational for the use of genistein-containing soy matrices in chemoprevention trials for breast and prostate cancer. J Cell Biochem 1995; 22(suppl):181–187.
10. Adlercreutz H, Honjo H, Higashi A, Fotsis T, Hamalainen E, Hasegawa T, Okada H. Urinary excretion of lignans and isoflavonoid phytoestrogens in Japanese men and women consuming a traditional Japanese diet. Am J Clin Nutr 1991; 54:1093–1100.
11. Chen Z, Zheng W, Custer LJ, Dai Q, Shu XO, Jin F, Franke AA. Usual dietary consumption of soy foods and its correlation with the excretion rate of isoflavonoids in overnight urine samples among Chinese women in Shanghai. Nutr Cancer 1999; 33:82–87.
12. Cancer Facts and Figures. Atlanta: American Cancer Society, 1994.
13. Lee HP, Gourley L, Duffy SW, Esteve J, Lee J, Day NE. Dietary effects of breast cancer risk in Singapore. Lancet 1991; 337:1197–1200.
14. Coward L, Barnes NC, Setchell KDR, Barnes S. Genistein, daidzein, and their β-glycoside conjugates: antitumor isoflavones in soybean foods from American and Asian diets. J Agr Food Chem 1993; 41:1961–1967.
15. Akiyama T, Ishida J, Nakagawa S, Ogawara H, Watanabe S-I, Itoh N, Shibuya M, Fukami Y. Genistein, a specific inhibitor of tyrosine-specific protein kinases. J Biol Chem 1987; 262:5592–5595.

16. Kelloff GJ, Boone CW, Crowell JA, Steele VE, Lubet RA, Doody LA, Malone WF, Hawk ET, Sigman CC. New agents for cancer chemoprevention. J Cell Biochem 1996; 26(suppl):1–28.

17. Kelloff GJ, Crowell JA, Hawk ET, Steele VE, Lubet RA, Boone CW, Covey JM, Doody LA, Omenn GS, Greenwald P, Hong WK, Parkinson DR, Bagheri D, Baxter GT, Blunden M, Doeltz MK, Eisenhauer KM, Johnson K, Knapp GG, Longfellow DG, Malone WF, Nayfield SG, Seifried HE, Seall Lm, Sigman CC. Strategy and planning for chemopreventive drug development: clinical development plans II. J Cell Biochem 1996; 26(suppl):54–71.

18. Wang HK. The therapeutic potential of flavonoids. Expert Opin Invest Drug 2000; 9:2103–2119.

19. This P, Magdelenat H. Phytoestrogens and adjuvant endocrine treatment of breast cancer. J Clin Oncol 2000; 18:2792.

20. Uckun FM, Narla RK, Zeren T, Yanishevski Y, Myers DE, Waurzyniak B, Ek O, Schneider E, Messinger Y, Chelstrom LM, Gunther R, Evans W. In vivo toxicity, pharmacokinetics, and anticancer activity of genistein kinked to recombinant human epidermal growth factor. Clin Cancer Res 1998; 4:1125–1134.

21. Mikisicek RJ. Estrogenic flavonoids: structural requirements for biological activity. Proc Soc Exp Biol Med 1995; 208:44–50.

22. Gescher A, Pastorino U, Plummer SM, Manson MM. Suppression of tumor development by substances derived from the diet—mechanisms and clinical implications. Br J Clin Phamacol 1998; 45:1–12.

23. Adlercreutz CH, Goldin BR, Gorbach SL, Hockerstedt KA, Watanabe S, Hamalainen EK, Markkanen MH, Makela TH, Whala KT, Hase TA, Fotsis T. Soybean phytoestrogen intake and cancer risk. J Nutr 1995; 125(suppl):757S–770S.

24. Markovits J, Linassier C, Fosse P, Couprie J, Pierre J, Jacquemin-Sablon A, Saucier JM, Le Pecq JB, Larsen AK. Inhibitory effects of the tyrosine kinase inhibitor genistein on mammalian DNA topoisomerase II. Cancer Res 1989; 49:5111–5117.

25. Fotsis T, Pepper M, Adlercreutz H, Fleischmann G, Hase T, Montesano R, Schweigerer L. Genistein, a dietary-derived inhibitor of in vitro angiogenesis. Proc Natl Acad Sci USA 1993; 90:2690–2694.

26. Naim M, Gestetner B, Bondi A, Birk Y. Antioxidative and antihemolytic activities of soybean isoflavone. J Agr Food Chem 1976; 24:1174–1177.

27. Hunakova L, Sedlak J, Klobusicka M, Duraj J, Chorvath B. Tyrosine kinase inhibitor-induced differentiation of K-562 cell: alterations of cell cycle and cell surface phenotype. Cancer Lett 1994; 81:81–87.

28. Azuma Y, Onishi Y, Sato Y, Kizaki H. Induction of mouse thymocytes apoptosis by inhibitors of tyrosine kinases is associated with dephosphorylation of nuclear proteins. Cell Immunol 1993; 152:271–278.

29. American Cancer Society; www.cancer.org/docroot/STT/stt_0_2002.asp.

30. GLOBOCAN 2000: Cancer incidence, mortality, and prevalence worldwide; www-dep.iarc.fr/globocan/globocan.html.

31. Green S, Walter P, Kumar V, Krust A, Bornert JM, Argos P, Chambon P. Human oestrogen receptor cDNA: sequence, expression and homology to v-erb-A. Nature 1986; 320:134–139.

32. Kuiper GGJM, Enmark E, Pelto-Huikko M, Nilsson S, Gustafsson JA. Cloning of a novo estrogen receptor expressed in rat prostate and ovary. Proc Natl Acad Sci USA 1996; 93:5925–5930.

33. Casanova M, You L, Gaido KW, Archibeque-Engle S, Janszen DB, Heck HA. Developmental effects of dietary phytoestrogens in Sprague-Dawley rats and interactions of genistein and daidzein with rat estrogen receptors α and β in vitro. Toxicol Sci 1999; 51:236–144.

34. Fioravanti L, Cappelletti V, Miodini P, Ronchi E, Brivio M, Fronzo G. Genistein in the control of breast cancer cell growth: insights into the mechanism of action in vitro. Cancer Lett 1998; 130:143–152.

35. Li Y, Upadhyay S, Bhuiyan M, Sarkar FH. Induction of apoptosis in breast cancer cells MDA-MB-231 by genistein. Oncogene 1999; 18:3166–3172.

36. Nakagawa H, Yamamoto D, Kiyozuka Y, Tsuta K, Uemura Y, Hioki K, Tsutsui Y, Tsubura A. Effects of genistein and synergistic action in combination with eicosapentaenoic acid on the growth of breast cancer cell lines. J Cancer Res Clin Oncol 2000; 126:448–454.

37. Xu J, Loo G. Different effects of genistein on molecular markers related to apoptosis in two phenotypically dissimilar breast cancer cell lines. J Cell Biochem 2001; 82: 78–88.

38. Frey RS, Li J, Singletary KW. Effects of genistein on cell proliferation and cell cycle arrest in non-neoplastic human mammary epithelial cells: involvement of Cdc2, $p21^{waf/cip1}$, $p27^{kip1}$, and Cdc25C expression. Biochem Pharmacol 2001; 61:979–989.

39. Shao ZM, Shen ZZ, Fontana JA, Barsky SH. Genistein's "ER-dependent and independent" actions are mediated through ER pathways in ER-positive breast carcinoma cell lines. Anticancer Res 2000; 20:2409–2416.

40. Allred CD, Allred KF, Ju YH, Virant SM, Helferich WG. Soy diets containing varying amounts of genistein stimulate growth of estrogen-dependent (MCF-7) tumors in a dose-dependent manner. Cancer Res 2001; 61:5045–5050.

41. De Lemos ML. Effects of soy phytoestrogens genistein and daidzein on breast cancer growth. Ann Pharmacother 2001; 35:1118–1121.

42. Allred CD, Ju YH, Allred KF, Chang J, Helferich WG. Dietary genistin stimulates growth of estrogen-dependent breast cancer tumors similar to that observed with genistein. Carcinogenesis 2001; 22:1667–1673.

43. El-Deiry WS, Tokino T, Velculescu VE, Levy DB, Parsons R, Trent JM, Lin D, Mercer WE, Kinzler KW, Vogelstein B. WAF1, a potential mediator of p53 tumor suppression. Cell 1993; 75:817–825.

44. Nigg EA. Cellular substates of $p34^{CDK2}$ and its companion cyclin-dependent kinases. Trend Cell Biol 1993; 3:296–301.

45. Miyashita T, Reed JC. Tumor suppressor p53 is a direct transcriptional activator of the human bax gene. Cell 1995; 80:293–299.

46. Harper JW, Adami GR, Wei N, Keyomarsi K, Elledge SJ. The p21 Cdk-interacting protein Cip1 is a potent inhibitor of G1 cyclin-dependent kinases. Cell 1993; 75: 805–816.

47. Griffiths K, Morton MS, Denis L. Certain aspects of molecular endocrinology that relate to the influence of dietary factors on pathogenesis of prostate cancer. Eur Urol 1999; 35:443–455.

48. Shen JC, Klein RD, Wei Q, Guan Y, Contois JH, Wang TT, Chang S, Hursting SD. Low-dose genistein induces cyclin-dependent kinase inhibitors and $G_1$ cell-cycle arrest in human prostate cancer cells. Mol Carcinogen 2000; 29:92–102.

49. Kumi-Diaka J, Sanderson NA, Hall A. The mediating role of caspase-3 protease in the intracellular mechanism of genistein-induced apoptosis in human prostatic carcinoma cell lines, DU145 and LANCaP. Biol Cell 2000; 92:595–604.

50. Choi YH, Lee WH, Park KY, Zhang L. p53-independent induction of p21$^{WAF1/CIP1}$, reduction of cyclin B1 and G2/M arrest by the isoflavone genistein in human prostate carcinoma cells. Jpn J Cancer Res 2000; 91:164–173.

51. Agarwal R. Cell signaling and regulator of cell cycle as molecular target for prostate cancer prevention by dietary agents. Biochem Pharmacol 2000; 60:1051–1059.

52. Mitchell JH, Duthie SJ, Collins AR. Effects of phytoestrogens on growth and DNA integrity in human prostate tumor cell lines: PC-3 and LNCaP. Nutr Cancer 2000; 38:223–228.

53. Knowles LM, Zigrossi DA, Tauber RA, Hightower C, Milner JA. Flavonoids suppress androgen-independent human prostate tumor proliferation. Nutr Cancer 2000; 38:116–122.

54. Li X, Marani M, Mannucci R, Kinsey B, Andriani F, Nicoletti I, Denner L, Marcelli M. Over-expression of BCL-$X_L$ underlies the molecular basis for resistance to staurosporine-induced apoptosis in PC-3 cells. Cancer Res 2001; 61:1699–1706.

55. Kobayashi T, Nakata T, Kuzumaki T. Effect of flavonoids on cell cycle progression in prostate cancer cells. Cancer Lett 2002; 176:17–23.

56. Davis JN, Kucuk O, Sarkar FH. Genistein inhibits NF-κB activation in prostate cancer cells. Nutr Cancer 1999; 35:167–174.

57. Piontek M, Hengels KJ, Porschen R, Strohmeyer G. Anti-proliferative effect of tyrosine kinase inhibitors in epidermal growth factor-stimulated growth of human gastric cancer cells. Anticancer Res 1993; 13:2119–2123.

58. Matsukawa Y, Marui N, Sakai Y, Yoshida M, Matsumoto K, Nishino H, Aoike A. Genistein arrests cell cycle progression at G2-M. Cancer Res 1993; 53:1328–1331.

59. Kuo SM, Leavitt PS. Genistein increases metallothionein expression in human intestinal cells, Caco-2. Biochem Cell Biol 1999; 77:79–88.

60. Kameoka S, Leavitt P, Chang C, Kuo SM. Expression of antioxidant proteins in human intestinal Caco-2 cells treated with dietary flavonoids. Cancer Lett 1999; 146:161–167.

61. Pitt BR, Schwarz M, Woo ES, Yee E, Wasserllos K, Tran S, Weng W, Mannix RJ, Watkins SA, Tyurina YY, Tyurin VA, Kagan VE, Lazo JS. Overexpression of metallothionein decreases sensitivity of pulmonary endothelial cells to oxidant injury. Am J Physiol 1997; 273:L856–L865.

62. Djuric Z, Chen G, Doerge DR, Heibrun LK, Kucuk O. Effect of soy isoflavone supplementation on markers of oxidative stress in men and women. Cancer Lett 2001; 172:1–6.

63. Iishi H, Tatsuta M, Baba M, Yano H, Sakai N, Akedo H. Genistein attenuates peritoneal metastasis of azoxymethan-induced intestinal adenocarcinomas in Wistar rats. Int J Cancer 2000; 86:416–420.

64. Arai N, Strom A, Rafter JJ, Gustafsson JA. Estrogen receptor β mRNA in colon cancer cells: growth effects of estrogen and genistein. Biochem Biophys Res Comm 2000; 270:425–431.

65. Salti GI, Grewal S, Mehta RR, Das Gupta TK, Boddie AW, Constantinou AI. Genistein induces apoptosis and topoisomerase II-mediated DNA breakage in colon cancer cells. Eur J Cancer 2000; 36:796–802.
66. Park JH, Oh EJ, Choi YH, Kim DK, Kang KI, Yoo MA. Synergistic effects of dexamethasone and genistein on the expression of Cdk inhibitor p21$^{WAF1/CIP1}$ in human hepatocellular and colorectal carcinoma cells. Int J Oncol 2001; 18:997–1002.
67. Lian F, Bhuiyan M, Li YW, Wall N, Kraut M, Sarkar FH. Genistein-induced G$_2$-M Arrest, p21$^{WAF1}$ upregulation, and apoptosis in a non-small-cell lung cancer cell line. Nutr Cancer 1998; 31:184–191.
68. Lian F, Li YW, Bhuiyan M, Sarkar FH. P53-independent apoptosis induced by genistein in lung cancer cells. Nutr Cancer 1999; 33:125–131.
69. Gonzalez-Espinosa C, Garcia-Sainz JA. Protein kinases and phosphatases modulated c-fos expression in rat hepatocytes. Effect of angiotensin II and phorbol myristate actate. Life Sci 1995; 56:723–728.
70. Mousavi Y, Adlercreutz H. Genistein is an effective stimulator of sex hormone-binding globulin production in heptocarcinoma human liver cancer cells and suppresses proliferation of these cells in culture. Steroids 1993; 58:301–304.
71. Adlercreutz H, Hockerstedt K, Bannwart C, Bloigu S, Hamalainen E, Fotsis T, Ollus A. Effect of dietary components, including lignans and phytoestrogens, on enterohepatic circulation and live metabolism of estrogens and on sex hormone binding globulin (SHBG). J Steroid Biochem 1987; 27:1135–1144.
72. Dangles V, Femenia F, Laine V, Berthelemy M, Le Rhun D, Poupon MF, Levy D, Schwartz-Cornil I. Two- and three-dimensional cell structures govern epidermal growth factor survival function in human bladder carcinoma cell lines. Cancer Res 1997; 57:3360–3364.
73. Saito S, Hata M, Fukuyama R, Sakai K, Kudoh J, Tazaki H, Shimizu N. Screening of H-ras gene point mutations in 50 cases of bladder carcinoma. Int J Urol 1997; 4: 178–185.
74. Chow NH, Liu HS, Lee EI, Chang CJ, Chan SH, Cheng HL, Tzai TS, Lin JS. Significance of urinary epidermal growth factor and its receptor expression in human bladder cancer. Anticancer Res 1997; 17:1293–1296.
75. Theodorescu D, Laderoute KR, Calaoagan JM, Gulding KM. Inhibition of human bladder cancer cell motility by genistein is dependent on epidermal growth factor receptor but not p21$^{ras}$ gene expression. Int J Cancer 1998; 78:775–782.
76. Zhou JR, Mukherjee P, Gugger ET, Tanaka T, Blackburn GL, Clinton SK. Inhibition of murine bladder tumorigenesis by soy isoflavones via alterations in the cell cycle, apoptosis, and angiogenesis. Cancer Res 1998:5231–5238.
77. Naraghi S, Khoshyomn S, DeMattia JA, Vane DW. Receptor tyrosine kinase inhibition suppresses growth of pediatric renal tumor cells in vitro. J Pediatr Surg 2000; 35:884–890.
78. Chen CL, Levine A, Rao A, O'Neill K, Messinger Y, Myers DE, Goldman F, Hurvitz C, Casper JT, Uckun FM. Clinical pharmacokinetics of the CD19 receptor-directed tyrosine kinase inhibitor B43-genistein in patients with B-lineage lymphoid malignancies. J Clin Pharmacol 1999; 39:1248–1255.
79. Myers DE, Jun X, Waddick KG, Forsyth C, Chelstrom LM, Gunther RL, Tumer NE, Bolen J, Uckun FM. Membrane-associated CD19-LYN complex is an endoge-

nous p53-independent and Bcl-2-2independent regulator of apoptosis in human B-lineage lymphoma cells. Proc Natl Acad Sci USA 1995; 92:9575–9579.

80. Wijetunge S, Lymn JS, Hughes AD. Effects of protein tyrosine kinase inhibitors on voltage-operated calcium channel currents in vascular smooth muscle cells and pp60$^{c-sac}$ kinase activity. Br J Pharmacol 2000; 129:1347–1354.

81. Mall M, Wissner A, Seydewitz HH, Hubner M, Kuehr J, Brandis M, Greger R, Kunzelmann K. Effect of genistein on native epithelial tissue from normal individuals and CF patients and on ion channels expressed in *Xenopus oocytes*. Br J Pharmacol 2000; 130:1884–1892.

82. Schnabel P, Mies F, Nohr T, Geisler M, Bohm M. Differential regulation of phospholipase C-β isozymes in cardiomyocyte. Biochem Biophys Res Comm 2000; 275: 1–6.

83. Thorburn J, Thorburn A. The tyrosine kinase inhibitor, genistein, prevents α-adrenergic-induced cardiac muscle cell hypertrophy by inhibiting activation of Ras-MAP kinase signaling pathway. Biochem Biophys Res Comm 1994; 202:1586–1591.

84. Chen CC, Jin YT, Liau YE, Huang CH, Liou JT, Wu LW, Huang W, Young KC, Lai MD, Liu HS, Shieh B, Li C. Profiling of gene expression in bladder tumor cells treated with genistein. J Biomed Sci 2001; 8:214–22.

85. Duggan DJ, Bittner M, Chen Y, Meltzer P, Trent JM. Expression profiling using cDNA microarrays. Nat Genet 1999; 21(suppl):10–14.

86. Lennon G, Auffray C, Polymeropoulos M, Soares MB. The I.M.A.G.E. Consortium: an integrated molecular analysis of genomes and their expression. Genomics 1996; 33:151–152.

87. Chen JJW, Wu R, Yang PC, Huang JY, Sher YP, Han MH, Kao WC, Lee PJ, Chiu TF, Chang F, Chu YW, Wu CW, Peck K. Profiling expression patterns and isolating differentially-expressed genes by cDNA microarray system with colorimetry detection. Genomics 1998; 51:313–324.

88. Nayak SK, O'Toole C, Price ZH. A cell line from an anaplastic transitional cell carcinoma of human urinary bladder. Br J Cancer 1977; 35:142–151.

89. Bubenik J, Baresova M, Viklicky V, Jakoubkova J, Sainerova H, Donner J. Established cell line of urinary bladder carcinoma (T24) containing tumor-specific antigen. Int J Cancer 1973; 11:765–773.

90. Yeh MY, Yu DS, Chen SC, Lin MS, Chang SY, Ma CP, Han SH. Establishment and characterization of a human urinary bladder carcinoma cell line (TSGH-8301). J Surg Oncol 1988; 37:177–184.

91. Eisen MB, Spellman PT, Brown PO, Botstein D. Cluster analysis and display of genome-wide expression patterns. Proc Natl Acad Sci USA 1998; 95:14863–14868.

92. van Straaten F, Muller R, Curran T, Van Beveren C, Verma IM. Complete nucleotide sequence of a human c-onc gene: deduced amino acid sequence of the human c-fos protein. Proc Natl Aca Sci USA 1983; 80:3183–3187.

93. Li C, Mitchell DH, Coleman DL. Analysis of Egr-1 protein induction in murine peritoneal macrophages treated with granulocyte-macrophage colony-stimulating factor. Yale J Biol Med 1994; 67:269–276.

94. Chen KS, DeLuca HF. Isolation and characterization of a novel cDNA from HL-60 cells treated with 1,25-dihydroxyvitamin D-3. Biochim Biophys Acta 1994; 1219: 26–32.

95. Nishiyama A, Matsui M, Iwata S, Hirota K, Masutani H, Nakamura H, Takagi Y, Sono H, Gon Y, Yodoi J. Identification of thioredoxin-binding protein-2/vitamin $D_3$ up-regulated protein 1 as a negative regulator of thioredoxin function and expression. J Biol Chem 1999; 274:21645–21650.
96. Junn E, Han SH, Im JY, Yang Y, Cho EW, Um HD, Kim DK, Lee KW, Han PL, Rhee SG, Choi I. Vitamin D3 up-regulated protein 1 mediates oxidative stress via suppressing the thioredoxin function. J Immunol 2000; 164:6287–6295.
97. Peterson G. Evaluation of the biochemical targets of genistein in tumor cell. J Nutr 1995; 125(suppl):784S–789S.
98. Miyazawa M, Sakano K, Nakamura SI, Kosaka H. Antimutagenic activity of isoflavones from soybean seeds (*Glycine max* Merrill). J Agr Food Chem 1999; 47: 1346–1349.

# 5

# Gene Regulatory Activity of *Ginkgo biloba* L.

**Gerald Rimbach**
*University of Reading, Reading, United Kingdom*

**Rainer Cermak and Siegfried Wolffram**
*Christian-Albrechts-University of Kiel, Kiel, Germany*

## I. CHEMICAL COMPOSITION

The leaves of *Ginkgo biloba* L. contain a variety of chemical compounds, some of which are unique to this ancient tree. Various substances including long-chain hydrocarbons and derivatives, alicyclic acids, cyclic compounds, carbohydrates and derivatives, lectins, carotenoids, isoprenoids (sterols, terpenoids), and flavonoids can be found in its leaves. Several of these compounds have antibiotic and antifungal properties or afford protection against herbivores, attributes that may very well have contributed to the long survival of this unique plant [1,2]. From a pharmacological perspective, flavonoids and terpenoids are probably the most interesting components. In dried *Ginkgo* leaves, the content of the flavonoid glycosides is around 0.5–1% w/w, whereas the content of terpenoids is often below 0.1% w/w [2]. The phenolic aglycon moiety of the flavonoid glycosides is made up predominantly of the flavonols quercetin or kaempferol, the occurence of isorhamnetin being significantly smaller. The flavonols myricetin, 3'-*O*-methylmyricetin, and tamarixetin and the flavones luteolin, apigenin, and 4'-*O*-methylapigenin are found only in small amounts (Fig. 1A). The glycoside moiety consists of mono-, di-, or triglycosides (mainly D-glucose and L-rhamnose units), which are connected to the aglycon at position 3, 3', or 7 via *O*-glycosidic bonds. Some of these glycosides are additionally acylated by *p*-coumaric acid (Fig. 1B) [3]. Since these latter compounds are distinctive to *Ginkgo biloba* leaves and extracts, they are considered to be lead substances in this species. Other compounds with

*A*

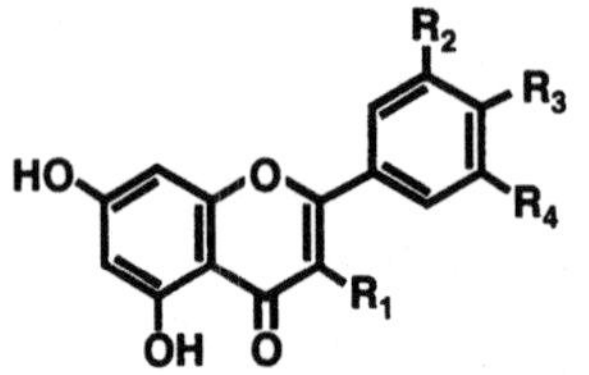

| Flavonoidaglycon | R₁ | R₂ | R₃ | R₄ |
|---|---|---|---|---|
| Quercetin | OH | OH | OH | H |
| Kaempferol | OH | H | OH | H |
| Isorhamnetin | OH | O-CH₃ | OH | H |
| Myricetin | OH | OH | OH | OH |
| 3'-O-Met-myricetin | OH | O-CH₃ | OH | OH |
| Tamarixetin | OH | OH | O-CH₃ | H |
| Luteolin | H | OH | OH | H |
| Apigenin | H | H | OH | H |
| 4'-O-Met-apigenin | H | H | O-CH₃ | H |

*B*

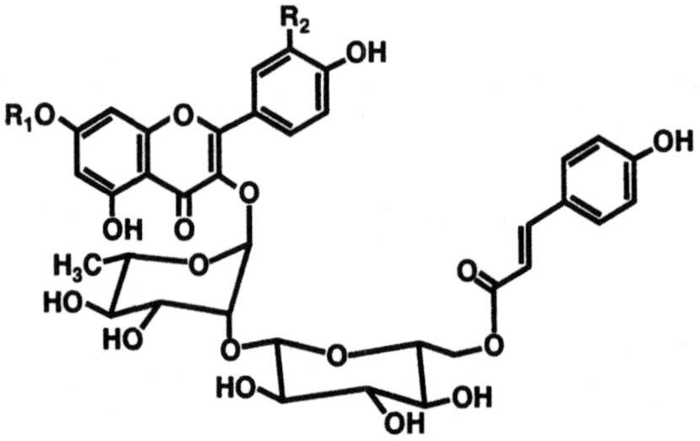

| Coumaroyl flavonol glycoside | R₁ | R₂ |
|---|---|---|
| Quercetin 3-O-α-L-[6-p-coumaroyl-(β-D)-glucosyl-(1,2)-rhamnoside] | H | OH |
| Kaempferol 3-O-α-L-[6-p-coumaroyl-(β-D)-glucosyl-(1,2)-rhamnoside] | H | H |
| Isorhamnetin 3-O-α-L-[6-p-coumaroyl-(β-D)-glucosyl-(1,2)-rhamnoside] | H | O-CH₃ |
| Quercetin 3-O-α-L-[6-p-coumaroyl-(β-D)-glucosyl-(1,2)-rhamnoside]-7-O-β-D-glucoside | glucoside | OH |
| Kaempferol 3-O-α-L-[6-p-coumaroyl-(β-D)-glucosyl-(1,2)-rhamnoside]-7-O-β-D-glucoside | glucoside | H |

**Figure 1** Structures of flavon- and flavonol aglycons found in leaves of *Ginkgo biloba* (*A*). Structures of some flavonol glycoside esters with coumaric acid, distinctive compounds of *Ginkgo biloba* (*B*). The coumaroyl esters of the quercetin-rhamnoglucoside and of the kaempferol-rhamnoglucoside are most common in *Ginkgo* leaves.

a flavonoid-like structure, such as nonglycosidic biflavonoids, catechins, and pro-anthocyanidins, have also been isolated from this plant [1,2,4].

In addition to flavonoids, compounds that belong to the group of terpenoids are of particular interest. This group contains the ginkgolides, with a diterpenoid structure, and the sesquiterpenoid bilobalide, these being unique to _Ginkgo biloba_ (Fig. 2). Five different ginkgolides have been isolated, of which ginkgolides A, B, C, and J are found in the leaves; the fifth, ginkgolid M, occurs solely in the roots. In the leaves, ginkgolides A, B, and C predominate, with ginkgolide J only found in smaller amounts. Ginkgolides are hexacyclic diterpenes possessing three lactone rings forming electrophilic cage-like structures that enables them to interact with cations and positively charged residues of other molecules. They are chemically very stable, being resistant to concentrated hydrochloric and sulfuric acids as well as to caustic soda solution. Bilobalide, a $C_{15}$-trilactone, is considered to be a degradation product of the ginkgolides. Its concentration in _Ginkgo biloba_ leaves is as high as that of all the ginkgolides [1,2,4]. A unique structural feature of all these terpene lactones is their tertiary butyl residue (Fig. 2). Besides these

**A**

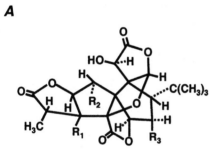

| Ginkgolide | R₁ | R₂ | R₃ |
|------------|-----|-----|-----|
| A | OH | H | H |
| B | OH | OH | H |
| C | OH | OH | OH |
| J | OH | H | OH |
| M | H | OH | OH |

**B**

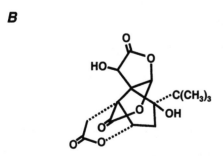

**Figure 2**  Structures of ginkgolides (_A_) and bilobalide (_B_) found in leaves of _Ginkgo biloba_. Ginkgolide M is present only in the roots.

distinct isoprenoids, *Ginkgo biloba* leaves also contain several acyclic polyprenoids and steroids.

In addition to the above substances, a large variety of polysaccharides, long-chain hydrocarbons, alcohols, and acids have been isolated, as well as (Z,Z)-4,4′-(1,4-pentadiene-1,5-diyl)diphenol and 6-hydroxykynurenic acid [1,2]. The cytotoxic, mutagenic, and allergenic potential of some constituents, especially anacardic or ginkgolic acids, limits the use of a crude leaf extract for therapeutical or prophylactic purposes in humans [5]. Thus, several extraction processes have been developed for removal of toxic substances and concomitant enrichment of desired compounds. The extract EGb 761 (formerly patented by Dr. W. Schwabe GmbH & Co, Karlsruhe, Germany) contains 22–27% flavone glycosides, 5–7% terpenoids (2.8–3.4% ginkgolides A, B, C and 2.6–3.2% bilobalide), and not more than 5 ppm ginkgolic acids [6]. This has been the most widely used extract of *Ginkgo biloba* leaves in clinical studies and is the basis for many commercially available medicinal products of *Ginkgo biloba*.

## II. BIOAVAILABILITY

The bioavailability of the constituents of EGb 761 was investigated by Moreau et al. in 1986 using an extract derived from plants grown in a medium containing $^{14}C$-acetate [7]. After intragastric application of this radiolabeled extract, activity in plasma peaked after $1\frac{1}{2}$ hours ($T_{max}$), indicating the absorption of at least some constituents in the upper gastrointestinal tract. At least 60% of the total radioactivity was found in body tissues, urine, or expired $CO_2$, suggesting a high bioavailabilty for *Ginkgo biloba* constituents. Interestingly, radioactivity accumulated in glandular and neuronal tissue, but this study neither identified the plant constituents that were absorbed nor determined whether they reached the systemic circulation in intact or in metabolized form [7].

More recently, the absorption of specific *Ginkgo* terpenoids has been verified. The occurrence of intact ginkgolides A, B and bilobalide in plasma after oral application of the *Ginkgo* extract has been demonstrated in both humans and rats [8,9]. Plasma concentrations peaked at 2–4 hours after application of the extract in humans and simultaneous application of a phospholipid complex enhanced the oral bioavailability of the terpene lactones [9]. One human study suggested that the absorption of these three terpene lactones after oral intake of the extract is nearly complete with bioavailability coefficients larger than 0.8 [10]. Interestingly, all of these studies detected only traces of ginkgolide C in plasma, although it was clearly present in the batches of *Ginkgo* extract used. Detailed knowledge of the absorption mechanism of these terpenoids is still missing.

Several studies have demonstrated the occurrence of the main *Ginkgo* flavonols quercetin, kaempferol, and isorhamnetin in plasma or urine of humans and mice after ingestion of EGb 761 [11–13]. However, intact flavonol glycosides are not present in the systemic circulation [14,15], a consequence of the intensive metabolic activity of the gastrointestinal tract. Flavonoid glycosides are deglycosylated by intestinal and/or bacterial glycosidases followed by methylation and conjugation with glucuronide and/or sulfate in the intestine and liver [16]. Thus, flavonoids are present mainly as conjugates in plasma. It remains a matter of debate whether certain quercetin glucosides can be absorbed *via* specific small intestinal carrier mechanisms (e.g., the sodium-dependent glucose transporter SGLT1) prior to deglycosylation by cytosolic β-glucosidases [17]. An alternative hypothesis presumes that β-glucosidases of the brush border membrane (e.g., lactase phlorizin hydrolase, LPH) deglycosylate the glycoside extracellularly before the lipophilic aglycon can enter the intestinal epithelium [18]. Only flavonol glucosides, which are substrates for LPH or cytosolic β-glucosidases, are already absorbed in the small intestine. Conjugates of those flavonols already appear in the systemic circulation within one hour after ingestion.

Flavonols from other glycosides like quercetin-3-glucorhamnoside (rutin), which is not a substrate for these enzymes, is only detected after several hours in the circulation, indicating the need for microbial deglycosylation in the terminal ileum or large intestine prior to absorption of the aglycon. The bioavailability of quercetin from rutin is relatively much smaller than from several other quercetin glucosides, which is probably due to ongoing degradation of the flavonol skeleton by microbial enzymes and/or a slow release of the aglycon. The bioavailability of flavonols from flavonol glycosides is, thus, dependent on the sugar moiety [15,19].

This bioavailability is further reduced by the secretion of flavonoid metabolites from the enterocytes as well as to their biliary excretion [20]. Data on the total bioavailability of single flavonoids in humans are scarce and contradictory, with estimates ranging from 0 to 50% [21]. In pigs, total bioavailability of quercetin was around 17% when all metabolites with an intact flavonol structure were taken into account [22]. This all suggests that the total bioavailability of the flavonoid glycosides present in *Ginkgo biloba* is much smaller than that of the terpene lactones. It is also important to note that in contrast to the situation for the terpenoids, the circulatory forms of the flavonoid glycosides are not the same as the original compounds. In vivo, the potentially bioactive flavonoids are mainly a mixture of glucuronidated derivatives of quercetin, kaempferol, and isorhamnetin.

Many of the positive effects of EGb 761, especially on central nervous system (CNS) functions, have been ascribed to the antioxidant properties of the extract [23]. Its flavonoid portion is able to protect cultured hippocampal cells against the toxicity induced by the nitric oxide radical [24]. The antioxidative potential of flavonoids has been demonstrated in numerous investigations, but most studies describe the antioxidative properties of the aglycones and do not take into account that circulating postabsorptive forms are different from the ingested parent substances [25].

## III.  NEUROMODULATORY EFFECTS

Supplementation of diets with plant extracts such as *Ginkgo biloba* (EGb 761) for health and prevention of degenerative diseases is popular. It is, however, often difficult to analyze the biological activities of plant extracts because of their complex natures and the occurrence of possible synergistic and/or antagonistic effects. Genome-wide expression monitoring with high-density oligonucleotide arrays offers one way to examine the molecular targets of plant extracts and may prove a valuable tool for evaluating therapeutic claims. Here, we briefly describe the results of work on the effect of EGb 761 on differential gene expression in relation to its potential neuromodulatory, anticancer, anti-inflammatory, and photo-protective properties.

EGb 761 is commonly used to combat a variety of neurological disturbances such as Alzheimer's disease or various common geriatric complaints including vertigo, depression, short-term memory loss, hearing loss, lack of attention or vigilance [26]. To gain further insights into the biochemical effects of *Ginkgo biloba*, the transcriptional effects of the extract on the brains of mice were profiled using oligonucleotide microarrays [12]. These microarrays represent all (~6000) sequences in the Mouse UniGene database that have been functionally characterized, as well as ~6000 expressed sequence tag (EST) clusters. The effects on gene transcription of EGb 761 were measured in the hippocampus and cerebral cortex of adult female mice ($n = 10$ per group) fed a diet with or without *Ginkgo biloba* extract (300 mg EGb 761/kg diet). After the 4-week diet regimen, the hippocampi and cortices were removed and pooled with respect to both tissue type and treatment. The concentrations of quercetin, kaempferol, and isorhamnetin (major constituents of EGb 761) in plasma samples were measured by HPLC [22]. Quercetin, kaempferol, and isorhamnetin concentrations of the EGb761-supplemented group were significantly higher than those of the control group, confirming the absorption of at least some of the major components of EGb 761 into the systemic circulation.

Of the 12,000 combined genes and ESTs represented on the array, only 10 changed in expression level by a factor of threefold or more, and all were upregulated. In the cortex, mRNAs for neuronal tyrosine/threonine phosphatase 1 and microtubule-associated tau were significantly enhanced. These proteins are both associated with formation/breakdown of intracellular neurofibrillary tangles, a hallmark lesion of Alzheimer's disease. Hyperphosphorylated tau has been found to be the major protein of these neurofibrillary tangles, possibly because of an imbalance in tau kinase and phosphatase activities in the affected neurons [27]. Hyperphosphorylated tau isolated from brains of patients with Alzheimer's disease has been shown to be efficiently dephosphorylated in vitro by protein phosphatases 1, 2A, and 2B. Additionally, selective inhibition of protein phosphatase 2A by okadaic acid in metabolically competent rat brain slices has been shown to induce a hyperphosphorylation and accumulation of tau like that in Alzheimer's disease [28]. Thus, upregulation of neuronal phosphatase 1 by EGb 761 could play a neuroprotective role in the brain.

The expression of α-amino-3-hydroxy-5-methyl-4-isoxazolepropionic-acid-2 (AMPA-2), calcium and chloride channels, prolactin, and growth hormone (GH), all of which are associated with brain function, was also upregulated. Within the past decade, studies have revealed that GH may exert significant effects on the central nervous system. Cognitive impairments are well-documented hallmark features of GH deficiency, and clinical studies have reported psychological improvements (in mood and well-being) and beneficial effects on certain functions including memory, mental alertness, motivation, and working capacity in adults receiving GH-replacement therapy. Moreover, GH therapy in children deficient in this protein has been reported to produce marked behavioral improvement [29].

In the hippocampus, only transthyretin mRNA was induced. Transthyretin plays a role in the transport of tyroxine and retinol-binding protein in the brain. Thyroid hormones regulate neuronal proliferation and differentiation in discrete regions of the brain during development and are necessary for normal cytoskeletal outgrowth [30]. Transthyretin has also been shown in vitro to sequester amyloid beta protein and prevent amyloid beta aggregation from arising in amyloid formation [31]. Moreover, transthyretin levels in cerebrospinal fluid have been found to be significantly decreased in Alzheimer's disease patients. Thus, one mechanism whereby EGb 761 may exert neurological effects is the modulation of transthyretin levels and, as a consequence, by either hormone transport or amyloid beta sequestration in the brain.

## IV. ANTI-INFLAMMATORY AND ANTIATHEROGENIC EFFECTS

Monocyte-derived macrophages are the principal inflammatory cell type in the atheromatic plaque environment. Macrophages, due to their phagocytic nature, remove noxious material that accumulates in the vessel wall during lesion formation [32]. In early stages of atherosclerotic lesion formation, macrophages and endothelial cells interplay, shifting from the normal homeostasis and possibly triggering processes that exacerbate endothelial dysfunction [33]. During macrophage activity the generation of reactive oxygen (ROS) and nitrogen species (RNS) induces a condition of oxidative-nitrosative stress. ROS and RNS may finally affect the antioxidant network, redox-sensitive transcription factors, and gene expression in the endothelial cell. Indeed, oxidative stress–induced injury has been observed in the environment of artheriosclerotic plaques, suggesting the importance of free radicals in the etiology of cardiovascular diseases [34,35].

Kobuchi et al. [36] studied the effect of *Ginkgo biloba* extract (EGb 761) on the synthesis of nitric oxide in RAW264.7 macrophages activated by interferon-$\gamma$ (IFN-$\gamma$) and lipopolysaccharide (LPS). EGb 761 inhibited nitrite and nitrate production in a concentration-dependent manner. Reverse transcription polymerase chain reaction (RT-PCR) analysis revealed that the effect of EGb 761 on NO production in macrophages is mediated by downregulation of inducible NO synthase (iNOS) gene expression. These data indicate that EGb 761 not only acts directly as an NO scavenger but also inhibits NO production through its gene regulatory activity.

NF-$\kappa$B is an ubiquitous transcription factor that plays a significant role in immune response by regulating the gene expression of multiple inflammatory and immune genes [37,38]. NF-$\kappa$B activation in endothelial cells has consequently been considered as critical event in the pathogenesis of atherosclerosis [39]. A recent study demonstrated activation of NF-$\kappa$B in cells within atherosclerotic lesions but not in cells of normal vessels [40]. Furthermore, NF-$\kappa$B activation has been demonstrated in an arterial injury model [41]. In endothelial cells NF-$\kappa$B regulates the inducible expression of genes encoding chemokines, adhesion molecules, and growth factors. Activated NF-$\kappa$B may also modulate the endothelial cell production of chemotactic substances such as monocyte chemoattractant protein (MCP-1) [42]. MCP-1 is a CC-chemokine consisting of 76 amino acids that has recently been suggested to play a key role in atherogenesis and tissue injury through its involvement in recruitment of monocytes into the arterial wall [43,44]. It has recently been demonstrated that co-culture between human endothelial cells (HUVEC) and macrophages induced a pro-oxidant environment and resulted in increased DNA binding and NF-$\kappa$B transactivation of these cells [45]. Activation of NF-$\kappa$B in endothelial cells was accompanied by an increase in the expression of mRNA encoding for the MCP-1 protein. Most importantly, pre-incubation of endothelial cells with EGb 761 significantly downregulated both NF-$\kappa$B transactivation (Fig. 3) as well as MCP-1 gene expression in HUVEC [46].

Overall, these regulatory effects of EGb 761 on nitric oxide metabolism, NF-$\kappa$B activity, and chemokine expression suggest that these may be implicated in the beneficial effects of *Ginkgo biloba* on inflammatory disorders and atherogenesis.

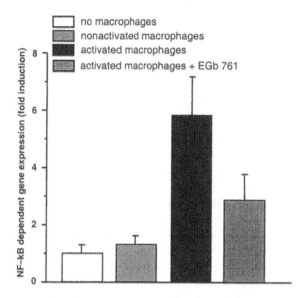

**Figure 3** NF-κB transactivation activity in human endothelial cells (HUVEC) co-incubated with RAW 264.7 macrophages treated with 10 U/ml IFN-γ. Twelve hours before exposure to macrophages, HUVEC are transiently transfected with two plasmids containing either Firefly or Renilla luciferases hooked up to a multiple NF-κB consensus sequence and to a simple thymidine kinase promoter, respectively. Twelve hours after co-incubation, HUVEC are collected and both Renilla and Firefly luciferase activities are measured in cell lysate. The effect of the preincubation with *Ginkgo biloba* extract EGb761 (100 μg/mL) is also shown. Results are expressed as the ratio of Renilla to Firefly luciferase activity in the cell lysate.

# V. ANTICARCINOGENIC PROPERTIES

Reactive oxygen and reactive nitrogen species are major contributors to chronic disorders such as cancer, photoaging, and neurodegenerative diseases. It is well established that *Ginkgo biloba* extract EGb 761 has potent antioxidant properties achieved by either scavenging free radicals or chelating pro-oxidant transition metals [47]. Since reactive oxygen species regulate the expression of a large number of genes, it is likely that EGb 761 modulates genes under the control of redox-sensitive transcription factors such as NF-κB, AP-1, and Nrf-1 [46,48]. Furthermore, flavonoids, major constituents of EGb 761, have been shown to target transcription processes independent of their antioxidant properties. It has been demonstrated that flavonoids inhibit cyclin-dependent kinases and topoisomerase, thereby controlling cellular proliferation, differentiation, and apoptosis [49,50]. Interestingly, terpenoids, also present in EGb 761, arrest cancer cells in the G1 phase and initiate apoptosis [51]. Gohil et al. [52] recently examined the mRNA expression profile of human T-24 bladder carcinoma cells in response to EGb 761 treatment (100 μg/mL for 72 h). T-24 cells offer an attractive in vitro model to test the transcriptional response of *Ginkgo biloba* since they express constitutively high levels of the oncogenic G-protein Ha-ras, and it is well established that a high expression of Ras increases reactive oxygen species and cellular proliferation [53]. The analysis of the transcriptional response by Affymetrix gene chip technology revealed a net activation of transcription due to EGb 761 treatment of T-24 cells; the expression of 139 genes was upregulated and the abundance of 16 mRNAs was downregulated. Functional classification of the affected mRNAs showed the largest changes in the abundance of mRNAs encoding for intracellular vesicular transport proteins, mitochondrial proteins, and transcription factors. Furthermore, several transcripts encoding proteins with antioxidant function were induced in T-24 cells in the presence of EGb 761; these included mRNAs for heme oxygenase and manganese-dependent mitochondrial superoxide dismutase (MnSOD). Heme oxygenase-1 catalyzes the oxidation of heme to carbon monoxide, iron, and biliverdin, a potent antioxidant. Heme oxygenase-1 activity has also been reported to result in growth arrest and an increased resistance to hyperoxia [54]. Interestingly, MnSOD is expressed at decreased levels in cancer cells, leading to the suggestion that overexpression of human MnSOD suppresses growth of cancer cells [55]. The incubation of T-24 cells with EGb 761 induced at least four genes encoding proteins essential for the repair and synthesis of DNA, including the transcripts for DNA mismatch repair enzyme, DNA ligase, the regulatory subunit of DNA polymerase, and topoisomerase-I. Furthermore $^3$H-thymidine incorporation in cellular DNA of T-24 cells incubated with EGb 761 was significantly decreased by up to 50%.

## VI. PHOTOPROTECTION

As the outermost barrier of the human body, skin is chronically exposed to environmental stressors such as ultraviolet (UV) radiation, ozone, and oxidizing chemicals. A variety of enzymatic and nonenzymatic antioxidants protect skin against damaging effects from reactive oxygen and nitrogen species. The cellular thiol tripeptide, glutathione (GSH), plays a pivotal role in the antioxidant network of the skin. Thus, it has been demonstrated that GSH depletion renders cultured keratinocytes more sensitive to UVA- and UVB-induced mutations and cell death [56]. Furthermore, decreased cellular GSH levels have been associated with immunosuppression and possibly skin aging [57]. The presence of sufficient amounts of intracellular GSH, therefore, seems to be essential for ensuring efficient cutaneous photoprotection. Several approaches have been made to boost GSH levels in skin. Exogenous GSH application and cysteine both failed to increase cutaneous GSH [58], but better results have been obtained with GSH-esters. Application of such esters in a millimolar range increased cellular GSH levels in human cultured cells up to 75%, while topical treatment of hairless mice with GSH-esters led to a significant rise in cutaneous GSH [59].

The effect of EGb 761 on GSH homeostasis in cultured human kerationcytes (HaCaT) has recently been studied [60]. HaCaT were incubated with EGb 761, some of its purified flavonoids (quercetin, kaempferol, rutin), or terpenoids (ging-kolides A, B, C, J, bilobalide) for up to 72 hours. Incubation of HaCaT with the purified flavonoids or terpenoids had no effect on cellular GSH levels. However, EGb 761 treatment (up to 200 µg/mL) resulted in dose-dependent increases in cellular GSH. *Ginkgo biloba* subfractions lacking terpenoids (CP205) or both terpenoids and flavonoids (HE215) caused a similar increase in cellular GSH to that of EGb 761, indicating that flavonoids and terpenes in the extract may not be responsible for augmentation of intracellular GSH.

Western blot analysis of extracts from cells treated with EGb 761 revealed increased levels of the catalytic subunit of γ-glutamylcysteinyl synthetase (γ-GCS), the rate-limiting enzyme in GSH synthesis. The abundance of mRNA for the catalytic subunit (assayed by RT-PCR) was also increased by the treatment with EGb 761.

The increase in GSH synthesis due to EGb 761 was accompanied by an increase in AP-1 and Nrf-1 DNA-binding activity. Increased levels of cellular GSH by EGb 761 were also observed in other cell lines, including those from human bladder and liver as well as murine macrophages, indicating that the induction of γ-GCS mRNA, protein, and GSH may be an ubiquitous effect of EGb 761 in mammalian cells [60].

While GSH-esters serve as direct precursors of cellular GSH, our data suggest that EGb 761 can augment intracellular GSH through transcriptional upregulation of γ-GCS gene expression. It seems, therefore, that EGb 761 displays antioxidant properties both by direct scavenging of ROS and RNS, as described previously, and by inducing gene expression and protein levels of γ-GCS. Further studies in vivo are needed to find out whether oral or topical applications of EGb 761 may have any practical implication in boosting cellular GSH, thereby preventing free radical–mediated damage to the skin.

## VII. CONCLUSION

A wide spectrum of beneficial activity to human health has been associated with *Ginkgo biloba* extract EGb 761. More recently the ability of EGb 761 to affect gene expression and cell response has been reported, thus providing a novel mechanistic perspective on its biological activity. The efficacy of *Ginkgo biloba* to positively influence the molecular mechanisms implicated in various diseases is becoming increasingly important. Differential changes in the expression of several groups of genes are key factors underlying the complex behavior of plant extracts. The determination of a global picture of the effect of EGb 761 on gene expression through genomic techniques will afford a better understanding of its action on the molecular level. Overall, gene chips and DNA microarrays in combination with proteomics represent new avenues in understanding gene regulation and signal-transduction pathways. This approach may yield better insights into the molecular mechanisms of plant extracts, thereby offering a novel strategy in *Ginkgo biloba* research. It also provides a model for more general future investigations.

## REFERENCES

1.  Hölzl J. Die Zusammensetzung von *Ginkgo biloba*. Pharm Unserer Zeit 1992; 21: 215–223.
2.  Sticher O. Quality of Ginkgo preparations. Planta Med 1993; 59:2–11.
3.  Tang Y, Lou F, Wang J, Li Y, Zhuang S. Coumaroyl flavonol glycosides from the leaves of *Ginkgo biloba*. Phytochemistry 2001; 58:1251–1256.
4.  Bedir E, Tatli II, Khan RA, Zhao J, Takamatsu S, Walker LA, Goldman P, Khan IA. Biologically active secondary metabolites from *Ginkgo biloba*. J Agric Food Chem 2002; 50:3150–3155.
5.  Hecker H, Johannisson R, Koch E, Siegers CP. In vitro evaluation of the cytotoxic potential of alkylphenols from *Ginkgo biloba*. L. Toxicology 2002; 177:167–177.
6.  Kressmann S, Müller WE, Blume HH. Pharmaceutical quality of different *Ginkgo biloba* brands. J Pharm Pharmacol 2002; 54:661–669.
7.  Moreau JP, Eck CR, McCabe J, Skinner S. Absorption, distribution and elimination of a labelled extract of *Ginkgo biloba* leaves in the rat. Presse Med 1986; 15:1458–1461.
8.  Biber A, Koch E. Bioavailability of ginkgolides and bilobalide from extracts of *Ginkgo biloba* using GC/MS. Planta Med 1999; 65:192–193.
9.  Mauri P, Simonetti P, Gardana C, Minoggio M, Morazzoni P, Bombardelli E, Pietta P. Liquid chromatography/atmospheric pressure chemical ionization mass spectrometry of terpene lactones in plasma of volunteers dosed with *Ginkgo biloba* L. extracts. Rapid Commun Mass Spectrom 2001; 15:929–934.
10. Fourtillan JB, Brisson AM, Girault J, Ingrand I, Decourt JP, Drieu K, Jouenne P, Biber A. Pharmacokinetic properties of bilobalide and ginkgolides A and B in healthy subjects after intravenous and oral administration of *Ginkgo biloba* extract (EGb 761). Therapie 1995; 50:137–144.
11. Wojcicki J, Gawronska-Szklarz B, Bieganowski W, Patalan M, Smulski HK, Samochowiec L, Zakrzewski J. Comparative pharmacokinetics and bioavailability of flavonoid glycosides of *Ginkgo biloba* after a single oral administration of three formulations to healthy volunteers. Mater Med Pol 1995; 27:141–146.

12. Watanabe CM, Wolffram S, Ader P, Rimbach G, Packer L, Maguire JJ, Schultz PG, Gohil K. The in vivo neuromodulatory effects of the herbal medicine *Ginkgo biloba*. Proc Natl Acad Sci USA 2001; 98:6577–6580.

13. Watson DG, Oliveira EJ. Solid-phase extraction and gas chromatography-mass spectrometry determination of kaempferol and quercetin in human urine after consumption of *Ginkgo biloba* tablets. J Chromatogr B 1999; 723:203–210.

14. Sesink ALA, O'Leary KA, Hollman PCH. Quercetin glucuronides but not glucosides are present in human plasma after consumption of quercetin-3-glucoside or quercetin-4'-glucoside. J Nutr 2001; 131:1938–1941.

15. Graefe EU, Wittig J, Mueller S, Riethling AK, Uehleke B, Drewelow B, Pforte H, Jacobasch G, Derendorf H, Veit M. Pharmacokinetics and bioavailability of quercetin glycosides in humans. J Clin Pharmacol 2001; 41:492–499.

16. Williamson G, Day AJ, Plumb GW, Couteau D. Human metabolic pathways of dietary flavonoids and cinnamates. Biochem Soc Trans 2000; 28:16–22.

17. Wolffram S, Blöck M, Ader P. Quercetin-3-glucoside is transported by the glucose carrier SGLT1 across the brush border membrane of rat small intestine. J Nutr 2002; 132:630–635.

18. Day AJ, Canada FJ, Diaz JC, Kroon PA, McLauchlan R, Faulds CB, Plumb GW, Morgan MRA, Williamson G. Dietary flavonoid and isoflavone glycosides are hydrolysed by the lactase site of lactase phlorizin hydrolase. FEBS Lett 2000; 468: 166–170.

19. Hollman PCH, Bijsman MNCP, van Gameren Y, Cnossen EPJ, Vries JHM, Katan MB. The sugar moiety is a major determinant of the absorption of dietary flavonoid glycosides in man. Free Radic Res 1999; 31:569–573.

20. Crespy V, Morand C, Manach C, Besson C, Demigne C, Remesy C. Part of quercetin absorbed in the small intestine is conjugated and further secreted in the intestinal lumen. Am J Physiol 1999; 277:G120–G126.

21. Graefe EU, Derendorf H, Veit M. Pharmacokinetics and bioavailability of the flavonol quercetin in humans. Int J Clin Pharmacol Ther 1999; 37:219–233.

22. Ader P, Wessman A, Wolffram S. Bioavailability and metabolism of the flavonols quercetin in the pig. Free Radic Biol Med 2000; 28:1056–1067.

23. Christen Y. Oxidative stress and Alzheimer disease. Am J Clin Nutr 2000; 71: 621S–629S.

24. Bastianetto S, Zheng WH, Quirion R. The *Ginkgo biloba* extract (EGb 761) protects and rescues hippocampal cells against nitric oxide-induced toxicity: involvement of its flavonoid constituents and protein kinase C. J Neurochem 2000; 74:2268–2277.

25. Rice-Evans C. Flavonoid antioxidants. Curr Med Chem 2001; 8:797–807.

26. Le Bars PL, Katz MM, Berman N, Itil TM, Freedman AM, Schatzberg AF. A placebo-controlled, double-blind, randomized trial of an extract of *Ginkgo biloba* for dementia. North American EGb Study Group. JAMA 1997; 278:1327–1332.

27. Iqbal K, Alonso AC, Gong CX, Khatoon S, Pei JJ, Wang JZ, Grundke-Iqbal I. Mechanisms of neurofibrillary degeneration and the formation of neurofibrillary tangles. J Neural Transm Suppl 1998; 53:169–180.

28. Gong CX, Lidsky T, Wegiel J, Zuck L, Grundke-Iqbal I, Iqbal K. Phosphorylation of microtubule-associated protein tau is regulated by protein phosphatase 2A in

mammalian brain. Implications for neurofibrillary degeneration in Alzheimer's disease. J Biol Chem 2000; 275:5535–5544.

29.  Nyberg F. Growth hormone in the brain: characteristics of specific brain targets for the hormone and their functional significance. Front Neuroendocrinol 2000; 21: 330–348.

30.  Porterfield SP. Thyroidal dysfunction and environmental chemicals-potential impact on brain development. Environ Health Perspect 2000; 108(suppl 3):433–438.

31.  Tsuzuki K, Fukatsu R, Yamaguchi H, Tateno M, Imai K, Fujii N, Yamauchi T. Transthyretin binds amyloid β-peptides, Aβ1–42 and Aβ1–40 to form complex in the autopsied human kidney—possible role of transthyretin for Aβ sequestration. Neurosci Lett 2000; 281:171–174.

32.  Brand K, Page S, Walli AK, Neumeier D, Baeuerle PA. Role of nuclear factor-κ B in atherogenesis. Exp Physiol 1997; 82:297–304.

33.  Tedesco F, Fischetti F, Pausa M, Dobrina A, Sim RB, Daha MR. Complement-endothelial cell interactions: pathophysiological implications. Mol Immunol 1999; 36:261–268.

34.  Zhu Y, Lin JHC, Liao HL, Friedli O, Verna L, Marten NW, Straus DS, Stemerman MB. LDL induces transcription factor activator protein-1 in human endothelial cells. Arterioscler Thromb Vasc Biol 1998; 18:473–480.

35.  Parry GCN, Mackman N. NF-KB mediated transcription in human monocytic cells and endothelial cells. Trends Cardiovascul Med 1998; 8:138–142.

36.  Kobuchi H, Droy-Lefaix MT, Christen Y, Packer L. *Ginkgo biloba* extract (EGb 761): inhibitory effect on nitric oxide production in the macrophage cell line RAW 264.7. Biochem Pharmacol 1997; 53:897–903.

37.  Baeuerle PA, Baltimore D. NF-κB: ten years after. Cell 1996; 87:13–20.

38.  Barnes PJ, Karin M. Nuclear factor-κB: a pivotal transcription factor in chronic inflammatory diseases. N Engl J Med 1997; 336:1066–1071.

39.  Navab M, Fogelman AM, Berliner JA, Territo MC, Demer LL, Frank JS, Watson AD, Edwards PA, Lusis AJ. Pathogenesis of atherosclerosis. Am J Cardiol 1995; 76:18C–23C.

40.  Brand K, Page S, Rogler G, Bartsch A, Brandl R, Knuechel R, Page M, Kaltschmidt C, Baeuerle PA, Neumeier D. Activated transcription factor nuclear factor-κB is present in the atherosclerotic lesion. J Clin Invest 1996; 97:1715–1722.

41.  Lindner V, Collins T. Expression of NF-κB and I κB-α by aortic endothelium in an arterial injury model. Am J Pathol 1996; 148:427–438.

42.  Widmer U, Manogue KR, Cerami A, Sherry B. Genomic cloning and promoter analysis of macrophage inflammatory protein (Mip)-2, Mip-1α, and Mip-1β, members of the chemokine superfamily of proinflammatory cytokines. J Immunol 1993; 150:4996–5012.

43.  Yoshimura T, Takeya M, Takahashi K, Kuratsu JI, Leonard EJ. Production and characterization of mouse monoclonal antibodies against human monocyte chemoattractant protein-1. J Immunol 1991; 147:2229–2233.

44.  Gu L, Okada Y, Clinton SK, Gerard C, Sukhova GK, Libby P, Rollins BJ. Absence of monocyte chemoattractant protein-1 reduces atherosclerosis in low density lipoprotein receptor-deficient mice. Mol Cell 1998; 2:275–281.

45. Rimbach G, Valacchi G, Canali R, Virgili F. Macrophages stimulated with IFN-γ activate NF-κB and induce MCP-1 gene expression in primary human endothelial cells. Mol Cell Biol Res Commun 2000; 3:238–242.

46. Rimbach G, Saliou C, Canali R, Virgili F. Interaction between cultured endothelial cells and macrophages: in vitro model for studying flavonoids in redox-dependent gene expression. Methods Enzymol 2001; 335:387–397.

47. Yoshikawa T, Naito Y, Kondo M. *Ginkgo biloba* leaf extract: review of biological actions and clinical applications. Antioxid Redox Signal 1999; 1:469–480.

48. Saliou C, Valacchi G, Rimbach G. Assessing bioflavonoids as regulators of NF-κB acitivity and gene expression in mammalian cells. Methods Enzymol 2001; 335: 380–387.

49. Brusselbach S, Nettelbeck DM, Sedlacek HH, Muller R. Cell cycle-independent induction of apoptosis by the anti-tumor drug flavopiridol in endothelial cells. Int J Cancer 1998; 77:146–152.

50. Zi X, Agarwal R. Silibinin decreases prostate-specific antigen with cell growth inhibition *via* G1 arrest, leading to differentiation of prostate carcinoma cells: implications for prostate cancer intervention. Proc Natl Acad Sci USA 1999; 96:7490–7495.

51. Elson CE, Peffley DM, Hentosh P, Mo H. Isoprenoid-mediated inhibition of mevalonate synthesis: potential application to cancer. Proc Soc Exp Biol Med 1999; 221: 294–311.

52. Gohil K, Moy RK, Farzin S, Maguire JJ, Packer L. mRNA expression profile of a human cancer cell line in response to *Ginkgo biloba* extract: induction of antioxidant response and the Golgi system. Free Radic Res 2000; 33:831–849.

53. Irani K, Xia Y, Zweier JL, Sollott SJ, Der CJ, Fearon ER, Sundaresan M, Finkel T, Goldschmidt-Clermont PJ. Mitogenic signaling mediated by oxidants in Ras-transformed fibroblasts. Science 1997; 275:1649–1652.

54. Lee PJ, Alam J, Wiegand GW, Choi AM. Overexpression of heme oxygenase-1 in human pulmonary epithelial cells results in cell growth arrest and increased resistance to hyperoxia. Proc Natl Acad Sci USA 1996; 93:10393–10398.

55. St Clair DK, Oberley LW. Manganese superoxide dismutase expression in human cancer cells: a possible role of mRNA processing. Free Radic Res Commun 1991; 12–13:771–778.

56. Tyrrell RM, Pidoux M. Correlation between endogenous glutathione content and sensitivity of cultured human skin cells to radiation at defined wavelengths in the solar ultraviolet range. Photochem Photobiol 1988; 47:405–412.

57. Keogh BP, Allen RG, Pignolo R, Horton J, Tresini M, Cristofalo VJ. Expression of hydrogen peroxide and glutathione metabolizing enzymes in human skin fibroblasts derived from donors of different ages. J Cell Physiol 1996; 167:512–522.

58. Meister A. Glutathione deficiency produced by inhibition of its synthesis, and its reversal; applications in research and therapy. Pharmacol Ther 1991; 51:155–194.

59. Steenvoorden DP, Hasselbaink DM, d Beijersbergen van Henegouwen GM. Protection against UV-induced reactive intermediates in human cells and mouse skin by glutathione precursors: a comparison of N-acetylcysteine and glutathione ethylester. Photochem Photobiol 1998; 67:651–656.

60. Rimbach G, Gohil K, Matsugo S, Moini H, Saliou C, Virgili F, Weber SU, Packer L. Induction of glutathione synthesis in human keratinocytes by *Ginkgo biloba* extract (EGb761). Biofactors 2001; 15:39–52.

# 6

# Cancer Chemoprevention with Sulforaphane, a Dietary Isothiocyanate

**Yuesheng Zhang**
*Roswell Park Cancer Institute, Buffalo, New York, U.S.A.*

## I. INTRODUCTION

Sulforaphane [1-isothiocyanato-4-(methylsulfinyl)-butane, SF] is a plant-derived aliphatic isothiocyanate (ITC), first isolated as an antimicrobial agent from the leaves of hoary cress and other plants by Procháska and coworkers in the late 1950s [1–4]. However, its cancer chemopreventive activity was not recognized until 1992, when it was found to be the principal ingredient of broccoli extracts exhibiting potent induction of phase 2 detoxification enzymes [5]. In plants, SF and other ITCs are synthesized and stored in cells as a glucosinolate (glucoraphanin in the case of sulforaphane) (Fig. 1), and a large of number of plants are known to contain glucoraphanin [6]. The anionic and nonelectrophilic glucoraphanin is highly water soluble and heat stable. In contrast, SF possesses an isothiocyanate, $-N = C = S$, group that is highly electrophilic, reacting with sulfur-, nitrogen- and oxygen-based nucleophiles, and heat labile (giving rise primarily to a thiourea derivative, whose biological properties are unknown) [7]. Glucoraphanin is but one of the more than 100 glucosinolates that are synthesized in plants [6,8]; many are of interest since anticarcinogenic activity has been observed in more than 20 ITCs [9–11]. Although intact glucosinolates are generally assumed to be biologically inert, the anticarcinogenic activity of intact glucoraphanin cannot be completely ruled out [12]. Conversion of glucosinolates to ITCs and other products occurs when plant cells are disrupted, such as by chewing or chopping, and is catalyzed by thioglucoside glucohydrolase (EC 3.2.3.1), commonly known as myrosinase. Myrosinase and the coexisting glucosinolates are physically segregated from one another in normal plant cells, but come into contact when the

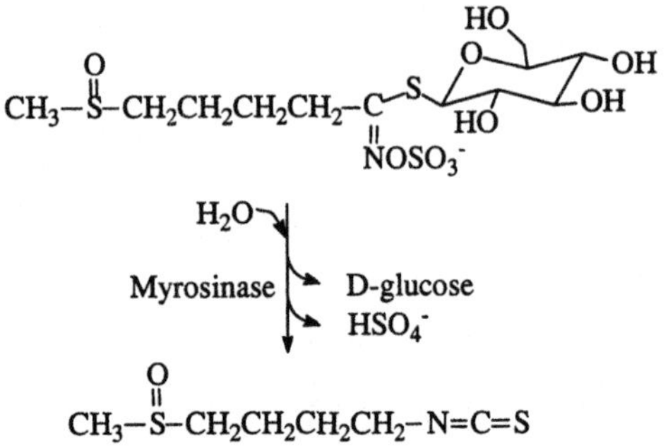

Figure 1   Chemical structures of glucoraphanin and sulforaphane and the myrosinase-catalyzed conversion

cells are damaged. In addition to ITCs, hydrolysis of glucosinolates by myrosinase may also give rise to thiocyanates, nitriles, and other products [13], and some (notably indoles) may undergo subsequent chemical reaction. In addition to the formation of SF, enzymatic hydrolysis of glucoraphanin has been shown to yield a nitrile [CH$_3$–(SO)–CH$_2$–CH$_2$–CH$_2$–CH$_2$–CN] as a minor product [14,15], which seems to lack anticarcinogenic activity [16]. Myrosinase is also present in enteric microflora, and a significant portion of orally ingested glucoraphanin (10–20%) in humans can be converted to SF in the intestine [17–19].

Since first shown to be a potent inducer of phase 2 detoxification enzymes in 1992 [5], SF has been extensively studied for its anticarcinogenic activities in both cultured cells and animal models. Accumulating evidence indicates that SF is effective against a variety of carcinogens and is without cell or tissue specificity. It is capable of interrupting several steps in the carcinogenic process, including protection of DNA by inhibiting carcinogen-activating enzymes and inducing carcinogen-detoxifying enzymes, reduction in proliferation of DNA-damaged cells (initiated cells) by inducing apoptosis and cell-cycle arrest, as well as other mechanisms to be discussed below. Initial pharmacokinetic study of SF in humans has been conducted. Moreover, plants superior to mature broccoli and perhaps other vegetables as dietary sources of glucoraphanin/SF have been developed. Such interdisciplinary investigations bring us closer to developing an SF-based strategy for cancer chemoprevention in humans.

## II.   INHIBITION OF CHEMICAL CARCINOGENESIS BY SF

SF has been shown to inhibit the genotoxicity of a number of chemical carcinogens. In mouse primary hepatocytes, $N$-nitrosodimethylamine (NDMA)–induced

DNA damage [unscheduled DNA synthesis (UDS)] was inhibited in a dose-dependent manner by SF at low $\mu$M concentrations, whereas the isothiocyanate itself did not induce UDS in the hepatocytes [20]. SF also inhibited both NDMA and 2-amino-3-methylimidazo[4,5-f] quinoline–induced DNA strand breaks (assessed in the Comet assay) in human liver THLE cells [21]. These effects appeared to result, in part at least, from inhibition of cytochrome P-450 enzymes, including P-450 2E1 and P-450 1A2 that activate the carcinogens [20,21]. In another study, sulforaphane potently inhibited benzo(a)pyrene (BP)- and 1,6-dinitropyrene (DNP)–DNA adduct formation in human breast epithelial cells (MCF-10F), BP and DNP adduct formation being reduced by 68–80% and 30–50% at 0.1–2 $\mu$M, respectively [22]. Moreover, BP-induced adduct formation was also significantly inhibited in human bronchial epithelial (BEAS-2B) cells [23]. In the latter experiments, it was unclear whether inhibition of any P-450 enzymes by SF contributed to the inhibition of DNA adduct formation; however, phase 2 carcinogen-detoxifying enzymes, including quinone reductase (QR) and glutathione S-transferase (GST), were significantly induced by SF in the MCF-10F cells.

When evaluated in animal models, sulforaphane strongly blocked chemical carcinogen-induced tumorigenesis. Using a 9,10-dimethyl-1,2-benzanthracene (DMBA)–induced mammary tumor model in female Sprague-Dawley rats, daily intragastric administration of 75 and 150 $\mu$mol SF for just 5 days was found to reduce tumor incidence by 48.6 and 61.3%, respectively. Tumor multiplicity also was inhibited 68.6 and 81.8% by 75 and 150 $\mu$mol SF [24]. Moreover, in rats that formed tumors, their development was significantly delayed by SF. In this study, sulforaphane was just administered once daily for 5 days at rat age day 46–50, each rat receiving 8 mg intragastric DMBA 3 hours after the last dose of SF. The experiment was terminated at approximately 5 months after the carcinogen administration. Although the tumor-inhibitory mechanism of SF was not investigated in this study, Gerhäuser and coworkers later found that a single intragastric dose of 17 $\mu$mol SF to each BALB/c female mouse induced both QR and GST 2.6- to 2.8-fold in the mammary glands [25]. These workers also found that sulforaphane inhibited the formation of DMBA-induced preneoplastic lesions in mouse mammary organ culture in a dose-dependent manner in a concentration range of 0.01–1 $\mu$M. These data suggest that induction of phase 2 detoxification enzymes plays a role in the inhibition of mammary tumorigenesis. In an azoxymethane (AOM)-induced colonic aberrant crypt foci (ACF) model in F344 rats, Chung and coworkers [26] found that SF in the dosing regimen described below reduced the formation of total ACF in both the initiation and postinitiation phases by 29–33%. AOM was administered subcutaneously (15 mg/kg body wt) once weekly for 2 weeks; the isothiocyanate was administered intragastrically either once daily for 3 days (20 and 50 $\mu$mol) before each AOM treatment (the last dose of SF was given 2 h before AOM dosing) or 3 times weekly for 8 weeks (5 and 20 $\mu$mol), beginning 2 days after the last dose of AOM. SF was effective even when given after carcinogen exposure, suggesting that it might exert its effect by preventing DNA-damaged cells (i.e., initiated cells) from progressing. More recently, Fahey and coworkers [27] reported that sulforaphane also prevented BP-induced stomach tumors in female ICR mice. When SF was added to the diet, 2.5 μmol/kg diet (approximately 7.5 μmol SF daily per mouse), starting one week before BP administration (120 mg/kg body wt intragastrically, once weekly for 4 weeks) and continued 2 days after the last BP dosing, a 39% reduction in tumor multiplicity was observed. Interestingly, sulforaphane had no effect on BP-induced stomach tumorigenesis in ICR mice unable to upregulate their phase 2 detoxification enzymes by SF due to deletion of a critical transcriptional factor (Nrf2), implying that induction of phase 2 detoxification enzymes by the compound plays a crucial role in this model system. These in vivo animal studies showed that SF prevented carcinogenesis, potentially through multiple mechanisms, and also revealed that its cancer-preventive activity was *not* species, organ, or carcinogen specific.

## III.  MECHANISM OF ACTION OF SF

### A.  A Novel Method for Measurement of SF and Its Dithiocarbamate Metabolites

Efforts to understand the cancer-preventive mechanism of SF and other ITCs have been facilitated by developments in analytical methodology. During the course of studies on SF, it was observed that ITCs, including SF, reacted with vicinal dithiols to form five-membered cyclic products and free amines [28]. Using 1,2-benzenedithiol as the vicinal dithiol, SF and all other reactive ITCs were converted to 1,3-benzodithiole-2-thione (Fig. 2). The reaction was stoichiometrically complete in a short time under mild conditions in the presence of excess 1,2-benzenedithiol. The resultant 1,3-benzodithiole-2-thione exhibits spectroscopic characteristics ($a_m = 23,000$ at 365 nm) that allow accurate measurement of as low as a few picomoles with a simple isocratic HPLC procedure [28,29]. Dithiocarbamates resulting from reaction of ITCs with a monothiol such as GSH and N-acetylcysteine also reacted quantitatively with 1,2-benzenedithiol to yield the same cyclic product (Fig. 2). This is especially significant for the study of ITCs in vivo because ITCs, including SF, are primarily metabolized in animals and humans to dithiocarbamates through the mercapturic acid pathway (Fig. 3). However, as all ITCs and dithiocarbamates form the same product on reaction with 1,2-benzenedithiol, this cyclocondensation assay can measure only the total amount of ITCs and their dithiocarbamate metabolites in a sample. Its use has, however, enabled the rapid and sensitive measurement of ITCs and/or

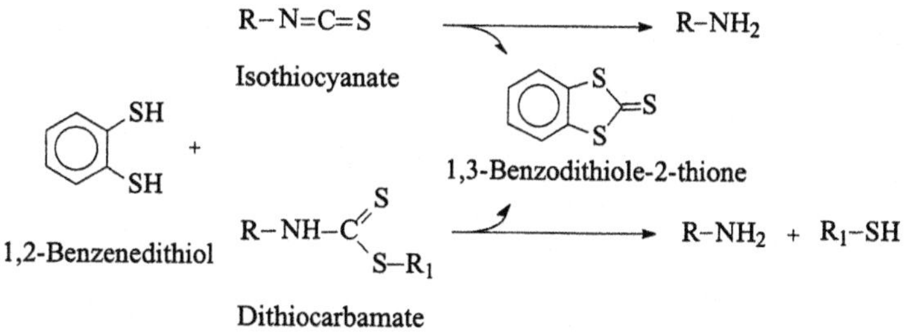

**Figure 2** Cyclocondensation reaction of isothiocyanates and dithiocarbamates with 1,2-benzenedithiol

dithiocarbamates in a variety of samples, including vegetable extracts, cell lysates, and human fluids, without recourse to radiolabeled compounds [17,18,30–34].

## B.  Induction of Phase 2 Enzymes

The induction of phase 2 enzymes was the first recognized cancer chemopreventive mechanism of SF [5]. Phase 2 enzymes [including glutathione S-transferase (GST), UDP-glucuronosyltransferase (UGT), quinone reductase (QR), and γ-glutamylcysteine synthetase (GCS)] are cellular enzymes able to reduce or eliminate the electrophilic character of chemical carcinogens and reactive oxygen spe-

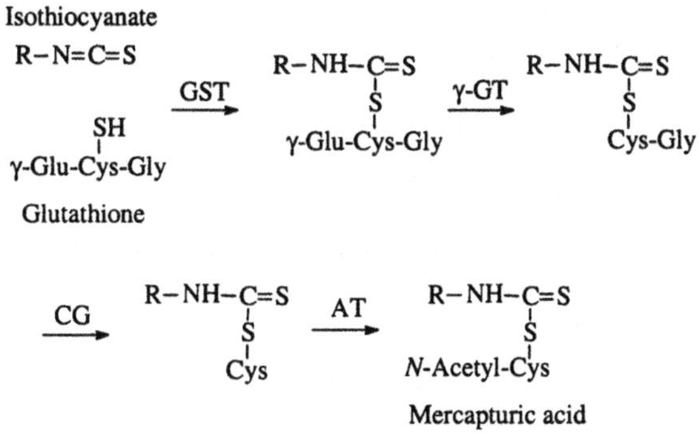

**Figure 3** Metabolism of isothiocyanates through the mercapturic acid pathway. Isothiocyanates are conjugated with glutathione, which is promoted by glutathione S-transferase (GST). The conjugates are metabolized sequentially by γ-glutamyltranspeptidase (γ-GT), cystinylglycinase (CG), and N-acetyltransferase (AT) to form mercapturic acids

cies, thereby affording protection to critical cellular targets such as DNA [35]. An enzyme that (1) catalyzes the conjugation of electrophilic compounds including carcinogens with endogenous ligands, resulting in less toxic and more readily excreted products, as demonstrated by GST and UGT, (2) catalyzes reactions that destroy the reactive centers of carcinogens or other toxic compounds, as demonstrated by QR, or (3) synthesizes and maintains cellular antioxidants, as demonstrated by GCS and QR, is broadly defined as a phase 2 enzyme.

Phase 2 enzymes, as well as the antioxidants they generate, exhibit low substrate specificity and therefore can detoxify a wide variety of carcinogens, reactive oxygen species, and other toxic chemicals. Moreover, many phase 2 enzymes are transcriptionally controlled by a regulatory DNA element, the antioxidant response element (ARE), that resides in the 5′-flanking region of their genes [36,37], leading to coordinate induction of these enzymes upon exposure to ARE-mediated inducers. A signaling system that appears to transmit an inducer signal to the ARE has been revealed [38]. Simultaneous induction of different phase 2 enzymes provides a powerful and versatile protection against carcinogens and other toxic chemicals in cells.

Sulforaphane is an ARE-mediated inducer that induces all of the above-described enzymes as well as other phase 2 enzymes [5,39–41]. The phase 2 enzyme-inducing activity of SF has been observed in a variety of cultured cells and rodent tissues [5,22,25,42,43], demonstrating that it is neither cell nor tissue specific and is active in vivo. Sulforaphane is one of the most potent inducers of phase 2 enzymes among ITCs, as screening of more than 40 laboratory-synthesized analogues has shown [5,44]. Incubation of cells in culture with only 0.2–5 μM SF for 24 hours led to a significant induction of phase 2 enzymes [5,32,43]. Because many phase 2 enzymes protect cells against carcinogens and detoxify reactive oxygen species, SF has also been suggested to be an indirect antioxidant [45]. For example, induction of phase 2 enzymes was found to be closely correlated with the degree of protection by sulforaphane against oxidative stressors (manadione, tert-butyl hydroperoxide, 4-hydroxynonenal, and peroxynitrite) in human retinal pigment epithelial cells, keratinocytes, and mouse leukemia cells [46].

In an effort to more fully understand the relation between structure and inducer activity, the results of modifying the structural features of SF (including its chirality, oxidation state of the methylthiol moiety, and the number and rigidity of methylene bridging units) were evaluated using QR as the marker of induction of phase 2 enzymes in murine hepatoma Hepa 1c1c7 cells [5,44]. Table 1 includes results for representative sulforaphane analogs from over 40 compounds examined. SF is chiral with the natural compound possessing the R configuration [5]: however, both (R)-SF and synthetic (R,S)-SF showed identical inducer potency. There also has been no report that the chirality of the molecule affects its other anticarcinogenic mechanisms. However, change of the oxidation state of the sulfur

**Table 1**   SF Analogs: Relation of Structure to Inducer Activity

| Analog | | Concentration required to double QR ($\mu M$)[a] |
|---|---|---|
| $\overset{\displaystyle O}{\underset{\displaystyle CH_3}{\overset{\|}{S}}}-(CH_2)_4-NCS$ | (Isolated) | 0.2 |
| $CH_3-\overset{O}{\overset{\|}{S}}-(CH_2)_4-NCS$ | (Synthetic) | 0.2 |
| $CH_3-S-(CH_2)_4-NCS$ | | 2.3 |
| $CH_3-\overset{O}{\underset{O}{\overset{\|}{\underset{\|}{S}}}}-(CH_2)_4-NCS$ | | 0.8 |
| $CH_3-\overset{O}{\overset{\|}{C}}-(CH_2)_4-NCS$ | | 0.2 |
| $CH_3-CH_2-(CH_2)_4-NCS$ | | 15 |
| $CH_3-\overset{O}{\overset{\|}{C}}$ ⬡ NCS | | 0.3 |
| ⬡ NCS  $CH_3-\overset{O}{\overset{\|}{C}}$ | | 0.3 |

[a] Concentration of the test compound required to double the QR activity in a standard microtiter plate assay in which murine hepatoma Hepa 1c1c7 cells were exposed to the compound for 48 hours [5,44].

atom in the methylthiol group from sulfoxide to sulfone reduced inducer activity fourfold, and the sulfide analog was more than 10 times less active. Moreover, if the sulfoxide group was replaced with a methylene group, the inducer activity was reduced 75-fold. Significantly, the sulfoxide group could be replaced with a carbonyl group without losing any inducer activity; this is important because the latter compound is more easily synthesized [44]. Change in the number of methylene units from 4 to 5 or 3 did not significantly affect inducer activity (results not shown), nor did the rigidity of the methylene bridge appear to have much effect on inducer activity, as shown by the finding that the norbonyl ITCs were almost equally active. It should be noted again, however, that the above-described structure-activity relationship analyses were only based on induction of QR in murine hepatoma Hepa 1c1c7 cells [5,44].

## C. Intracellular SF Accumulation

In the above cyclocondensation assay system, sulforaphane, as well as other ITCs, was observed to be rapidly accumulated in all cell lines tested, with intracellular accumulation reaching millimolar levels [32,43,47,48]. SF appeared to enter cells freely, but was subsequently almost entirely conjugated with glutathione (GSH) (Fig. 4) [48,49]. Preliminary evidence suggests that this conjugate may be further metabolized, for example, by binding to cellular proteins and the formation of other dithiocarbamate metabolites. As expected, exposure of cells to high concentrations of SF led to depletion of cellular GSH [49]. These results revealed that cellular GSH was the principal driving force for accumulation, while cellular GST may further enhance such accumulation [47]. This is consistent with sulforaphane undergoing spontaneous conjugation with GSH under mild conditions to form the corresponding dithiocarbamate (GS-SF) and SF being a substrate of GST [50].

Intracellular accumulation of sulforaphane appeared to be critical for its phase 2 enzyme-inducing activity; for example, when compared with eight other ITCs, the potent activity of SF in inducing both QR and GST in murine hepatoma Hepa 1c1c7 cells was closely related to its overall intracellular accumulation (as measured by the area under time-concentration curve, AUC) [32]. This phenomenon was not cell-specific, as other cell lines including murine keratinocytes (PE cells) and human HepG2 hepatic cells also showed similar responses [43]. Moreover, when utilizing cells stably transfected with an ARE-driven reporter construct, the intracellular AUC-dependent induction of phase 2 enzymes by SF and other ITCs was found to be mediated by the ARE [43]. This finding further shows that the intracellular AUC-dependent induction of phase 2 enzymes is likely to be a common mechanism across a variety of cell types, because ARE-regulated induction of phase 2 enzymes appears to exist across cell types. These observations suggest that the overall intracellular accumulation of ITCs (AUCs) may serve as useful markers of their activity in elevating cellular detoxification capac-

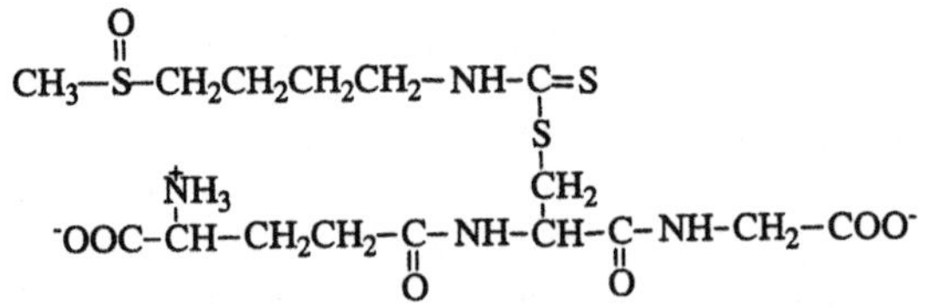

SF conjugated with GSH (GS-SF)

**Figure 4**   The chemical structure of sulforaphane conjugated with glutathione

ity. If confirmed by in vivo experimentation, such markers may provide valuable guidance in developing ITCs as cancer chemopreventive agents in humans. However, it was subsequently found that the accumulated GS-SF was rapidly exported by membrane transporters including multidrug resistance–associated protein-1 (MRP-1) and P-glycoprotein-1 (Pgp-1) [48]; this indicated that continuous uptake of SF was necessary to sustain the intracellular accumulation levels of sulforaphane required to offset the rapid export of GS-SF [48]. Because intracellular accumulation of SF is likely to be important for other anticarcinogenic mechanisms, discussed below, this finding could have important implications for the development of an effective dosing regimen of this compound for cancer prevention.

## D. Inhibition of Carcinogen-Activating Cytochrome P-450 Enzymes

P-450 enzymes (CYP) are a family of membrane-bound, heme-containing monooxygenases, for which over 100 genes have been identified in the mammalian system. These enzymes belong to the phase 1 enzyme family in drug metabolism and are critical for metabolic processing of numerous endogenous and exogenous compounds. However, some P-450 enzymes also can activate chemical carcinogens (procarcinogens); thus, many carcinogens, including benzo(a)pyrene, nitrosamines, 2-amino-1-methyl-6-phenyl-imidazo(4,5-b)pyridine (PhIP), and 4-(methylnitrosamino)-1-(3-pyridyl)-1-butanone (NNK), only become carcinogenic after metabolism by P-450 enzymes. In both cultured cells and animals models, inhibition of specific carcinogen-activating P-450 enzyme(s) effectively blocked carcinogen-induced DNA damage and tumorigenesis. Many investigators, therefore, view the development of drugs that inhibit specific P-450 enzymes as an important element of cancer chemoprevention; it should be stressed that the potential adverse effects of such inhibition on normal cellular functions have not been adequately addressed. Barceló and coworkers showed SF to be a potent competitive inhibitor of CYP2E1 ($K$i of $37.0 \pm 4.5$ $\mu$M), which is responsible for the activation of several carcinogens, including $N$-nitrosodimethylamine [20,21]. Sulforaphane also inhibited CYP1A2 and CYP3A4, and possibly CYP1A1 and CYP2B1/2, in human cells [21,51]. In the case of CYP3A4, the major P-450 enzyme in the human liver, SF (5–25 $\mu$M for 48–72 h) inhibited both its enzyme activity and expression in human hepatocytes [51]. Interestingly, when this isothiocyanate was directly incubated with P-450 enzymes under cell-free conditions, such as enzymes expressed on bacterial membranes or rat microsomes, levels of 40–100 $\mu$M did not significantly inhibit the catalytic activities of CYP 1A1, 1A2, 1B1, 2B1, and 3A4 [52,53]. The reason for this discrepancy between the results of cell-based and cell-free experiments is not entirely clear but might relate to cellular accumulation of SF, leading to much higher intracellular SF concentrations than occur in the extracellular space.

## E.   Induction of Apoptosis and Inhibition of Cell Growth

Although blocking carcinogens from damaging cellular targets by inhibiting carcinogen-activating P-450 enzymes and/or inducing carcinogen-detoxifying enzymes has proven effective in preventing carcinogenesis in cultured cells and animal models, there is ample evidence that the induction of apoptosis and inhibition of proliferation are also important if DNA damage occurs. During the transformation of a normal cell to an invasive and ultimately metastatic state, increased apoptosis and reduced proliferation of transformed cells is considered pivotal in halting or delaying cancer formation and progression. Drugs that induce apoptosis and/or inhibit cell proliferation also inhibit carcinogen-induced tumorigenesis in animal models [54,55].

Sulforaphane has been shown to induce both apoptosis and cell cycle arrest in several cell lines, including HT29 human colon cancer cells [56], LNCaP human prostate cancer cells [57], and Jurkat human T-leukemia cells [58]. Incubation of such cells with 3–30 μM SF for 24–48 hours resulted in a dose-dependent induction of apoptosis and cell cycle arrest, showing that this compound can induce apoptosis and arrest cell growth and suggesting that this activity is unlikely to be cell type specific. There is also evidence that the concentration of SF required to induce apoptosis and cell cycle arrest may be much lower in undifferentiated cancer cells than in normal cells [56], offering the possibility that this isothiocyanate may be selectively toxic to malignant cells. Because p53 expression was unaffected by SF in HT29 cells [56], induction of apoptosis by SF in these cells might be independent of p53, suggesting possible induction of cell death in p53-defective cancer cells. In contrast, however, p53 expression in Jurkat cells was markedly increased by SF [58]. Interestingly, sulforaphane appeared to arrest different cells in different phases; thus, it arrested LNCaP cells in G1 phase, but both HT29 cells and Jurkat cells in G2-M phase. Details of the regulatory mechanism of the cell cycle by SF is still largely unclear, and further work is being directed towards its elucidation.

## F.   Other Effects of SF That May Also Contribute to Its Anticarcinogenic Activity

In addition to the mechanisms described above, sulforaphane has recently been shown to possess other activities that may contribute to its protection against cancer. Payen et al. [59] showed that incubation of either primary human or rat hepatocytes with 20–50 μM SF for 48–72 hours significantly induced multidrug resistance–associated protein 2 (MRP2), a membrane drug transporter involved in removal of toxic intracellular chemicals. MRP2 and other drug transporters play a key role in developing drug resistance in cancer cells, but the same mechanism may also protect normal cells, offering the possibility that increased expression of these pumps might contribute to the prevention of cancer. Currently, it is not known whether SF induces other membrane drug transporters or if the induction is cell specific.

Sulforaphane also appears to possess anti-inflammatory activity since treatment of Raw 264.7 murine macrophages with SF (1–10 μM for 24 h) caused a dose-dependent inhibition of lipopolysaccharide (LPS)-induced secretion of pro-inflammatory and pro-carcinogenic signaling factors, i.e., nitric oxide, prostaglandin $E_2$, and tumor necrosis factor α (TNF-α) [60]. It was found that SF selectively reduced DNA binding of NF-κB, a critical transcription factor in LPS-stimulated inflammatory response. Ornithine decarboxylase (ODC) is a rate-limiting enzyme in the synthesis of polyamines, with increased expression of ODC being linked to tumor promotion. SF was shown to inhibit 12-*O*-tetradecanoylphorbol-13-acetate (TPA)–induced ODC activity in mouse epidermal cells (ME308) ($IC_{50}$, 6.8 μM) [23]. Moreover, while its antimicrobial activity has been long recognized [3,4], Fahey et al. [27] recently showed that SF (approximately 20 μM) potently inhibited extracellular, intracellular, and antibiotic-resistant strains of *Helicobacter pylori*, a significant risk factor for human gastric cancer, suggesting that it may be especially useful against stomach carcinogenesis.

## IV.  EDIBLE PLANTS AS CARRIERS OF SF

Because sulforaphane occurs in common vegetables such as broccoli, it is likely that many humans have been exposed to this compound through vegetable-derived diets. Given its cancer chemopreventive activity and the absence of any significant toxicities in humans, it seems reasonable to recommend the regular consumption of vegetables rich in SF before the pure compound is permitted for use in humans. As mentioned earlier, SF is synthesized and stored in plant cells as the glucosinolate glucoraphanin and is formed when this thioglucoside is hydrolyzed by myrosinase. Although glucoraphanin is known to occur in a variety of plants, including many cruciferous vegetables [6,18], broccoli may be one of the most abundant sources of this compound in edible plants (Table 2). However, the content of glucoraphanin in plants is likely to vary significantly even within the same species. For example, different samples of broccoli (both frozen and fresh) sold in supermarkets have been found to differ in glucoraphanin content by as much as ninefold, unrelated to their physical appearance or whether grown under conventional or organic conditions [12]. It is thus impossible to estimate how much glucorapha-

**Table 2**  Glucosinolate Profile in Mature Broccoli (*cv* Saga) and 3-Day-Old Broccoli Sprouts

| Glucosinolate | μmol/g fresh weight | |
|---|---|---|
| | 3-Day sprouts | Mature broccoli |
| Glucoraphanin[a] | 16.6 | 1.08 |
| Glucoerucin | 5.41 | Not detected |
| 4-Hydroxyglucobrassicin | 0.71 | Not detected |
| Glucobrassicin | Not detected | 1.67 |
| Neoglucobrassicin | Not detected | 0.62 |

[a] Glucoiberin, the glucosinolate presursor of iberin, had the same retention time as glucoraphanin in HPLC analysis; hence, there might be a small amount of glucoiberin in both mature broccoli and broccoli sprouts.

nin is present in a particular vegetable without the actual determination of its content.

The degree of conversion of glucoraphanin to SF may also be significantly affected by the method of preparation and the condition of enteric microflora. Steaming or boiling is likely to reduce conversion by inactivating endogenous myrosinase and destroying the heat-labile SF [7]. Consumption of fresh and raw vegetables is, therefore, preferable but inconsistent with the taste preferences of the majority of consumers. Any glucoraphanin not hydrolyzed by vegetable myrosinase could be converted to SF by enzymes in the enteric flora, but it has been demonstrated that the growth inhibition of enteric microflora significantly influences the conversion of glucosinolates in the intestine since use of antibiotics known to inhibit the microflora reduced this conversion [18]. It is presently unknown how much SF should be given to a human in order to achieve cancer prevention and, indeed, whether glucoraphanin itself is an anticarcinogen.

Research has shown that all glucoraphanin in mature broccoli might already have been synthesized in the seeds and that no significant synthesis of this compound occurred thereafter. Thus, as broccoli sprouts grow, their glucoraphanin content per unit plant weight declines in an exponential manner from its maximal content in the seeds [12]. Thus, there is approximately 15 times more glucoraphanin in 2- to 3-day-old broccoli sprouts (*cv* Saga) than in the heads of the mature cultivar [12]. However, whether the same phenomenon exists in other plants is largely unknown. Although Faulkner et al. [61] have reported that glucoraphanin content in mature broccoli could be increased 10-fold by crossing broccoli cultivars with selected wild taxa of the *Brassica oleracea*, the exploitation of broccoli sprouts may offer another advantage. Investigations have revealed that whereas indole glucosinolates (4-hydroxyglucobrassin, glucobrassicin, and neoglucobrassicin) comprised 68% of the total in mature broccoli (*cv* Saga), this proportion fell to 3% in the sprouts [12] (Table 2). Similar results were obtained in sprouts grown from other varieties of broccoli seeds [62].

Hydrolysis of indole glucosinolates by myrosinase yields unstable ITCs that spontaneously decompose to compounds including indole-3-carbinol, indole-3-acetonitrile, thiocyanate ions, and 3,3′-diindolylmethane that are potentially tumor-promotive [63,64]. Indole-3-carbinol may also condense under the acidic conditions of the stomach to form compounds that closely resemble 2,3,7,8-tetrachlorodibenzo-*p*-dioxin (TCDD, or dioxin) in structure, toxicity, and carcinogenicity [65]. Thus, it is considered desirable to consume vegetables that are low in indole glucosinolates even though other studies have shown that indole-3-carbinol may also prevent carcinogenesis [66,67] and suggest that broccoli sprouts or sprout extracts, carefully analyzed for the content of glucoraphanin or SF, may be the most suitable dietary source of these compounds.

In addition to glucoraphanin/SF, broccoli sprouts also appear to contain two minor glucosinolates that give rise to erucin $[CH_3-S-CH_2-CH_2-CH_2-CH_2-N=C=S]$ and iberin $[CH_3-S(O)-CH_2-CH_2-CH_2-N=C=S]$; both isothiocyanates are known inducers of phase 2 enzymes [5], suggesting potential tumor-inhibitory activity, but neither has been examined for other anticarcinogenic properties.

## V. HUMAN METABOLISM AND DISPOSITION OF SF

ITCs are metabolized in mammals principally by the mercapturic acid pathway, initial conjugation with GSH promoted by GST giving rise to the corresponding conjugates (68–70). These undergo sequential enzymatic modifications to form cycteinylglycine, cysteine, and *N*-acetylcysteine conjugates (mercapturic acids), which are excreted in the urine (Fig. 3). Collectively, these metabolites are called dithiocarbamates, since they contain the characteristic $R-NH-C(-S-R_1)=S$ skeleton. Many dithiocarbamates, including those derived from SF, have been shown to be equally, if not more, active cancer chemoprotectors than their parent ITCs [55,57,71,72] because of their ability to dissociate to ITCs. By gradually releasing ITCs, they may more effectively deliver these bioactive molecules. When human volunteers were fed a broccoli sprout extract preparation containing 100–200 μmol ITCs (but no glucosinolates), the majority of which was SF, nearly 90% of the ingested ITCs were excreted in the urine as ITC/dithiocarbamates in 72 hours as measured by the cyclocondensation assay, with 58% being excreted in the first 8 hours [33,62]. This surprisingly high recovery provides further evidence that urinary excretion is the principal disposal route of ITCs. However, feeding fresh or steamed mature broccoli to humans only resulted in detection of 32.3% and 10.2% of the administered total ITC/dithiocarbamates, respectively, in the urine samples 24 hours after consuming [17]. These much reduced recoveries were probably a reflection of incomplete conversion of glucosinolates to ITCs.

Pharmacokinetic studies have been conducted in human volunteers who received single doses of 200 μmol of ITCs contained in the extract of 3-day old broccoli sprouts, cv DeCicco (containing no intact glucosinolates) [33,62]. The precise content of sulforaphane in the extract was unknown, but analysis showed that SF may comprise up to 79% of the total ITCs, the remainder being erucin and iberin [33]. Concentrations of total ITC/dithiocarbamates in plasma, erythrocytes, and urine were determined with the cyclocondensation assay, as described earlier. The pharmacokinetic behavior of the ITCs was very similar, all subjects showing rapid absorption of ITCs, reaching peak concentrations of 0.94–2.27 μM in plasma and erythrocytes at 1 hour after feeding, and declining with first-order kinetics (half-life 1.77 ± 0.13 h). Serum contents were very similar to those of plasma. The cumulative urinary excretion of dithiocarbamates at 8 hours was 58.3 ± 2.8% of the dose. Not only was the plasma concentration of ITC/dithiocarbamate low and rapidly declining after feeding of the ITCs, but the ITC/dithiocarbamates circulating in the blood appeared unable to accumulate in the cells. The intracellular concentrations of ITC/dithiocarbamates in erythrocytes were slightly lower than those in the plasma. If freshly isolated human erythrocytes were incubated with sulforaphane, however, there was a rapid accumulation of intracellular SF. This reached 333 μM (a 67-fold increase) when the cells were incubated with 5 μM SF for 30 minutes at 37°C.

Since it had been previously shown [49] that cells did not accumulate intact dithiocarbamates (e.g., SF conjugates of GSH and N-acetylcysteine), it seemed highly likely that the majority, if not all, of the plasma-circulating ITC/dithiocarbamate were present in the form of dithiocarbamates. Therefore, while in animal experiments a high dose of ITC (e.g., 600 μmol SF/kg body weight) may lead to high concentrations of plasma ITC/dithiocarbamate, resulting in intracellular levels sufficient for anticarcinogenic activity, similarly effective intracellular concentrations of ITC/dithiocarbamates may be difficult to achieve in humans, since the amounts ingested by humans are typically much lower. For example, 200 μmol of broccoli sprout ITCs represent only 2.8 μmol/kg body weight (assuming an average body weight of 70 kg), corresponding to just 5% of the dose given to rodents.

Thus, a significantly higher dose of SF than contained in regular diets appears necessary to achieve effective intracellular concentrations of SF and/ or its dithiocarbamate metabolites in human tissues and cells. Indeed, broccoli supplementation in the human diet [lyophilized tablets, the equivalent of 30 g of fresh broccoli per day per person for 14 days, representing approximately 0.4 g/ kg body weight (assuming an average body weight of 70 kg, ITC content unknown)] did not affect the expression of GST in both lymphocytes and colon mucosa, whereas a single dose of the tablet given to ICR mice (equivalent of 5 g of fresh broccoli/kg body weight) significantly increased colonic GST activity [73]. However, a number of dietary questionnaires have shown consumption of broccoli or other ITC-containing vegetables to be correlated with lower tumor incidence (colorectal adenomas and lung cancer) in GST-deficient individuals (GST M1-null and/or GST P1-null) [34,74–76], suggesting that SF and other anticarcinogenic ITCs may be more effective in these individuals. The mechanism, however, remains to be elucidated, and more quantitative clinical studies are obviously needed to confirm these findings.

Because most of the broccoli ITCs ingested by humans are excreted in urine and storage of urine in the bladder may allow a degree of dissociation of dithiocarbamates to free ITCs, high intracellular accumulation of ITCs in bladder cells may be more readily achieved. Since the transitional epithelia that line the inner surface of the bladder are the site where the majority of the bladder cancers originate and these cells are directly exposed to urine, it seems likely that the bladder may be the most important target organ for SF for cancer prevention. If (as described above) 58.3% of the ingested 200 μmol sprout ITCs is excreted in urine within 8 hours after dosing, and assuming 1 L of urine is produced during that period, urinary ITC may reach 116 μM providing that all dithiocarbamates are converted to ITCs. However, sulforaphane conjugated with $N$-acetylcysteine appeared to comprise the majority of its urinary dithiocarbamates [77], which is slower in releasing SF than other dithiocarbamates [78]; thus, the amount of free ITCs that will form from dithiocarbamates in urine stored in the bladder remains to be determined. Nevertheless, urine collected from individuals who consumed DiCicco broccoli sprout extracts markedly induced a marker phase 2 detoxification enzyme (QR) in murine hepatoma Hepa 1c1c7 cells grown on microtiter plates (a widely used bioassay system for the detection of inducers of phase 2 detoxification enzymes) [62]. Based on these analyses, studies have recently been initiated aimed at determining the preventive effects of SF or broccoli sprouts on bladder carcinogenesis.

## VI. CONCLUDING REMARKS

The isothiocyante sulforaphane is a potent cancer chemopreventive agent capable of interrupting a number of steps in the carcinogenic process through anticarcinogenic mechanisms that include inhibition of carcinogen-activating P-450 enzymes, induction of carcinogen-detoxifying phase 2 enzymes, and induction of apoptosis and cell cycle arrest. More recent observations suggest that it may also induce membrane transporters to remove carcinogens from cells, inhibit inflammation, downregulate ornithine decarboxylase, and eliminate *Helicobacter pylori*, which is linked to stomach cancer. All these mechanisms suggest sulforaphane to be a highly promising cancer chemopreventive agent in humans. Moreover, the anticarcinogenic activity of SF does not appear to be cell or organ specific, nor is it carcinogen specific.

Nevertheless, preliminary studies raise the question whether orally consumed SF from a regular diet would be sufficient to achieve the effective intracellular accumulation in cells and tissues that appears critical for at least some of its anticarcinogenic mechanisms. While SF is rapidly and efficiently absorbed after oral ingestion, absorbed SF is also rapidly excreted in urine as dithiocarbamates. As a result, effective intracellular concentrations of SF may be difficult to achieve. Evidence also suggests that the majority of SF that circulates in the blood may be in the form of dithiocarbamates that do not accumulate in cells. However, because ingested SF is almost exclusively excreted and concentrated in urine as dithiocarbamates, which may release SF when stored in the bladder, the urinary bladder may be an ideal target site for SF.

Broccoli sprouts appear to be an excellent source of glucoraphanin/SF. Standard methods that have been developed for selecting the seeds, growing the sprouts, and analyzing the content of glucosinolates/ITCs [12,29,79,80] hold out the possibility of sprouts with well-defined glucoraphanin/sulforaphane content being made available to the public.

## ACKNOWLEDGMENTS

Studies in the author's laboratory were supported in part by U.S. Public Service Grant CA80962. The author wishes to thank Mr. Bradley C. Helbing for editorial assistance in the preparation of this manuscript.

## REFERENCES

1. Procháska Z. Isolation of sulforaphane from hoary cress (*Lepidium draba L.*). Collet Czech Chem Commun 1959; 24:2429–2430.
2. Procháska Z, Sanda V, Jrousek L. Isothiocyanate im Wirsing- und Rosenkohl. Collet Czech Chem Commun 1959; 24:3606–3610.
3. Procháska Z, Komersová I. Isolation of sulforaphane from *Cardaria draba* and its antimicrobial effect. Cesk Farm 1959; 8:373–376.
4. Dornberger K, Bökel V, Heyer J, Schönfeld C, Tonew M, Tonew E. Untersuchungen über die isothiocyanate crysolin und sulforaphan aus *Cardaria draba L.* Pharmazie 1975; 30:792–796.
5. Zhang Y, Talalay P, Cho C-G, Posner GH. A major inducer of anticarcinogenic protective enzymes from broccoli: isolation and elucidation of structure. Proc Natl Acad Sci USA 1992; 89:2399–2403.
6. Fahey JW, Zalcmann AT, Talalay P. The chemical diversity and distribution of glucosinolates and isothiocyanates among plants. Phytochemistry 2001; 56:5–51.

7. Jin Y, Wang M, Rosen RT, Ho C-T. Thermal degradation of sulforaphane in an aqueous solution. J Agric Chem 1999; 47:3121–3123.
8. Fenwick GR, Heaney RK, Mullin WJ. Glucosinolates and their breakdown products in food and food plants. CRC Crit Rev Food Sci Nutr 1983; 18:123–201.
9. Zhang Y, Talalay P. Anticarcinogenic activities of organic isothiocyanates: chemistry and mechanisms. Cancer Res 1994; 54(suppl):1976s–1981s.
10. Hecht SS. Chemoprevention by isothiocyanates. J Cell Biochem 1995; 22(suppl): 195–209.
11. Hecht SS. Chemoprevention of cancer by isothiocyanates, modifiers of carcinogen metabolism. J Nutr 1999; 129(suppl):768s–774s.
12. Fahey JW, Zhang Y, Talalay P. Broccoli sprouts: an exceptionally rich source of inducers of enzymes that protect against chemical carcinogenesis. Proc Natl Acad Sci USA 1997; 94:10367–10372.
13. Pessina A, Thomas RM, Palmieri S, Luisi PL. An improved method for the purification of myrosinase and its physicochemical characterization. Arch Biochem Biophys 1990; 280:383–389.
14. Chiang WCK, Pusateri DJ, Leitz REA. Gas chromatography/mass spectrometry method for the determination of sulforaphane and sulforaphane nitrile in broccoli. J Agric Food Chem 1998; 46:1018–1021.
15. Matusheski NV, Wallig MA, Juvik JA, Klein BP, Kushad MM, Jeffery EH. Preparative HPLC method for the purification of sulforaphane and sulforaphane nitrile from *Brassica oleracea*. J Agric Food Chem 2001; 49:1867–1872.
16. Basten GP, Bao Y, Williamson G. Sulforaphane and its glutathione conjugate but not sulforaphane nitrile induce UDP-glucuronosyl transferase (UGT1A1) and glutathione transferase (GSTA1) in cultured cells. Carcinog 2002; 23:1399–1404.
17. Conaway CC, Getahun SM, Liebes LL, Pusateri DJ, Topham DKW, Botero-Omary M, Chung F-L. Disposition of glucosinolates and sulforaphane in humans after ingestion of steamed and fresh broccoli. Nutr Cancer 2000; 38:168–178.
18. Shapiro TA, Fahey JW, Wade KL, Stephenson KK, Talalay P. Human metabolism and excretion of cancer chemoprotective glucosinolates and isothiocyanates of cruciferous vegetables. Cancer Epidemiol Biomark Prev 1998; 7:1091–1100.
19. Getahun SM, Chung F-L. Conversion of glucosinolates to isothiocyanates in humans after ingestion of cooked watercress. Cancer Epidemiol Biomark Prev 1999; 8: 447–451.
20. Barceló S, Gardiner JM, Gescher A, Chipman JK. CYP2E1-mediated mechanism of anti-genotoxicity of the broccoli constituent sulforaphane. Carcinogenesis 1996; 17:277–282.
21. Barceló S, Macé K, Pfeifer AMA, Chipman JK. Production of DNA strand breaks by N-nitrosodimethylamine and 2-amino-3-methylimidazo[4,5-f]quinoline in THLE cells expressing human CYP isoenzymes and inhibition by sulforaphane. Mutat Res 1998; 402:111–120.
22. Singletary K, MacDonald C. Inhibition of benzo[a]pyrene- and 1,6-dinitropyrene-DNA adduct formation in human mammary epithelial cells by dibenzoylmethane and sulforaphane. Cancer Lett 2000; 155:47–54.
23. Lee SK, Song L, Mata-Greenwood E, Kelloff GJ, Steele VE, Pezzuto JM. Modulation of in vitro biomarkers of the carcinogenic process by chemopreventive agents. Anticancer Res 1999; 19:35–44.

24. Zhang Y, Kensler TW, Cho C-G, Posner GH, Talalay P. Anticarcinogenic activities of sulforaphane and structurally related synthetic norbornyl isothiocyanates. Proc Natl Acad Sci USA 1994; 91:3147–3150.

25. Gerhäuser C, You M, Liu J, Moriarty RM, Hawthorne M, Mehta RG, Moon RC, Pezzuto JM. Cancer chemopreventive potential of sulforamate, a novel analogue of sulforaphane that induces Phase 2 drug-metabolizing enzymes. Cancer Res 1997; 57:272–278.

26. Chung F-L, Conaway CC, Rao CV, Reddy BS. Chemoprevention of colonic aberrant crypt foci in Fischer rats by sulforaphane and phenethyl isothiocyanate. Carcinogenesis 2000; 21:2287–2291.

27. Fahey JW, Haristoy X, Dolan PM, Kensler TW, Scholtus I, Stephenson KK, Talalay P, Lozniewski A. Sulforaphane inhibits extracellular, intracellular, and antibiotic-resistant strains of *Helicobacter pylori* and prevents benzo[a]pyrene-induced stomach tumors. Proc Natl Acad Sci USA 2002; 99:7610–7615.

28. Zhang Y, Cho C-G, Posner GH, Talalay P. Spectroscopic quantitation of organic isothiocyanates by cyclocondensation with vicinal dithiols. Anal Biochem 1992; 205: 100–107.

29. Zhang Y, Wade KL, Prestera T, Talalay P. Quantitative determination of isothiocyanates, dithiocarbamates, carbon disulfide, and related thiocarbonyl compounds by cyclocondensation with 1,2-benzenedithiol. Anal Biochem 1996; 239:160–167.

30. Chung F-L, Jiao D, Getahun M, Yu MC. A urinary biomarker for uptake of dietary isothiocyanates in humans. Cancer Epidemiol Biomark Prev 1998; 7:103–108.

31. Seow A, Shi C-Y, Chung F-L, Jiao D, Hankin JH, Lee H-P, Coetzee GA, Yu MC. Urinary total isothiocyanate (ITC) in a population-based sample of middle-aged and older Chinese in Singapore: relationship with dietary total ITC and glutathione S-transferase M1/TI/P1 genotypes. Cancer Epidemiol Biomark Prev 1998; 7:775–781.

32. Zhang Y, Talalay P. Mechanism of differential potencies of isothiocyanates as inducers of anticarcinogenic phase 2 enzymes. Cancer Res 1998; 58:4632–4639.

33. Ye L, Dinkova-Kostova AT, Wade KL, Zhang Y, Shapiro TA, Talalay P. Quantitative determination of dithiocarbamates in human plasma, serum, erythrocytes and urine: pharmacokinetics of broccoli sprout isothiocyanates in humans. Clin Chim Acta 2002; 316:43–53.

34. London SJ, Yuan J-M, Chung F-L, Gao Y-T, Coetzee GA, Ross RK, Yu MC. Isothiocyanates, glutathione S-transferase M1 and T1 polymorphisms, and lung-cancer risk: a prospective study of men in Shanghai, China. Lancet 2000; 356: 724–729.

35. Talalay P, Fahey JW, Holtzclaw WD, Prestera T, Zhang Y. Chemoprotection against cancer by phase 2 enzyme induction. Proceedings of the VII International Congress of Toxicology. WA: Seattle, 1995.

36. Hayes JD, McLellan LI. Glutathione and glutathione-dependent enzymes represent a coordinately regulated defence against oxidative stress. Free Radic Res 1999; 31: 273–300.

37. Jaiswal AK. Antioxidant response element. Biochem Pharmacol 1994; 48:439–444.

38. Dinkova-Kostova AT, Holtzclaw WD, Cole RN, Itoh K, Wakabayashi N, Katoh Y, Yamamoto M, Talalay P. Direct evidence that sulfhydryl groups of Keap1 and the

sensors regulating induction of phase 2 enzymes that protect against carcinogens and oxidants. Proc Natl Acad Sci USA 2002; 99:11908–11913.

39. Kashfi K, Zhang Y, Yang EK, Talalay P, Dannenberg AJ. Anticarcinogenic organic isothiocyanates induce UDP-glucuronosyltransferase. FASEB J 1995; 9:A868.

40. Prestera T, Holtzclaw WD, Zhang Y, Talalay P. Chemical and molecular regulation of enzymes that detoxify carcinogens. Proc Natl Acad Sci USA 1993; 90:2965–2969.

41. Prestera T, Talalay P, Alam J, Ahn YI, Lee PJ, Choi AM. Parallel induction of heme oxygenase-1 and chemoprotective phase 2 enzymes by electrophiles and antioxidants: regulation by upstream antioxidant-responsive elements (ARE). Mol Med 1995; 1: 827–837.

42. Brooks JD, Paton VG, Vidanes G. Potent induction of phase 2 enzymes in human prostate cells by sulforaphane. Cancer Epidemiol Biomark Prev 2001; 10:949–954.

43. Ye L, Zhang Y. Total intracellular accumulation levels of dietary isothiocyanates determine their activity in elevation of cellular glutathione and induction of phase 2 detoxification enzymes. Carcinogenesis 2001; 22:1987–1992.

44. Posner GH, Cho C-G, Green JV, Zhang Y, Talalay P. Design and synthesis of bifunctional isothiocyanate analogs of sulforaphane: correlation between structure and potency as inducers of anticarcinogenic detoxification enzymes. J Med Chem 1994; 37:170–176.

45. Fahey JW, Talalay P. Antioxidant functions of sulforaphane: a potent inducer of phase II detoxication enzymes. Food Chem Toxicol 1999; 37:973–979.

46. Gao X, Dinkova-Kostova AT, Talalay P. Powerful and prolonged protection of human retinal pigment epithelial cells, keratinocytes, and mouse leukemia cells against oxidative damage: the indirect antioxidant effects of sulforaphane. Proc Natl Acad Sci USA 2001; 98:15221–15226.

47. Zhang Y. Molecular mechanism of rapid cellular accumulation of anticarcinogenic isothiocyanates. Carcinogenesis 2001; 22:425–431.

48. Zhang Y, Callaway EC. High cellular accumulation of sulforaphane, a dietary anticarcinogen, is followed by rapid transporter-mediated export as a glutathione conjugate. Biochem J 2002; 364:301–307.

49. Zhang Y. Role of glutathione in the accumulation of anticarcinogenic isothiocyanates and their glutathione conjugates by murine hepatoma cells. Carcinogenesis 2000; 21:1175–1182.

50. Kolm RH, Danielson UH, Zhang Y, Talalay P, Mannervik B. Isothiocyanates as substrates for human glutathione transferases: structure-activity studies. Biochem J 1995; 311:453–459.

51. Mahéo K, Morel F, Langouët S, Kramer H, Ferrec EL, Ketterer B, Guillouzo A. Inhibition of cytochromes P-450 and induction of glutathione S-transferases by sulforaphane in primary human and rat hepatocytes. Cancer Res 1997; 57:3649–3652.

52. Langouët S, Furge LL, Kerriguy N, Nakamura K, Guillouzo A, Guengerich FP. Inhibition of human cytochrome P450 enzymes by 1,2-dithiole-3-thione, oltipraz and its derivatives, and sulforaphane. Chem Res Toxicol 2000; 13:245–252.

53. Conaway CC, Jiao D, Chung F-L. Inhibition of rat liver cytochrome P-450 isozymes by isothiocyanates and their conjugates: a structure-activity relationship study. Carcinogenesis 1996; 17:2423–2427.

54. Piazza GA, Thompson WJ, Pamukcu R, Alila HW, Whitehead CM, Liu L, Fetter JR, Gresh WE, Klein-Szanto AJ, Farnell DR, Eto I, Grubbs CJ. Exisulind, a novel proapoptotic drug, inhibits rat urinary bladder tumorigenesis. Cancer Res 2001; 61: 3961–3968.

55. Yang YM, Conaway CC, Chiao JW, Wang CX, Amin S, Whysner J, Dai W, Reinhardt J, Chung FL. Inhibition of benzo(a)pyrene-induced lung tumorigenesis in A/J mice by dietary N-acetylcysteine conjugates of benzyl and phenethyl isothiocyanates during the postinitiation phase is associated with activation of mitogen-activated protein kinases and p53 activity and induction of apoptosis. Cancer Res 2002; 62: 2–7.

56. Gamet-Payrastre L, Li P, Lumeau S, Cassar G, Dupont M-A, Chevolleau S, Gasc N, Tulliez J, Tercé F. Sulforaphane, a naturally occurring isothiocyanate, induces cell cycle arrest and apoptosis in HT29 human colon cancer cells. Cancer Res 2000; 60:1426–1433.

57. Chiao JW, Chung F-L, Kancherla R, Ahmed T, Mittelman A, Conaway CC. Sulforaphane and its metabolite mediate growth arrest and apoptosis in human prostate cancer cells. Int J Oncol 2002; 20:631–636.

58. Fimognari C, Nüsse M, Cesari R, Iori R, Cantelli-Forti G, Hrelia P. Growth inhibition, cell-cycle arrest and apoptosis in human T-cell leukemia by the isothiocyanate sulforaphane. Carcinogenesis 2002; 23:581–586.

59. Payen L, Courtois A, Loewert M, Guillouzo A, Fardel O. Reactive oxygen species-related induction of multidrug resistance-associated protein 2 expression in primary hepatocytes exposed to sulforaphane. Biochem Biophy Res Commu 2001; 282: 257–263.

60. Heiss E, Herhaus C, Klimo K, Bartsch H, Gerhäuser C. Nuclear factor κB is a molecular target for sulforaphane-mediated anti-inflammatory mechanisms. J Biol Chem 2001; 276:32008–32015.

61. Faulkner K, Mithen R, Williamson G. Selective increase of the potential anticarcinogen 4-methylsulphinylbutyl glucosinolate in broccoli. Carcinogenesis 1998; 19: 605–609.

62. Shapiro TA, Fahey JW, Wade KL, Stephenson KK, Talalay P. Chemoprotective glucosinolates and isothiocyanates of broccoli sprouts: metabolism and excretion in humans. Cancer Epidemiol Biomark Prev 2001; 10:501–508.

63. Kim DJ, Han BS, Ahn B, Hasegawa R, Shirai T, Ito N, Tsuda H. Enhancement by indole-3-carbinol of liver and thyroid gland neoplastic development in a rat medium-term multiorgan carcinogenesis model. Carcinogenesis 1997; 18:377–381.

64. Ge X, Yannai S, Rennert G, Gruener N, Fares FA. 3,3′-Diindolylmethane induces apoptosis in human cancer cells. Biochem Biophys Res Commun 1996; 228: 153–158.

65. Bjeldanes LF, Kim JY, Grose KR, Bartholomew JC, Bradfield CA. Aromatic hydrocarbon responsiveness-receptor agonists generated from indole-3-carbinol in vitro and in vivo: comparisons with 2,3,7,8-tetrachlorodibenzo-p-dioxin. Proc Natl Acad Sci USA 1991; 88:9543–9547.

66. Bradlow HL, Michnovicz J, Telang NT, Oxborne MP. Effects of dietary indole-3-carbinol on estradiol metabolism and spontaneous mammary tumors in mice. Carcinogenesis 1991; 12:1571–1574.

67. Guo D, Schut HAJ, Davis CD, Snyderwine EG, Bailey GS, Dashwood RH. Protection by chlorophylin and indole-3-carbinol against 2-amino-1-methyl-6-phenylimidazo-[4,5-b]pyridine (PhIP)-induced DNA adducts and colonic abberant crypts in the F344 rat. Carcinogenesis 1995; 16:2931–2937.
68. Brüsewitz G, Cameron BD, Chasseaud LF, Görler K, Hawkins DR, Koch H, Mennicke WH. The metabolism of benzyl isothiocyanate and its cysteine conjugate. Biochem J 1977; 162:99–107.
69. Mennicke WH, Görler K, Krumbiegel G. Metabolism of some naturally occurring isothiocyanates in the rat. Xenobiotica 1983; 13:203–207.
70. Mennicke WH, Görler K, Krumbiegel G, Lorenz D, Rittmann N. Studies on the metabolism and excretion of benzyl isothiocyanate in man. Xenobiotica 1988; 18: 441–447.
71. Jiao D, Smith TJ, Yang CS, Pittman B, Desai D, Amin S, Chung F-L. Chemopreventive activity of thiol conjugates of isothiocyanates for lung tumorigenesis. Carcinogenesis 1997; 18:2143–2147.
72. Xu K, Thornalley PJ. Studies on the mechanism of the inhibition of human leukaemia cell growth by dietary isothiocyanates and their cysteine adducts in vitro. Biochem Pharm 2000; 60:221–231.
73. Clapper ML, Szarka CE, Pfeiffer GR, Graham TA, Balshem AM, Litwin S, Goosenberg EB, Frucht H, Engstrom PF. Preclinical and clinical evaluation of broccoli supplements as inducers of glutathione S-transferase activity. Clin Cancer Res 1997; 3:25–30.
74. Lin HJ, Probst-Hensch NM, Louie AD, Kau IH, Witte JS, Ingles SA, Frankl HD, Lee ER, Haile RW. Glutathione transferase null genotype, broccoli, and lower prevalence of colorectal adenomas. Cancer Epidemiol Biomark Pre 1998; 7:647–652.
75. Spitz MR, Duphorne CM, Detry MA, Pillow PC, Amos CI, Lei L, de Andrade M, Gu X, Hong WK, Wu X. Dietary intake of isothiocyanates: evidence of a joint effect with glutathione S-transferase polymorphisms in lung cancer risk. Cancer Epidemiol Biomark Prev 2000; 9:1017–1020.
76. Lin HJ, Zhou H, Dai A, Huang H-F, Lin JH, Frankl HD, Lee ER, Haile RW. Glutathione transferase GSTT1, broccoli, and prevalence of colorectal adenomas. Pharmacogenetics 2002; 12:175–179.
77. Kassahun K, Davis M, Hu P, Martin B, Baillie T. Biotransformation of the naturally occurring isothiocyanate sulforaphane in the rat: identification of phase 1 metabolites and glutathione conjugates. Chem Res Toxicol 1997; 10:1228–1233.
78. Conaway CC, Krzeminski J, Amin S, Chung F-L. Decomposition rates of isothiocyanate conjugates determine their activity as inhibitors of cytochrome P450 enzymes. Chem Res Toxicol 2001; 14:1170–1176.
79. Prestera T, Fahey JW, Holtzclaw WD, Abeygunawardana C, Kachinski JL, Talalay P. Comprehensive chromatographic and spectroscopic methods for the separation and identification of intact glucosinolates. Anal Biochem 1996; 239:168–179.
80. Troyer JK, Stephenson KK, Fahey JW. Analysis of glucosinolates from broccoli and other cruciferous vegetables by hydrophilic interaction liquid chromatography. J Chromatogr A 2001; 991:299–304.

# 7

## Phytochemicals Protect Against Heterocyclic Amine–Induced DNA Adduct Formation

**Yongping Bao and James R. Bacon**
*Institute of Food Research, Colney, Norwich, United Kingdom*

## I. INTRODUCTION

In the 1970s, Sugimura and colleagues isolated potent mutagens from cooked meat and fish, thereby stimulating a major international effort on the study of heterocyclic amines (HCAs) and the search for protective compounds [1–3]. Since then, more than 200 reports have been published on dietary compounds that protect against HCA-induced mutagenicity and carcinogenicity. Schwab et al. [2] noted that some 600 compounds and complex mixtures that attenuate the mutagenicity of HCAs have been identified, including fatty acids, amino acids, pigments, flavorings, biogenic amines, vitamins, dietary fiber, and dietary phytochemicals. Dashwood [4] also compiled a list of more than 150 natural or synthetic phytochemicals, micronutrients, and antioxidants able to modulate HCA-induced mutagenicity and carcinogenicity. These include curcumin [5], resveratrol [6], isoflavones [7–9], soy saponins [10], carotenoids [11,12], organosulfur compounds [9,13–15], oltipraz [9,16,17], glucosinolate breakdown products, including isothiocyanates (ITCs) [3,9,18–20] and various polyphenols [9,21–26].

During the last two decades, much research on phytochemicals has been focused on the bioactivities of ITCs and polyphenols [27–32]. ITCs can inhibit the development of tumors in many of the experimental models studied and have been investigated as possible chemopreventive agents for specific human cancers [27,33]. Sulforaphane, an isothiocyanate derived from the glucosinolate glucoraphanin, has been considered as one of the 40 most promising anticancer agents

[34]. The mechanism of the protective role of ITCs in cancer prevention is believed to involve the modulation of carcinogen metabolism by induction of phase II detoxification enzymes and inhibition of phase I carcinogen-activating enzymes [32,35–37]. Recent work showed that sulforaphane is a potent inducer of thioredoxin reductase 1 (TrxR1) and that the interactions of sulforaphane and selenium resulted in a synergistic effect on expression of TrxR1 in cultured human HepG2 cells [38]. ITCs, including sulforaphane, also induce apoptosis in precancerous cells and tumor cells [39–41]. It has been proposed that low doses of ITCs can induce antioxidant response element (ARE)–mediated gene expression and that high doses of ITCs can activate apoptosis [42]. Other ITCs, such as benzyl ITC (BITC) and phenethyl ITC (PEITC), can also induce phase II metabolism and apoptosis at similar concentrations to sulforaphane [43,44].

Flavonoids are widely distributed in plant foods, and more than 5000 different compounds have been described [45]. In a dietary context, the most significant class of flavonoids are the flavonols, which include quercetin and kaempferol found in tea, onion, and apple [28]. Some studies support a protective effect of flavonoid consumption in cardiovascular disease and cancer, though a few studies demonstrate no effects [45]. Flavonoids possess both a wide range of antioxidant properties in vitro, such as inhibition of lipid and LDL oxidation, and also antiproliferative activity in human cancer cell lines [46–48].

HCAs are widely distributed in normally prepared foods, being found principally in wellcooked meat and fish and at lower levels in some other foods and beverages. Estimates of human exposure to 2-amino-1-methyl-6-phenylimidazo[4,5-b] pyridine (PhIP) and 2-amino-3,8-dimethylimidazo[4,5-f]quinoxaline (MeIQx), the two most abundant HCAs, range from ng/person/day to a few μg/person/day [49,50]. There are, of course, several ways to diminish the formation of HCAs in food, the most obvious being to avoid prolonged cooking of meat at high temperatures [50], but it is impossible to eliminate them completely from the human diet. A strategy to minimize the risks of HCAs in DNA damage and cancer initiation must, therefore, take into account the roles of the many dietary compounds, including phytochemicals, that have been shown to modulate the in vivo metabolism of HCAs. This chapter focuses on the protective effect of phytochemicals; following short overviews on DNA adducts as biomarkers of DNA damage and the risk of cancer and the application of accelerator mass spectrometry (AMS), recent results are presented on the effects of sulforaphane and quercetin on PhIP-DNA adduct formation.

## II. DNA ADDUCTS: BIOMARKERS OF DNA DAMAGE AND RISK OF CANCER

It is well documented that diet plays a crucial role in the etiology of human cancer [51]. Food is a complex mixture of compounds, and while some of them meet nutritional demands, others may constitute a health risk. Dietary carcinogens include nitrosamines, benzo[a]pyrene diol-epoxide, mycotoxins, oxidative agents, and HCAs [50,52,53]; these, together with endogenous reactive oxygen species, continuously assault DNA in which important genetic information is coded. The reaction of carcinogens with DNA appears to be one of the early events in the initiation phase of cancer. Cancer cells are produced through the accumulation of gene alterations in normal cells, and usually more than several genetic alterations in a single cell are required for its malignant transformation. DNA adducts, such as 8-oxo-7,8-dihydroguananine (8-OH-dG), malondialdehyde-deoxyguanosine (M1-dG), and aflatoxin-B1-DNA adducts, have been characterized as markers of DNA damage [54–56]. These DNA adducts can be base- and position-specific, are affected by sequence context, and are repaired at different rates depending on whether or not they are on the transcribed or nontranscribed strand of DNA [57]. In general, if a mutation occurs in a gene specifying a DNA repair enzyme, and as a result DNA repair lacks fidelity, then a cascade of new mutations will ensue leading to further cell dysregulation and dysfunction such as aberrant protein expression, altered gene switching, and changes in cell cycle control [58]. Tumors produced by HCAs in rodents show alterations in cancer-related genes—H-*ras*, K-*ras*, *p53*, and *Apc* [59–61]. Therefore, more mutations can result from a single genetic event, such as adduct formation, if it is not eliminated. The mutations induced by PhIP may include GC→TA transversion and also GC→AT transition mutations [62]. PhIP also induced GC-to-TA dominant mutation in mismatch repair (MMR)–deficient cells [63]. The measurement of total genomic DNA reaction of carcinogens is, however, only the first step in dissecting out which are the critical lesions for cancer initiation. There is increasing evidence that some types of DNA adduct can be considered as early markers of biologically effective dose and as markers of risk [64]. It is generally believed that DNA adduct formation is a requisite step in the process of chemical carcinogenesis [65]. Persistence of certain DNA adducts has been related to increased susceptibility, while inhibition of DNA adduct formation has been associated with inhibition of tumor induction [66].

Biomarkers of risk, such as DNA adducts, are required that can measure the effects of exposure to a potential carcinogen and the modulating effects of protective factors such as dietary nutrients and phytochemicals. Biomarkers can thus be used to establish the role of phytochemicals in protecting against DNA damage and determine the optimal intake of certain phytochemicals. Application of knowledge of the risk factors involved in cancer formation and also of the role of factors that can protect against cancer in the diet and environment is likely to be instrumental in chemoprevention.

## III. HCAs, PHIP, AND PHIP-DNA ADDUCTS

HCAs formed during high-temperature cooking of meat have been implicated in human colon cancer. An association between the intake of fried, broiled, or roasted (red) meat and the development of cancer has been observed in some studies [67]. Approximately 20 HCAs have been identified in cooked food [68]. Human exposure to HCAs depends on the eating habits of the individual, including the type, cooking methods, and quantities of foods consumed. The levels of some HCAs can vary by at least 5000-fold in moderately cooked meat and flame-grilled chicken [69]. The average intake of HCAs in Germany was estimated using a questionnaire [70]; median dietary intake of total HCAs was 103 ng/day including PhIP (63 ng), MeIQx (34 ng), and a small amount of 2-amino-3,4,8-trimethylimidazo[4,5-f]quinoxaline (DiMeIQx) (2 ng). The most abundant form of HCAs in well-done meat is PhIP, followed by MeIQx. PhIP has been detected in many laboratories in cooked pork, beef, fish, and chicken, at levels ranging from a few to hundreds of ng/g [68]. Apart from cooked meat, PhIP has also been detected in some beverages and cigarette smoke; for example, PhIP was found to be present in all brands of beer and also in wine at concentrations of 8–25 ng (mean = 14.1 ng/L) and ND–83 ng (mean = 30.4 ng/L), respectively [71], whereas smokers may be exposed to PhIP at up to 23 ng per cigarette [68,72].

PhIP has been shown to be carcinogenic in animal studies, and the intake of cooked meat has been related to formation of cancer in the colon [73,74], prostate, and breast [75–77]. This is supported by the finding that DNA adducts were formed in colon tissue of cancer patients undergoing surgery who received dietary relevant doses of $^{14}$C-labeled PhIP [78]. PhIP itself does not readily form adducts with DNA and requires activation in two stages to form reactive species that can form DNA adducts [79,80]. Normally, colonic cells do not express the key enzyme, cytochrome P450(CYP)1A2, required for the first step of this bioactivation, which is believed to occur in the liver [81]. CYP1A2 converts PhIP to N-OH-PhIP, which can be further activated, in either the liver or target tissues such as the colon, by either N-acetyl-transferase 2 (NAT2) or sulfotransferase to form the immediate carcinogens, N-acetoxy-PhIP or PhIP-N-sulfate, respectively. Both species can readily form adducts with DNA, more specifically with guanine at the C8 of 2'-deoxyguanosine, to form the DNA adduct, N-(deoxyguanosin-8-yl)-2-amino-1-methyl-6-phenylimidazo[4,5-b] pyridine (dG-C8-PhIP) [82–84]. If not repaired, these adducts may lead to mutations such as GC→TA transversions. DNA adducts formed with HCAs can be rapidly removed or repaired, although the mechanisms involved are not well understood [85]. Mammalian cells are also able to detoxify PhIP, through metabolism by the phase II enzymes, principally UDP-glucuronosyltransferase (UGT) 1A1 and glutathione-S-transferase (GST), to form less toxic and readily excretable metabolites [79,86]. There are other pathways, for example, involving CYP 1A1 and 1B1 to form 4'-OH-PhIP, by which PhIP may also be detoxified [87,88]. The metabolic activation and detoxification pathways have been summarized in Fig. 1.

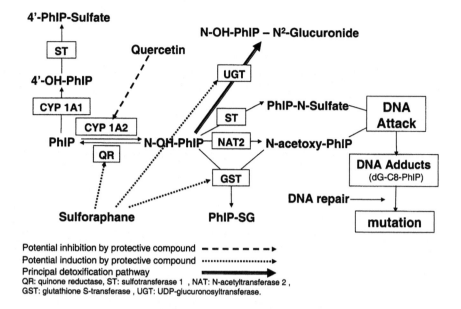

**Figure 1**  Bioactivation and detoxification of PhIP in human liver.


In determining the risk of PhIP and its carcinogenic activity, it is critical to assess the balance between enzyme pathways that catalyze activation and detoxification. The advantage in selecting PhIP-DNA adducts as an intermediate end-point biomarker is that they can be used to measure the overall effects of the processes of PhIP activation and detoxification and DNA repair. DNA adduct approaches have been demonstrated as biomarkers for exposure of genotoxic agents and risk of cancer [64]. They can provide new insights into the causes of cancer and the mechanisms of carcinogenesis and afford a useful intermediate endpoint in chemoprevention trials [52].


## IV.  APPLICATION OF ACCELERATOR MASS SPECTROMETRY

A number of different methods have been developed to detect and measure DNA adducts. These include physicochemical methods such as mass spectrometry, $^{32}$P-post-labeling, fluorescence, and AMS. The latter has been used to measure DNA adducts in human tissue after patients have ingested trace quantities of the food mutagens PhIP and MeIQx [85]. Such studies can assist in assessing the risks associated with low-level exposure to food genotoxins.

The number of PhIP-DNA adducts formed in vivo by the levels of PhIP normally found in plasma following ingestion of a portion of cooked meat or by the treatment of cultured cells in vitro by a similar physiologically-relevant dose of PhIP is proportionally very small (only 1 in $10^9$ bases may form an adduct). For AMS measurements, cells, animals or humans are exposed to low levels of isotopically labelled HCAs ($^{14}$C– or $^3$H–). DNA is extracted from cells or tissue using standard protocols for the isolation of genomic DNA, such as Qiagen DNA genomic tips, and graphitized. The high energy (4.5 million V) of a tandem Van de Graaff accelerator is used to generate positive ions of elements contained in the sample. Isotopes such as carbon ($^{14}$C, $^{13}$C, and $^{12}$C) can then be separated and quantified according to their differing mass to charge ratios. AMS can quantify isotope ratios to parts per quadrillion (that is, to attomole levels, $10^{-18}$ molar) [89]. Thus, it is particularly useful for measurement of levels of PhIP-DNA adducts at low doses when levels of adducts are correspondingly low. A sensitivity of detection can be attained equivalent to 0.01 dpm/mL or, with sufficient sample DNA, of the order of 1 adduct in $10^{12}$ bases [90]. However, one major disadvantage is that AMS is restricted by cost and access to most researchers. However, the increasing trend towards networking and collaboration may, to some extent, open up opportunities to a wider research community.

## A. The Use of AMS in In Vivo Studies

Doses of putative food mutagens can be administered to patients awaiting surgery and tissue samples subsequently taken during the procedure. The sensitivity of AMS is such that $^{14}$C-PhIP can be given safely at doses equivalent to the minimal levels found in the diet and still be detectable in subsequent tissue biopsies [78,91]. Such studies are not practicable using $^{32}$P-postlabeling detection. AMS requires the administration of radioisotopes, but this can be within ethically acceptable limits. Such in vivo studies are ideal for identifying levels of adduct formation in target organs, interindividual variation and genotypic effects, bioavailability, and metabolism.

## B. The Use of AMS in In Vitro Studies

In vitro studies to investigate the formation of DNA adducts, using either primary cells or immortal cell lines in culture, allow the investigation of the interactions of compounds like PhIP with other dietary components, such as phytochemicals, vitamins, and minerals. Measurements of DNA adduct formation in cells treated in culture can be compared with mechanistic studies to investigate effects of cotreatments on expression and activity of enzymes involved in the metabolism of PhIP and repair of DNA adducts. Some phytochemicals, such as the flavonoid quercetin, have been shown in animal studies to be able to reduce the formation of DNA adducts, while ITCs such as sulforaphane can induce phase II enzymes, such as UGT and GST, that have a detoxifying role in PhIP metabolism [31,86]. The pathway of PhIP metabolism in rodents is, however, different from that in humans, and the doses used in most studies are in excess of normal dietary levels; these factors make the results of animal studies difficult to extrapolate to humans [69,86]. Metabolically competent cells such as human hepatoma HepG2 cells [2] and human primary hepatocytes have been used to study the formation of PhIP-DNA adducts [92]. Liver cells were treated with $^{14}$C-PhIP in a range of concentrations including treatments at or below levels that might be expected in human serum after the consumption of a cooked meat meal. The results are described below.

## V. PHIP-DNA ADDUCT FORMATION

## A. Dose Response

Using AMS, it is possible to detect the formation of PhIP-DNA adducts in cultured cells treated with, for example, 100 pM $^{14}$C-PhIP. Figure 2 shows log/log plots of the effects of treatments of cells with a dilution series of PhIP over 10 million–fold dose range on levels of adducts formed in human HepG2 cells. 100 pM $^{14}$C-PhIP generated 2 adducts/$10^{10}$ bases, and the response was linear up to 20 μM of PhIP. Detection of this level of adduct formation is beyond the sensitivity of other methods; in one report [93] PhIP-DNA adducts were not detected using $^{32}$P-postlabeling analysis in cultured human mammary cells treated with μM concentrations of PhIP. This was probably a combination of the cell type lacking the necessary activation enzymes and the low detection limit of $^{32}$P-postlabeling. Although there is no information available on the human plasma level of PhIP, it may be estimated that intake from a typical meal of well-done meat (containing 0.2 μmol PhIP/100 g portion) could result in a PhIP concentration in human plasma of the range of 1–10 nM (Felton, personal communication). This is consistent with another HCA, MeIQx, which was found to form DNA adducts in human tissue at doses equivalent to normal human exposure levels [50].

The level of adducts formed in human hepatocytes was higher than in HepG2 cells and showed significant (20-fold) interindividual variation, for example, after treatment with 1 μM PhIP for 24 h, the level of adducts were found between 332 and 6892 adducts/$10^9$ bases in different subjects ($n = 6$). HepG2 cells, in comparison, gave $92 \pm 5$ adducts/$10^9$ bases under the same conditions [92]. Differences in the levels of adducts obtained between cell types can principally be attributed to levels of PhIP activation by enzymes such as CYP 1A2. It

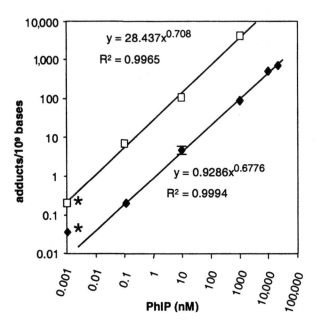

**Figure 2** $^{14}$C-PhIP DNA adduct formation in HepG2 cells (closed diamonds) and human hepatocytes (open squares) after 24-h treatment with $^{14}$C-PhIP. Error bars (10 nM PhIP treatment, HepG2) indicate mean $\pm$ SD (4.72 $\pm$ 0.39) of six treatments at different passage numbers. *Below detection limits <5 adducts/$10^{11}$ bases for HepG2 cells; <5 adducts/$10^9$ bases for hepatocytes.

has been reported that HepG2 cells showed only 10–20% of the cytochrome P-450 activity of the human hepatocytes [94]. Considering the differences in individual human exposure, rates of metabolism, and DNA repair, the difference in human cancer risk for HCAs could be more than 1000-fold [69].

## B.   Protective Effects of Sulforaphane and Quercetin

There are several reports on ITCs and flavonoids acting as chemopreventive agents against HCA-induced DNA adduct formation and carcinogenesis in rats [4,9,95]. Dietary phytochemicals can inhibit phase I activation and induce phase II detoxification enzymes and, therefore, attenuate the effect of PhIP on DNA adduct formation. ITCs, such as BITC, and quercetin strongly inhibit PhIP *N*-hydroxylation in rat liver microsomes and protect against PhIP-DNA adduct formation in rat colon [9,14].

Sulforaphane has also been shown to attenuate the induction of DNA strand breaks by *N*-nitrosodimethylamine (NDMA) and 2-amino-3-methylimidazo[4,5-f]quinoxaline (IQ) in T-antigen–immortalized human liver cells [96]. Moreover, sulforaphane induces levels of multidrug resistance–associated protein 2 (MRP2), an efflux pump contributing to biliary secretion of xenobiotics [97]. MRP2 protects against PhIP by biliary excretion of the parent compound and all major phase II metabolites [98,99]. Furthermore, in two separate human studies, consumption of cruciferous vegetables, including Brussels sprouts and broccoli, with cooked meat with known contents of HCAs affected both the rate of HCA metabolic excretion and the metabolic products [3,100].

When human HepG2 cells were co-treated with 10 nM PhIP and with either sulforaphane or quercetin, which have reported modulating effects on the phase I and II enzymes, both phytochemicals showed dose-response effects on the formation of PhIP-DNA adducts [92]. Sulforaphane and quercetin can reduce the number of DNA adducts formed at concentrations as low as 1 and 5 µM, respectively. The proportion of adduct formation inhibited is dependant on the concentration of PhIP treatment. At low levels of PhIP treatment, for example, 100 pM, co-treatment with sulforaphane or quercetin can reduce PhIP adducts to a level that is not distinguishable from background. At higher (supra-dietary) treatment levels of PhIP, e.g., 1 µM, however, the reduction in levels of adduct formation is proportionally smaller but still significant, 30% in the case of quercetin and 10% for sulforaphane (Fig. 3). It is interesting to notice that the threshold effects of PhIP on DNA adduct formation can be modulated by sulforaphane and quercetin.

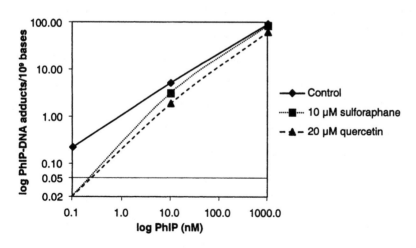

**Figure 3**   Effect of sulforaphane and quercetin on PhIP-DNA adduct formation in HepG2 cells at different [14]C-PhIP doses. (Control at 10 nM PhIP is one of six replicates described in Fig. 2). Detection limit < 5 adducts/$10^{11}$ bases.

In a comparison experiment, the possibility that vitamin C and β-carotene could affect PhIP-DNA adduct formation has been examined. Previous reports have already demonstrated the protective role of vitamin C in certain types of cancer and low intake of foods rich in vitamin C is associated with an increased risk of gastric cancer [100]. Vitamin C prevents DNA adduct formation in mice treated with the mycotoxins ochratoxin A and zearalenone [101]. Wu et al. [102] reported that vitamin C inhibited arylamine N-acetyltransferase (NAT) activity in human bladder tumor cells. This is one possible mechanism by which vitamin C can attenuate the formation of DNA adducts. It this study, 100 μM of vitamin C inhibited DNA adduct formation by about 30%. In contrast, quercetin is much more potent than vitamin C in protecting against PhIP-DNA adduct formation (Fig. 4). The role of β-carotene in 8-OH-dG formation is controversial. Some studies indicate that β-carotene can decrease adduct levels, whereas other studies do not support this claim [103]. Uehara et al. [11] reported that β-carotene decreased levels of IQ-DNA adducts in rats. However, it has also been reported that plasma β-carotene and α-tocopherol were not likely to influence the level of DNA adducts in lymphocytes [104]. In this study, it seems β-carotene did not modulate the level of DNA adduct induced by PhIP in human HepG2 cells (Fig. 4).

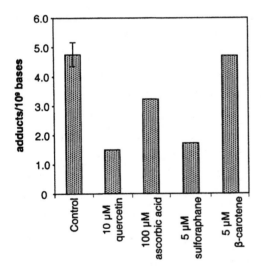

**Figure 4**   Effect of sulforaphane, quercetin, vitamin C, and β-carotene on PhIP-DNA adduct formation in HepG2 cells co-treated with 10 nM PhIP for 24 hours. Control is the average of six replicates treated with 10 nM PhIP.

As for the mechanisms of the protection, quercetin, acting as a blocking agent, can reduce the formation of PhIP-DNA adducts through the inhibition of PhIP bioactivation [3,14]. ITCs can induce phase II enzymes, such as UGT and GST, which may increase the rate of detoxification of PhIP [105]. The bioactivation and detoxification of PhIP and the interactions with phytochemicals have been summarized in Fig. 1.

## VI. SEARCH FOR OTHER PROTECTIVE PHYTOCHEMICALS

The limitation of many in vitro studies is the extrapolation of data from experimental models to the human situation. A major obstacle is the use of concentrations of carcinogens or protective compounds that are not physiologically relevant to normal exposure levels in humans and hence offer little or no insight regarding dose-response effects and interactions [106]. This study, using AMS, demonstrated the use of PhIP concentrations of 100 pM, which is lower than PhIP exposure from a normal diet. The formation of PhIP-DNA adducts was also dose-dependent.

When this measure of DNA adducts is combined with data from mechanistic and metabolic studies, it allows the investigation of the potential role of phytochemicals and micronutrients on the metabolism of PhIP and other foodborne genotoxins not only in liver cells, but also in other target tissues, such as breast and colon. The application of AMS in HCA-induced DNA adduct formation is applicable in human study. The level of DNA adducts is an intermediate endpoint biomarker, which can be used to measure the overall interactions and effects of PhIP and protective agents *via* activation, detoxification and repair [93,107].

The ultimate goal is to understand the relationship between HCAs and human cancer and to search for various measures to reduce the risk of, or to prevent, cancer. Food contains a complex mixture of nutrients, vitamins, additives, contaminants, and phytochemicals. The simplest way to reduce risk is to reduce the intakes of HCAs where possible through modified methods of cooking and, at the same time, to encourage the consumption of fruits and vegetables, especially of the cruciferous family. The results presented here support the epidemiological evidence that dietary phytochemicals can protect against cancer [108,109].

In searching for putative protective compounds, the following mechanisms of action should be considered: the protective compounds in the food matrix may act by one or more of the mechanisms that include (1) the inhibition of formation of HCA during food preparation, (2) inhibition of bioactivation through the CYP P450s, (3) induction of phase II detoxification enzyme expression, (4) induction of multidrug resistance transporters such as MRP2 [96,110], and (5) induction of DNA repair enzymes. The latter, the effect of phytochemicals on DNA repair, is the least studied of these mechanisms. More than 130 DNA repair genes have now been described [111], and modulation of DNA repair should lead to advanced understanding of the anticancer mechanism and cellular effects of phytochemicals. Further studies are required to identify potent dietary modulators of DNA repair.

## VII.  CONCLUSION

HCAs are widely distributed in our environment and humans are continuously exposed to HCAs in daily life at ng to μg per day per person. Even a PhIP exposure at pM level can generate DNA adducts in human cells and adducts could be involved in cancer development. More than one third of cancers are preventable by feasible dietary means, such as increasing vegetable intake and reducing the occurrence of dietary carcinogens including foodborne mutagens. Reducing these intakes is important, as "the dose makes the poison" (Paracelsus, 1493–1541), and this is relevant for dietary exposure to HCAs. From this study, the dose-dependent (linear) response and threshold effects of sulforaphane and quercetin on PhIP-DNA adduct formation have been demonstrated. The results support the promotion of higher intakes of fruits and vegetables.

Maintaining the integrity of DNA structure is of paramount importance to the cell. DNA damage, through adduct formation, increases the chance of mutation and hence cancer. Conversely, decreasing the amount of DNA damage should protect against or prevent cancer. The goal of understanding and reducing cancer through diet requires the understanding of the tumor initiation and controlling mechanisms and the influence of dietary protective phytochemicals. The detection of PhIP-DNA adducts using AMS is a very useful model for both the basic and applied research in diet and health.

## ACKNOWLEDGMENT

The authors would like to thank the UK Food Standards Agency for funding.

## REFERENCES

1.  Sugimura T, Kawachi T, Nagao M, Yahagi T, Seino Y, Okamoto T, Shudo K, Kosuge T, Tsuji K, Wakabayashi K, Iitaka Y, Itai A. Mutagenic principle(s) in tryptophan and phenylalanine pyrolysis products. Proc Jpn Acad 1977; 53:58–61.

2. Schwab CE, Huber WW, Parzefall W, Hietsch G, Kassie F, Schulte-Hermann R, Knasmuller S. Search for compounds that inhibit the genotoxic and carcinogenic effects of heterocyclic aromatic amines. Crit Rev Toxicol 2000; 30:1–69.

3. Knize MG, Kulp KS, Salmon CP, Keating GA, Felton JS. Factors affecting human heterocyclic amine intake and the metabolism of PhIP. Mutat Res 2002; 506–507: 153–162.

4. Dashwood RH. Modulation of heterocyclic amine-induced mutagenicity and carcinogenicity: an 'A-to-Z' guide to chemopreventive agents, promoters, and transgenic models. Mutat Res 2002; 511:89–112.

5. Collett GP, Robson CN, Mathers JC, Campbell FC. Curcumin modifies $Apc^{min}$ apoptosis resistance and inhibits 2-amino-1-methyl-6-phenylimidazo[4,5-b]pyridine (PhIP) induced tumour formation in $Apc^{min}$ mice. Carcinogenesis 2001; 22: 821–825.

6. Uenobe F, Nakamura S, Miyazawa M. Antimutagenic effect of resveratrol against Trp-P-1. Mutat Res 1997; 373:197–200.

7. Weisburger JH, Dolan L, Pittman B. Inhibition of PhIP mutagenicity by caffeine, lycopene, daidzein, and genistein. Mutat Res 1998; 416:125–128.

8. Ohta T, Nakatsugi S, Watanabe K, Kawamori T, Ishikawa F, Morotomi M, Sugie S, Toda T, Sugimura T, Wakabayashi K. Inhibitory effects of bifidobacterium-fermented soy milk on 2-amino-1-methyl-6-phenylimidazo[4,5-b]pyridine-induced rat mammary carcinogenesis, with a partial contribution of its component isoflavones. Carcinogenesis 2000; 21:937–941.

9. Hammons GJ, Fletcher JV, Stepps KR, Smith EA, Balentine DA, Harbowy ME, Kadlubar FF. Effects of chemoprotective agents on the metabolic activation of the carcinogenic arylamines PhIP and 4-aminobiphenyl in human and rat liver microsomes. Nutri Cancer 1999; 33:46–52.

10. Plewa MJ, Wagner ED, Berhow MA, Conway A, Rayburn AL, Anderson D. Antimutagenic activity of chemical fractions isolated from a commercial soybean processing by-product. Teratogen Carcinogen Mutagen 1999; 19:121–135.

11. Uehara N, Iwahori Y, Asamoto M, Baba-Toriyama H, Iigo M, Ochiai M, Nagao M, Nakayama M, Degawa M, Matsumoto K, Hirono I, Beppu H, Fujita K, Tsuda H. Decreased levels of 2-amino-3-methylimidazo[4,5-f]quinoline–DNA adducts in rats treated with β-carotene, α-tocopherol and freeze-dried aloe. Jpn J Cancer Res 1996; 87:342–348.

12. Rauscher R, Edenharder R, Platt KL. In vitro antimutagenic and in vivo anticlastogenic effects of carotenoids and solvent extracts from fruits and vegetables rich in carotenoids. Mutat Res 1998; 413:129–142.

13. Suzui N, Sugie S, Rahman KM, Ohnishi M, Yoshimi N, Wakabayashi K, Mori H. Inhibitory effects of diallyl disulfide or aspirin on 2-amino-1-methyl-6-phenylimidazo[4,5-b]pyridine-induced mammary carcinogenesis in rats. Jpn J Cancer Res 1997; 88:705–711.

14. Huber WW, McDaniel LP, Kaderlik KR, Teitel CH, Lang NP, Kadlubar FF. Chemoprotection against the formation of colon DNA adducts from the food-borne carcinogen 2-amino-1-methyl-6-phenylimidazo[4,5-b]pyridine (PhIP) in the rat. Mutat Res 1997; 376:115–122.

15. Guyonnet D, Belloir C, Suschetet M, Siess MH, Le Bon AM. Liver subcellular fractions from rats treated by organosulfur compounds from *Allium* modulate mutagen activation. Mutat Res 2000; 466:17–26.
16. Langouet S, Furge LL, Kerriguy N, Nakamura K, Guillouzo A, Guengerich FP. Inhibition of human cytochrome P450 enzymes by 1,2-dithiole-3-thione, oltipraz and its derivatives, and sulforaphane. Chem Res Toxicol 2000; 13:245–252.
17. Rao CV, Rivenson A, Zang E, Steele V, Kelloff G, Reddy BS. Inhibition of 2-amino-1-methyl-6-phenylimidazo[4,5-*b*]pyridine-induced lymphoma formation by oltipraz. Cancer Res 1996; 56:3395–3398.
18. Kinae N, Masuda H, Shin IS, Furugori M, Shimoi K. Functional properties of wasabi and horseradish. Biofactors 2000; 13:265–269.
19. Knasmuller S, Friesen MD, Holme JA, Alexander J, Sanyal R, Kassie F, Bartsch H. Effects of phenethyl isothiocyanate on metabolism and on genotoxicity of dimethylnitrosamine and 2-amino-1-methyl-6-phenylimidazo[4,5b]pyridine (PhIP). Mutat Res 1996; 350:93–102.
20. Kassie F, Laky B, Gminski R, Mersch-Sundermann V, Scharf G, Lhoste E, Kansmuller S. Effects of garden and watercress juices and their constituents, benzyl and phenethyl isothiocyanates, towards benzo(a)pyrene-induced DNA damage: a model study with the single cell gel electrophoresis/Hep G2 assay. Chem Biol Interact 2003; 142:285–296.
21. Bear WL, Teel RW. Effects of citrus flavonoids on the mutagenicity of heterocyclic amines and on cytochrome P4501A2 activity. Anticancer Res 2000; 20:3609–3614.
22. Oguri A, Suda M, Totsuka Y, Sugimura T, Wakabayashi K. Inhibitory effects of antioxidants on formation of heterocyclic amines. Mutat Res 1998; 402:237–245.
23. Kanazawa K, Yamashita T, Ashida H, Danno G. Antimutagenicity of flavones and flavonols to heterocyclic amines by specific and strong inhibition of the cytochrome P4501A family. Biosci Biotech Biochem 1998; 62:970–977.
24. Lee H, Wang HW, Su HY, Hao NJ. The structure-activity relationships of flavonoids as inhibitors of cytochrome P450 enzymes in rat liver microsomes and the mutagenicity of 2-amino-3-methylimidazo[4,5-*f*]quinoline. Mutagenesis 1994; 9:101–106.
25. Edenharder R, Keller G, Platt KL, Unger KK. Isolation and characterization of structurally novel antimutagenic flavonoids from spinach (*Spinacia oleracea*). J Agric Food Chem 2001; 49:2767–2773.
26. Hirose M, Futakuchi M, Tanaka H, Orita SI, Ito T, Miki T, Shirai T. Prevention by antioxidants of heterocyclic amine-induced carcinogenesis in a rat medium-term liver bioassay: results of extended and combination treatment experiments. Eur J Cancer Prev 1998; 7:61–67.
27. Fenwick GR, Heaney RK, Mullin WJ. Glucosinolates and their breakdown products in food and food plants. CRC Crit Rev Food Sci Nutr 1983; 18:123–201.
28. Williamson G, Faulkner K, Plumb GW. Glucosinolates and phenolics as antioxidants from plant foods. Eur J Cancer Prev 1998; 7:17–21.
29. Depeint F, Gee JM, Williamson G, Johnson IT. Evidence for consistent patterns between flavonoid structures and cellular activities. Proc Nutr Soc 2002; 61:97–103.
30. Day AJ, Bao Y, Morgan MR, Williamson G. Conjugation position of quercetin glucuronides and effect on biological activity. Free Rad Biol Med 2000; 29: 1234–1243.

31. Basten GP, Bao YP, Williamson G. Sulforaphane and its glutathione conjugate but not sulforaphane nitrile induce UDP-glucuronosyl transferase (UGT1A1) and glutathione transferase (GSTA1) in cultured cells. Carcinogenesis 2002; 23: 1399–1404.

32. Rose P, Faulkner K, Williamson G, Mithen R. 7-Methylsulfinylheptyl- and 8-methylsulfinyloctyl isothiocyanates from watercress are potent inducers of phase II enzymes. Carcinogenesis 2000; 21:1983–1988.

33. Conaway CC, Yang YM, Chung FL. Isothiocyanates as cancer chemopreventive agents: their biological activities and metabolism in rodents and humans. Curr Drug Metab 2002; 3:233–255.

34. Kelloff GJ, Crowell JA, Steele VE, Lubet RA, Malone WA, Boone CW, Kopelovich L, Hawk ET, Lieberman R, Lawrence JA, Ali I, Viner JL, Sigman CC. Progress in cancer chemoprevention: development of diet-derived chemopreventive agents. J Nutr 2000; 130:467S–471S.

35. Shapiro TA, Fahey JW, Wade KL, Stephenson KK, Talalay P. Human metabolism and excretion of cancer chemoprotective glucosinolates and isothiocyanates of cruciferous vegetables. Cancer Epidemiol Biomarkers Prev 1998; 7:1091–1100.

36. van Poppel G, Verhoeven DT, Verhagen H, Goldbohm RA. Brassica vegetables and cancer prevention: epidemiology and mechanisms. Adv Exp Med Biol 1999; 72:159–168.

37. Faulkner K, Mithen R, Williamson G. Selective increase of the potential anticarcinogen 4-methylsulphinylbutyl glucosinolate in broccoli. Carcinogenesis 1998; 19: 605–609.

38. Zhang JS, Svehlikova V, Bao YP, Howie AF, Beckett GJ, Williamson G. Synergy between sulforaphane and selenium in the induction of thioredoxin reductase 1 requires both transcriptional and translational modulation. Carcinogenesis 2003; 24:497–503.

39. Thornalley PJ. Isothiocyanates: mechanism of cancer chemopreventive action. Anti-Cancer Drugs 2002; 13:331–338.

40. Gamet-Payrastre L, Li P, Lumeau S, Cassar G, Dupont MA, Chevolleau S, Gasc N, Tulliez J, Terce F. Sulforaphane, a naturally occurring isothiocyanate, induces cell cycle arrest and apoptosis in HT29 human colon cancer cells. Cancer Res 2000; 60:1426–1433.

41. Chiao JW, Chung FL, Kancherla R, Ahmed T, Mittelman A, Conaway CC. Sulforaphane and its metabolite mediate growth arrest and apoptosis in human prostate cancer cells. Int J Oncol 2002; 20:631–636.

42. Kong AN, Yu R, Hebbar V, Chen C, Owuor E, Hu R, Ee R, Mandlekar S. Signal transduction events elicited by cancer prevention compounds. Mutat Res 2001; 480: 231–241.

43. Xu K, Thornalley PJ. Studies on the mechanism of the inhibition of human leukaemia cell growth by dietary isothiocyanates and their cysteine adducts in vitro. Biochem Pharmacol 2000; 60:221–231.

44. Bonnesen C, Eggleston IM, Hayes JD. Dietary indoles and isothiocyanates that are generated from cruciferous vegetables can both stimulate apoptosis and confer protection against DNA damage in human colon cell lines. Cancer Res 2001; 61: 6120–6130.

45. Ross JA, Kasum CM. Dietary flavonoids: bioavailability, metabolic effects, and safety. Ann Rev Nutr 2002; 22:19–34.

46. Aviram M, Fuhrman B. Wine flavonoids protect against LDL oxidation and atherosclerosis. Ann NY Acad Sci 2002; 957:146–161.

47. Kostyuk VA, Kraemer T, Sies H, Schewe T. Myeloperoxidase/nitrite-mediated lipid peroxidation of low-density lipoprotein as modulated by flavonoids. FEBS Lett 2003; 537:146–150.

48. Kuntz S, Wenzel U, Daniel H. Comparative analysis of the effects of flavonoids on proliferation, cytotoxicity, and apoptosis in human colon cancer cell lines. Eur J Nutr 1999; 38:133–142.

49. Turteltaub KW, Mauthe RJ, Dingley KH, Vogel JS, Frantz CE, Garner RC, Shen N. MeIQx-DNA adduct formation in rodent and human tissues at low doses. Mutat Res 1997; 376:243–252.

50. Nagao M, Sugimura T. Food Borne Carcinogens: Heterocyclic Amines. New York: John Wiley and Sons Ltd., 2000.

51. Knasmuller S, Steinkellner H, Majer BJ, Nobis EC, Scharf G, Kassie F. Search for dietary antimutagens and anticarcinogens: methodological aspects and extrapolation problems. Food Chem Toxicol 2002; 40:1051–1062.

52. Bartsch H. Studies on biomarkers in cancer etiology and prevention: a summary and challenge of 20 years of interdisciplinary research. Mutat Res 2000; 462:255–279.

53. Sugimura T. Nutrition and dietary carcinogens. Carcinogenesis 2000; 21:387–395.

54. Shen HM, Ong CN, Lee BL, Shi CY. Aflatoxin B1-induced 8-hydroxydeoxyguanosine formation in rat hepatic DNA. Carcinogenesis 1995; 16:419–422.

55. Shuker DE. The enemy at the gates? DNA adducts as biomarkers of exposure to exogenous and endogenous genotoxic agents. Toxicol Lett 2002; 134:51–56.

56. Leuratti C, Singh R, Lagneau C, Farmer PB, Plastaras JP, Marnett LJ, Shuker DE. Determination of malondialdehyde-induced DNA damage in human tissues using an immunoslot blot assay. Carcinogenesis 1998; 19:1919–1924.

57. Garner RC. The role of DNA adducts in chemical carcinogenesis. Mutat Res 1998; 402:67–75.

58. Bohr VA. DNA repair fine structure and its relations to genomic instability. Carcinogenesis 1995; 16:2885:2892.

59. Greenblatt MS, Bennett WP, Hollstein M, Harris CC. Mutations in the p53 tumor suppressor gene: clues to cancer etiology and molecular pathogenesis. Cancer Res 1994; 54:4855–4878.

60. Nagao M, Ushijima T, Toyota M, Inoue R, Sugimura T. Genetic changes induced by heterocyclic amines. Mutat Res 1997; 376:161–167.

61. Kakiuchi H, Watanabe M, Ushijima T, Toyota M, Imai K, Weisburger JH, Sugimura T, Nagao M. Specific 5′-GGGA-3′→5′-GGA-3′ mutation of the APC gene in rat colon tumors induced by 2-amino-1-methyl-6-phenylimidazo[4,5-b]pyridine. Proc Natl Acad Sci USA 1995; 92:910–914.

62. Shibutani S, Fernandes A, Suzuki N, Zhou L, Johnson F, Grollman AP. Mutagenesis of the N-(deoxyguanosin-8-yl)-2-amino-1-methyl-6-phenylimidazo[4,5-b]pyridine DNA adduct in mammalian cells. J Biol Chem 1999; 274:27433–27438.

63. Glaab WE, Kort KL, Skopek TR. Specificity of mutations induced by the food-associated heterocyclic amine 2-amino-1-methyl-6-phenylimidazo-[4,5-b]-pyridine

in colon cancer cell lines defective in mismatch repair. Cancer Res 2000; 60: 4921–4925.

64. Vineis P, Perera F. DNA adducts as markers of exposure to carcinogens and risk of cancer. Int J Cancer 2000; 88:325–328.

65. Hemminki K. DNA adducts, mutations and cancer. Carcinogenesis 1993; 14: 2007–2012.

66. Schut HA, Cummings DA, Smale MH, Josyula S, Friesen MD. DNA adducts of heterocyclic amines: formation, removal and inhibition by dietary components. Mutat Res 1997; 376:185–194.

67. Skog K, Solyakov A. Heterocyclic amines in poultry products: a literature review. Food Chem Toxicol 2002; 40:1231–1221.

68. Felton JS, Jägerstad M, Knize MG, Skog K, Wakabayshi K. Contents in food, beverages and tobacco. In Nagao M, Sugimura T, Eds. Food Borne Carcinogens: Heterocyclic Amines. New York: John Wiley & Sons Ltd, 2002:31–72.

69. Felton JS, Malfatti MA, Knize MG, Salmon CP, Hopmans EC, Wu RW. Health risks of heterocyclic amines. Mutat Res 1997; 376:37–41.

70. Rohrmann S, Becker N. Development of a short questionnaire to assess the dietary intake of heterocyclic aromatic amines. Public Health Nutr 2002; 5:699–705.

71. Manabe S, Suzuki H, Wada O, Ueki A. Detection of the carcinogen 2-amino-1-methyl-6-phenylimidazo[4,5-b]pyridine (PhIP) in beer and wine. Carcinogenesis 1993; 14:899–901.

72. Manabe S, Tohyama K, Wada O, Aramaki T. Detection of a carcinogen, 2-amino-1-methyl-6-phenylimidazo[4,5-b]pyridine (PhIP), in cigarette smoke condensate. Carcinogenesis 1991; 12:1945–1947.

73. Zheng W, Gustafson DR, Sinha R, Cerhan JR, Moore D, Hong CP, Anderson KE, Kushi LH, Sellers TA, Folsom AR. Well-done meat intake and the risk of breast cancer. J Natl Cancer Inst 1998; 90:1724–1729.

74. Sinha R, Rothman N. Role of well-done, grilled red meat, heterocyclic amines (HCAs) in the etiology of human cancer. Cancer Lett 1999; 143:189–194.

75. Ochiai M, Watanabe M, Kushida H, Wakabayashi K, Sugimura T, Nagao M. DNA adduct formation, cell proliferation and aberrant crypt focus formation induced by PhIP in male and female rat colon with relevance to carcinogenesis. Carcinogenesis 1996; 17:95–98.

76. Shirai T, Sano M, Tamano S, Takahashi S, Hirose M, Futakuchi M, Hasegawa R, Imaida K, Matsumoto K, Wakabayashi K, Sugimura T, Ito N. The prostate: a target for carcinogenicity of 2-amino-1-methyl-6-phenylimidazo4,5-b]pyridine (PhIP) derived from cooked foods. Cancer Res 1997; 57:195–198.

77. Sinha R, Gustafson DR, Kulldorff M, Wen WQ, Cerhan JR, Zheng W. 2-amino-1-methyl-6-phenylimidazo[4,5-b]pyridine, a carcinogen in high-temperature-cooked meat, and breast cancer risk. J Natl Cancer Inst 2000; 92:1352–1354.

78. Dingley KH, Curtis KD, Nowell S, Felton JS, Lang NP, Turteltaub KW. DNA and protein adduct formation in the colon and blood of humans after exposure to a dietary-relevant dose of 2-amino-1-methyl-6-phenylimidazo[4,5-b]pyridine. Cancer Epidemiol Biomarkers Prev 1999; 8:507–512.

79. Malfatti MA, Kulp KS, Knize MG, Davis C, Massengill JP, Williams S, Nowell S, MacLeod S, Dingley KH, Turteltaub KW, Lang NP, Felton JS. The identification

of [2-(14)C]2-amino-1-methyl-6-phenylimidazo[4,5-b]pyridine metabolites in humans. Carcinogenesis 1999; 20:705–713.

80. Schut HAJ, Sniderwine EG. DNA adducts of heterocyclic amine food mutagens: implications for mutagenesis and carcinogenesis. Carcinogenesis 1999; 20: 353–368.

81. Turesky RJ, Lang NP, Butler MA, Teitel CH, Kadlubar FF. Metabolic-activation of carcinogenic heterocyclic amines by human liver and colon. Carcinogenesis 1991; 12:1839–1845.

82. Frandson H. Excretion of DNA adducts of 2-amino-1-methyl-6-phenylimidazo[4,5-b]pyridine and 2-amino-3,4,8-trimethylimidazo[4,5-f]quinoxaline, PhIP-dG, PhIP-DNA and DiMeIQx-DNA from the rat. Carcinogenesis 1997; 18:1555–1560.

83. Lin D, Kaderlik KR, Tureky RJ, Miller DW, Lay JO, Kadlubar FF. Identification of N-(deoxyguanosin-8-yl)-2-amino-1-methyl-6-phenylimidazo[4,5-b]pyridine as the major adduct formed by the food-borne carcinogen 2-amino-1-methyl-6-phenylimidazo[4,5-b]pyridine, with DNA. Chem Res Toxicol 1992; 5:691–697.

84. Brown K, Hingerty BE, Guenther EA, Krishnan VV, Broyde S, Turteltaub KW, Cosman M. Solution structure of the 2-amino-1- methyl-6-phenylimidazo[4,5-b]pyridine C8-deoxyguanosine adduct in duplex DNA. Proc Natl Acad Sci USA 2001; 98:8507–8512.

85. Garner RC, Lightfoot TJ, Cupid BC, Russell D, Coxhead JM, Kutschera W, Priller A, Rom W, Steier P, Alexander DJ, Leveson SH, Dingley KH, Mauthe RJ, Turteltaub KW. Comparative biotransformation studies of MeIQx and PhIP in animal models and humans. Cancer Lett 1999; 143:161–165.

86. Langouet S, Paehler A, Welti DH, Kerriguy N, Guillouzo A, Turesky RJ. Differential metabolism of 2-amino-1-methyl-6-phenylimidazo[4,5-b]pyridine in rat and human hepatocytes. Carcinogenesis 2002; 23:115–122.

87. Wallin H, Mikalsen A, Guengerich FP, Ingelman-Sundberg M, Solberg KE, Rossland OJ, Alexander J. Differential rates of metabolic activation and detoxification of the food mutagen 2-amino-1-methyl-6-phenylimidazo[4,5-b]pyridine by different cytochrome P450 enzymes. Carcinogenesis 1990; 11:489–492.

88. Crofts FG, Strickland PT, Hayes CL, Sutter TR. Metabolism of 2-amino-1-methyl-6-phenylimidazo[4,5-b]pyridine (PhIP) by human cytochrome P4501B1. Carcinogenesis 1997; 18:1793–1798.

89. Vogel JS, Turteltaub KW, Finkel R, Nelson DE. Accelerator mass spectrometry—isotope quantification at attomole sensitivity. Anal Chem 1995; 67: 353A–359A.

90. Poirier MC, Santella RM, Weston A. Carcinogen macromolecular adducts and their measurement. Carcinogenesis 2000; 21:353–359.

91. Lightfoot TJ, Coxhead JM, Cupid BC, Nicholson S, Garner RC. Analysis of DNA adducts by accelerator mass spectrometry in human breast tissue after administration of 2-amino-1-methyl-6-phenylimidazo[4,5-b]pyridine and benzo[a]pyrene. Mutat Res 2000; 472:119–127.

92. Bacon JR, Williamson G, Garner RC, Lappin G, Langouet S, Bao YP. Sulforaphane and quercetin modulate PhIP-DNA adduct formation in human HepG2 cells and hepatocytes. Carcinogenesis 2003; 24:1903–1911.

93. Fan L, Schut HA, Snyderwine EG. Cytotoxicity, DNA adduct formation and DNA repair induced by 2-hydroxyamino-3-methylimidazo[4,5-f]quinoline and 2-hydroxyamino-1-methyl-6-phenylimidazo[4,5-b]pyridine in cultured human mammary epithelial cells. Carcinogenesis 1995; 16:775–779.

94. Doostdar H, Duthie SJ, Burke MD, Melvin WT, Grant MH. The influence of culture medium composition on drug metabolising enzyme activities of the human liver derived Hep G2 cell line. FEBS Lett 1988; 241:15–18.

95. Hirose M, Takahashi S, Ogawa K, Futakuchi M, Shirai T. Phenolics: Blocking agents for heterocyclic amine-induced carcinogenesis. Food Chem Toxicol 1999; 9–10:985–992.

96. Barcelo S, Mace K, Pfeifer AM, Chipman JK. Production of DNA strand breaks by N-nitrosodimethylamine and 2-amino-3-methylimidazo[4,5-f]quinoline in THLE cells expressing human CYP isoenzymes and inhibition by sulforaphane. Mutat Res 1998; 402:111–120.

97. Payen L, Courtois A, Loewert M, Guillouzo A, Fardel O. Reactive oxygen species-related induction of multidrug resistance-associated protein 2 expression in primary hepatocytes exposed to sulforaphane. Biochem Biophys Res Commun 2001; 282: 257–263.

98. Dietrich CG, de Waart DR, Ottenhoff R, Schoots IG, Elferink RP. Increased bioavailability of the food-derived carcinogen 2-amino-1-methyl-6-phenylimidazo[4,5-b]pyridine in MRP2-deficient rats. Mol Pharmacol 2001; 59:974–980.

99. Dietrich CG, de Waart DR, Ottenhoff R, Bootsma AH, van Gennip AH, Elferink RP. Mrp2-deficiency in the rat impairs biliary and intestinal excretion and influences metabolism and disposition of the food-derived carcinogen 2-amino-1-methyl-6-phenylimidazo[4,5-b]pyridine (PhIP). Carcinogenesis 2001; 22:805–811.

100. Webb PM, Bates CJ, Palli D, Forman D. Gastric cancer, gastritis and plasma vitamin C: results from an international correlation and cross-sectional study. The Eurogast Study Group. Int J Cancer 1997; 73:684–689.

101. Grosse Y, Chekir-Ghedira L, Huc A, Obrecht-Pflumio S, Dirheimer G, Bacha H, Pfohl-Leszkowicz A. Retinol, ascorbic acid and alpha-tocopherol prevent DNA adduct formation in mice treated with the mycotoxins ochratoxin A and zearalenone. Cancer Lett 1997; 114:225–229.

102. Wu HC, Lu HF, Hung CF, Chung JG. Inhibition by vitamin C of DNA adduct formation and arylamine N-acetyltransferase activity in human bladder tumor cells. Urol Res 2000; 28:235–240.

103. Jacobson JS, Begg MD, Wang LW, Wang Q, Agarwal M, Norkus E, Singh VN, Young TL, Yang D, Santella RM. Effects of a 6-month vitamin intervention on DNA damage in heavy smokers. Cancer Epidemiol Biomarkers Prev 2000; 9:1303–1311.

104. Wang Y, Ichiba M, Oishi H, Iyadomi M, Shono N, Tomokuni K. Relationship between plasma concentrations of beta-carotene and alpha-tocopherol and life-style factors and levels of DNA adducts in lymphocytes. Nutr Cancer 1997; 27:69–73.

105. Murray S, Lake BG, Gray S, Edwards AJ, Springall C, Bowey EA, Williamson G, Boobis AR, Gooderham NJ. Effect of cruciferous vegetable consumption on heterocyclic aromatic amine metabolism in man. Carcinogenesis 2001; 22: 1413–1420.

106. Knasmuller S, Steinkellner H, Majer BJ, Nobis EC, Scharf G, Kassie F. Search for dietary antimutagens and anticarcinogens: methodological aspects and extrapolation problems. Food Chem Toxicol 2002; 40:1051–1062.

107. Alexander J, Reistad R, Hegstad S, Frandsen H, Ingebrigtsen K, Paulsen JE, Becher G. Biomarkers of exposure to heterocyclic amines: approaches to improve the exposure assessment. Food Chem Toxicol 2002; 40:1131–1137.

108. Verhoeven DTH, Verhagen H, Goldbohm RA, van den Brandt PA, van Poppel G. A review of mechanisms underlying anticarcinogenicity by brassica vegetables. Chem Biol Interact 1997; 103:79–129.

109. Block G, Patterson B, Subar A. Fruit, vegetables, and cancer prevention: a review of the epidemiological evidence. Nutr Cancer 1992; 18:1–29.

110. Payen L, Sparfel L, Courtois A, Vernhet L, Guillouzo A, Fardel O. The drug efflux pump MRP2: regulation of expression in physiopathological situations and by endogenous and exogenous compounds. Cell Biol Toxicol 2002; 18:221–233.

111. Wood RD, Mitchell M, Sgouros J, Lindahl T. Human DNA repair genes. Science 2001; 291:1284–1289.

# 8

# Organosulfur-Garlic Compounds and Cancer Prevention

**John Milner**
*National Cancer Institute, National Institutes of Health, Rockville, Maryland, U.S.A.*

## I. INTRODUCTION

Garlic (*Allium sativum*) is cherished worldwide as part of a healthy diet, not only because of its savory characteristics but also for its perceived medicinal properties. It is fascinating to note that, historically, individuals who have not come in contact with each other have reached many of the same conclusions about the interrelationship between garlic and disease prevention [1,2]; such similarities in belief emphasize that folk wisdom should not simply be ignored, since it may provide valuable clues for discovering truth.

Increasing evidence undeniably points to the ability of garlic to alter several physiological processes that may influence heart disease and cancer risk [3–11]. Overall, these data continue to intrigue scientists, legislators, and consumers worldwide and offer support for earlier contentions. Although not as well examined, evidence is emerging that garlic and related sulfur constituents can influence immunocompetence [12–14] and, possibly, also learning/mental functions [15,16]. Thus, it is certainly conceivable that garlic is associated with a multitude of health benefits for both health promotion and disease risk reduction. Although the mechanism(s) accounting for many of the proposed health benefits remain largely unexplored, garlic continues to receive a reverence that is associated with few other foods.

Garlic is not simply a spice, herb, or vegetable, but a combination of all three. Together with onions, shallots, leeks, and chives, it is one of the major *Allium* foods consumed by humans. The garlic bulb consists of several individual bulblets or cloves, each weighing about 3 g. Nevertheless, absolute intakes are not known with any degree of certainty. The range of intake has been reported to vary from region to region and from individual to individual; average intakes in the United States have been approximated to be about 0.6 g per week or less [17], while those in parts of China may exceed 30 times this figure [18]. A recent study by Hsing et al. [19] provided evidence that health benefits were linked to daily intakes of more than 10 g compared to those consuming 2.2 g per day or less; it is not clear, however, that massive intakes (above 3 g/d) are really necessary to influence physiological processes associated with health.

While exaggerated intakes such as those in China may occur without adverse consequences, not all individuals would be expected to consume large quantities of garlic without complications. A spectrum of adverse allergic reactions, albeit of low incidence, can occur following contact with garlic [20]. In a recent U.S. study involving 132 children, 58% reported experiencing allergic reactions to food during a 2-year period. The offending food was identifiedin 34 of 41 reactions; milk was associated with 11 cases (32%) while wheat, celery, mango, or garlic were each linked with a single case (3%) [21]. Thus, the degree of allergic reaction to garlic is relatively low in the United States, and presumably this is also the situation in several other areas of the world.

Garlic is recognized to cause bleeding abnormalities in some individuals, depending on the quantity consumed and their blood-clotting capabilities; although this may not be an issue in most normal circumstances, it may pose a postoperative risk [22]. Animal studies provide additional evidence that some of the allyl sulfur compounds can foster hemolysis by enhanced free radical generation [23]. Although there appear to be few, if any, reports of hemolysis in humans following consumption of *Allium* vegetables, the potent hemolytic activity of their trisulfides and tetrasulfides in rodents indicates that they should be used with caution [24].

The garlic crop predominates in China, which alone represents about 65% of global production; South Korea and India are the next most important producers, each contributing about 5% to the world's supply, followed by the United States with about 3%. Approximately 250 million pounds of garlic are produced annually in the United States alone, with 80–90% originating in California. While some 50 million pounds is marketed annually as fresh garlic, the remainder is dehydrated and formulated as flakes and salt or used in packaged foods.

The U.S. Department of Agriculture (USDA) reports that on any typical day, about 18% of Americans consume at least one food containing garlic. Garlic has continued to be one of the top-selling herbs in the United States; in 1998 garlic sales in the United States were approximately $230 million [25], representing a 10% increase over the previous year. In countries such as Germany, the sale of garlic preparations ranks with those of the leading prescription drugs [26].

## II. CHEMISTRY OF GARLIC

Garlic usage is typically associated with its characteristic flavor and odor, which arise as a result of its high sulfhur content, about 1% dry weight [3,4,27]. While garlic is not a major source of essential nutrients, it can contribute to a variety of different dietary factors associated with health (Table 1): thus carbohydrates and protein constitute about 33% and 6.4% of its dry weight, respectively. Interestingly, the carbohydrate fraction contains a significant amount of oligosaccharides, which may influence gastrointestinal flora or function [28]. Most of the health benefits of garlic have been attributed to its sulfur compounds, but the matrix may also be important; moreover, it is noteworthy that garlic can be a relatively rich source of the amino acid arginine. Arginine breakdown may provide nitric oxide, involved in signal transduction of multiple pathways involved with human health [29], and this may account for some of the reported health benefits. Additionally, the presence of several other factors, including selenium and flavonoids, may influence the magnitude of the response to garlic [5,30,31].

The primary sulfur-containing constituents in garlic bulbs are γ-glutamyl-S-alk(en)yl-L-cysteines and S-alk(en)yl-L-cysteine sulfoxides. The content of S-alk(en)ylcysteine sulfoxide in garlic has been reported to range between 0.53 to 1.3% fresh weight, with alliin, S-allylcysteine sulfoxide, being the largest contributor [32]. Alliin concentrations can increase during storage as a result of the transformation of γ-glutamylcysteines. In addition to alliin, garlic bulbs contain small amounts of ( + )-S-methyl-L-cysteine sulfoxide (methiin) and ( + )-S-(*trans*-

**Table 1**  Content of Selected Components in Edible Garlic

| Component | Amount/100 g |
| --- | --- |
| Water, g | 58.6 |
| Protein, g | 6.4 |
| Lipid (fat), g | 0.5 |
| Carbohydrates, g | 33.1 |
| Fiber, total dietary, g | 2.1 |
| Calcium, mg | 181.0 |
| Magnesium, mg | 25.0 |
| Phosphorus, mg | 153.0 |
| Potassium, mg | 401.0 |
| Selenium, μg | 14.2 |
| Vitamin C, mg | 31.2 |

*Source*: USDA Nutrient Database for Standard Reference, Release 13 (November 1999).

1-propenyl)-L-cysteine sulfoxide, S-(2-carboxypropyl)glutathione, γ-glutamyl-S-allyl-L-cysteine, γ-glutamyl-S-(trans-1-propenyl)-L-cysteine, and γ-glutamyl-S-allylmercapto-L-cysteine [3,4,33].

The method used to process garlic can also influence the nature and levels of the sulfur compounds that predominate; some of the compounds found in various garlic preparations are shown in Table 2. The new analytical technique described by Arnault et al. [27] may assist in characterizing the impact of various processing methods on the content of the individual allyl sulfur compounds.

It is already recognized that allicin, allyl-2-propenethiosulfinate or diallyl thiosulfinate, is the major thiosulfinate occurring in garlic and its aqueous extracts. The characteristic odor of garlic arises from allicin and oil-soluble sulfur compounds formed when the bulb is crushed, damaged, or chopped. When this occurs, alliinase, an enzyme present in garlic, is activated and acts on alliin (found in the intact garlic) to produce odiferous alkyl alkane-thiosulfinates, including allicin. Since allicin is unstable, it further decomposes to yield sulfides, ajoene, and dithiins [26,34,35]. Tamaki and Sonoki [35] reported that strong garlic flavor and odor were linked to a higher content of volatile sulfur. Heating garlic is associated with a denaturing of alliinase, a reduction in allyl mercaptan, methyl mercaptan, and allyl methyl sulfide, and a corresponding reduction in odor [35].

Steam distillation of garlic produces an oil that also has several different allyl sulfur constituents; such garlic oils can contain about diallyl disulfide (DADS, ~26%), diallyl trisulfide (DATS, 19%), allyl methyl trisulfide (15%), allyl methyl disulfide (13%), diallyl tetrasulfide (8%), allyl methyl tetrasulfide (6%), dimethyl pentasulfide (4%), dimethyl trisulfide (3%), and dimethyl hexasulfide (1%). The oil derived from macerated garlic contains vinyldithiins and ajoenes. Storage of garlic in ethanol for several months produces so-called aged garlic extract (AGE); this process substantially reduces the amount of allicin and increases the amounts of S-allylcysteine (SAC), S-allylmercaptocysteine, and allixin.

**Table 2**   Sulfur Compounds in Garlic with Potential Health Benefits

| | |
|---|---|
| E-Ajoene | S-Allylmercaptocysteine |
| Z-Ajoene | S-Allylcysteine |
| Allicin | Diallyl disulfide |
| Allixin | Diallyl sulfide |
| Allyl mercaptan | Diallyl trisulfide |
| Allyl methyl sulfide | Dimethyl trisulfide |
| Allyl methyl trisulfide | 3-Vinyl-4-H-1,3-dithiin |
| Allyl methyl disulfide | |

The pharmacokinetics of allyl sulfur compounds have not been extensively examined, but it is considered unlikely that allicin occurs to any significant extent once garlic is processed and consumed. If it does exist, it will be rapidly transformed in the liver to diallyl disulfide (DADS) and allyl mercaptan, as suggested by studies by Egen-Schwind et al. [36]. Teyssier et al. [37] concluded that, in tissues, diallyl disulfide might be reconverted to allicin principally by oxidation involving cytochrome P450 monooxygenases and, to a limited extent, flavin-containing monooxygenases. Liver monooxygenases are also probably responsible for the oxidization of $S$-allylcysteine and many other sulfur compounds [38]. However, the extent to which this occurs in humans remains to be determined. Since DADS is known to cause an autocatalytic destruction of cytochrome P450 IIE1, it is unclear to what extent allicin would be formed under physiological conditions. Recently Germain et al. [39] provided evidence that DADS is absorbed and transformed into allyl mercaptan, allyl methyl sulfide, allyl methyl sulfoxide (AMSO), and allyl methyl sulfone (AMSO); while the latter predominated in tissues, both the mixed sulfoxide and sulfone were eliminated in the urine.

## III. IMPLICATION IN HEALTH

The literature contains numerous studies that demonstrate the ability of garlic and several of its associated allyl sulfur compounds to alter cellular processes that would logically be associated with improved health benefits. Notable among these are the abilities to serve as antimicrobial, anticarcinogenic, and antilipidemic agents. While long-term intervention studies are generally lacking, a variety of preclinical and epidemiological investigations support the contention that key molecular targets involved in the risk of several diseases can be affected by compounds arising from garlic consumption.

Scientists, legislators, and consumers are becoming increasingly aware that several foods, including garlic, may contribute to health, such as by reducing cancer risk [40–43]. Although major limitations exist in defining the precise role that garlic has in the overall cancer process, the likelihood of its significance is underscored by a relatively large number of epidemiological and preclinical studies [5,43,44]. Preclinical studies using several different types of cancer models undeniably provide some of the most convincing evidence that garlic and its associated sulfur components can suppress cancer risk and alter the biological behavior of tumors. A brief account follows of some of the current understanding of possible mechanisms by which these effects occur; reference is particularly made to cancer prevention.

## A.   Antimicrobial Effects and Cancer Prevention

Garlic is one of many plants with demonstrated antimicrobial activity; foods that are rich in tannins, terpenoids, alkaloids, flavonoids, and sulfur compounds are commonly reported to be particularly effective in reducing or preventing microbial growth. Garlic has been long recognized for its antimicrobial activity against several spoilage and pathogenic bacteria [45–47], and a range of gram-negative and gram-positive bacteria are known to be inhibited by garlic extracts [47–49]. The fate of *Salmonella*, *Escherichia coli* O157:H7, and *Listeria monocytogenes* preparations has recently been found to be influenced by the presence of garlic extracts [47]; addition of garlic to butter being observed to enhance the rates of inactivation of these three pathogens when incubated at 21 and 37°C.

The antimicrobial properties associated with garlic likely arise from several of its allyl sulfur components. In addition to allicin, compounds including diallyl sulfide, diallyl disulfide, E-ajoene, Z-ajoene, E-4,5,9-trithiadeca-1,6-diene-9-oxide (E-10-devinylajoene, E-10-DA), and E-4,5,9-trithiadeca-1,7-diene-9-oxide (*iso*-E-10-devinylajoene, iso-E-10-DA) have been proposed to contribute to these antimicrobial properties [6,50]. Although differences in efficacy among these compounds exist, relatively small amounts appear to be effective deterrents of microbial growth. Nevertheless, not all microorganisms are equally sensitive to garlic extracts or allyl sulfur compounds [47,51]. It is not known whether rates of uptake or metabolism within an organism determine the degree of the overall response to the individual allyl sulfur compounds. Such variation in sensitivity is unsurprising since mammalian cells have also been reported to vary in their response to allyl sulfur compounds. A more comprehensive examination of gene expression changes across various mammalian and prokaryocytic cells may provide important clues as to the molecular target(s) for these allyl sulfur compounds.

*Helicobacter pylori* colonization of the gastric mucosa is increasingly recognized as a factor leading to gastritis and likely gastric cancer. Since gastric cancer remains the second leading cause of cancer death worldwide, it is of major societal interest; high rates tend to occur in Japan, China, Central and South America, eastern Europe, and parts of the Middle East. Lower rates are found in North America, Australia and New Zealand, northern Europe, and India. While it is clear that the risk of stomach cancer for people infected with *H. pylori* is rather low and that it is not the sole factor involved, it is thought that the incidence may depend on specific genotypic polymorphisms of both the bacterium and its host. Changing several environmental risk factors by eliminating smoking, reducing salt intake, and increasing antioxidant intake appear to interfere with *H. pylori* and, thereby, modify risk. As Hooper et al. [52] have pointed out, it is becoming increasingly apparent that significant interactions exist between diet, microflora, and gastrointestinal genomics.

Cellini et al. [53] provided rather convincing evidence that aqueous garlic extracts (2–5 mg/mL) can inhibit *H. pylori* proliferation; reduced effectiveness occurred when the garlic was heated prior to extraction, suggesting that allicin or a breakdown product was likely to be responsible for the antiproliferative effects. Since both DAS and DADS have been shown to elicit a dose-dependent depression in *H. pylori* proliferation [54], heating may have reduced their formation. Canizares et al. [55] examined four solvents (water, acetone, ethanol, and hexane) for their efficacy to inhibit *Helicobacter*; while all provided some protection, acetone and ethanol were the most effective. The efficacy of various garlic preparations to inhibit *H. pylori* would be expected since Jonkers et al. [56] have reported the minimum inhibitory concentration (MIC) to range between 10 to 17.5 mg/mL for both raw garlic extracts and three commercially available garlic tablets. This sensitivity/selectivity is also demonstrated in studies by Sivam et al. [51] in which a MIC of 40 μg thiosulfinate/mL was found for *H. pylori* but no influence was observed on the proliferation of *Staphylococcus aureus*. Unfortunately, few clinical studies have been undertaken with garlic or individual allyl sulfides; until this is accomplished the physiological importance of such data will remain a focus of considerable debate and discussion.

Many results attest to the fact that garlic and its components can also serve as antifungal agents [46,49,57]. Lemar et al. [57] observed fresh garlic extracts to be more effective in retarding growth of *Candida albicans* than an extract of garlic powder. Similarly, garlic extracts appear to be as effective as some antibiotics [48]. Ajoene is also recognized in both in vitro and in vivo studies for its antimycotic activity. A fungal infection of the skin, tinea corporis, commonly known as ringworm, can also be influenced by sulfur compounds found in garlic. Ledezma et al. [58] reported that treatment with ajoene (0.4% as a gel) was as effective as terbinafine (1% as cream) in healing tinea corporis and tinea cruris in soldiers with dermatophytosis. Since ajoene can be readily prepared from garlic, it may be particularly useful as a public health strategy in developing countries

The antimicrobial effects of allyl sulfur compounds may reflect alterations in thiols occurring in various enzymes and/or a change in overall redox state. Similarly, changes in thiol homeostasis have been proposed as one potential mechanism by which garlic and associated sulfur compounds may suppress tumor proliferation [59]. There is evidence that both alliin and allicin possess antioxidant properties in a $H_2O_2$-Fe(II), Fenton oxygen-radical generating system [10,60]. Diallyl disulfide, but not diallyl sulfide, dipropyl sulfide, or dipropyl disulfide, has been found to inhibit liver microsomal lipid peroxidation induced by NADPH, ascorbate, and doxorubicin [61]. Water-soluble SAC has also been shown to possess antioxidant activity [62–64]. Hence the various compounds arising from garlic processing may contribute to its overall antimicrobial properties; while a change in redox status is a logical explanation for these effects, the data proving this mechanism or testing others are again woefully inadequate.

## B.  Influence on Multiple Tissues and Processes That Relate to Cancer

A number of studies involving animal models provide evidence that garlic and its associated components can reduce the incidence of cancers of the breast, colon, skin, uterine, esophagus, and lung [30,31,65–70]. The ability to inhibit tumors arising from different inducing agents and in different tissues indicates a generalized cellular response rather than a change in tissue specificity. Collectively, the protection appears to relate to fluctuations in several processes associated with cancer including the formation and bioactivation of carcinogens, DNA repair, tumor cell proliferation, and/or apoptosis. It is likely that several of these cellular processes are modified simultaneously, but clarification is needed in respect to the dose of allyl sulfur compound(s) needed and its temporality.

### 1.  Nitrosamine Formation and Metabolism

Suppressed nitrosamine formation has regularly been proposed as one of the likely mechanisms by which garlic may retard cancer incidence, and several studies suggest that allyl sulfur compounds can retard spontaneous and bacterial-mediated nitrosamine formation [50,71–74]. Since many nitrosamines are considered to be suspect carcinogens in a variety of biological systems, including humans [75,76], this response may be particular important physiologically. Chung et al. [77] found garlic extract to be more effective in blocking chemical nitrosation in vitro than an extract of strawberry or kale. Dion et al. [50] demonstrated that all allyl sulfur compounds were not equal in retarding nitrosamine formation: the ability of S-allylcysteine and S-propylcysteine to retard the formation of N-nitroso compounds, and the absence of any such effect with diallyl disulfide, dipropyl disulfide, and diallyl sulfide, reveal the critical role that the cysteine residue has in this inhibition.

The reduction in nitrosamine formation may actually arise secondarily to increased formation of nitrosothiols. Williams suggested almost 20 years ago [78] that several sulfur compounds may reduce nitrite availability for nitrosamine formation by enhancing the formation of nitrosothiols. Since the allyl sulfur content among garlic preparations can vary enormously, it is reasonable that all commercial preparation will not be equivalent in their ability to retard nitrosamine formation. While S-nitrosylation is known to influence health and disease [29], it is unclear how garlic influences this process across various cell types.

Some of the most compelling evidence that garlic depresses nitrosamine formation in humans is derived from studies conducted almost 15 years ago by Mei et al. [79]; these demonstrated that a daily dose of 5 g garlic completely blocked the enhanced urinary excretion of nitrosoproline that occurred as a result of exaggerated intake of nitrate and proline. The significance of this observation lies in the predictive value that nitrosoproline has for the synthesis of other potential carcinogenic nitrosamines [80]. Evidence that the effect of garlic occurs with nitrosamines other than those excreted in urine comes from the studies of Lin et al. [81] and Dion et al. [50].

The anticancer benefits attributed to garlic have also been associated with suppressed nitrosamine bioactivation. Evidence from a number of sources point to the effectiveness of garlic in blocking DNA alkylation, an initial step in nitrosamine carcinogenesis [81–83]. Consistent with such a reduction in bioactivation, Dion et al. [50] found that water-soluble $S$-allyl sulfide and lipid-soluble diallyl disulfide both retarded nitrosomorpholine mutagenicity in *Salmonella typhimurium* TA100. Aqueous extracts of garlic have also been shown to reduce the mutagenicity of ionizing radiation, peroxides, adriamycin and $N$-methyl-$N'$-nitronitrosoguanidine [84]. Reduced nitrosamine bioactivation may arise from changes in phase I enzymatic activity; in particular cytochrome P450 2E1 (CYP2E1) appears to be involved with this depression [12,85–87]. An autocatalytic destruction of CYP2E1 may account for some of the chemoprotective effects of diallyl sulfide and, possibly, other allyl sulfur compounds [88]. Variation in the content and overall activity of P450 2E1 may be an important variable in the degree of protection afforded by garlic and associated allyl sulfur components.

## 2. Bioactivation and Response to Other Carcinogens

Garlic and several of its allyl sulfur compounds can also effectively block the bioactivation and carcinogenicity of a host of carcinogens (Table 3). This protection, which encompasses a diverse array of compounds and cancers occurring in several tissues, again suggests an overarching biological response. Since metabolic activation is required for many of the carcinogens used in these studies, there is the likelihood that phase I and II enzymes are involved. Interestingly, little change in cytochrome P-450 1A1, 1A2, 2B1, or 3A4 activities has been observed following treatment with garlic or related sulfur compounds [89–91]. However, this lack of response may relate to the quantity and duration of exposure, the quantity of carcinogen administered, or the methods used to assess cytochrome content or activity. Wu et al. [92], using immunoblot assays, found the protein content of cytochrome P450 1A1, 2B1, and 3A1 to be increased by garlic oil and each of several isolated disulfides. These data demonstrated that the number of sulfur atoms in the allyl compound was inversely related to depression of these cytochromes. Thus, changes in phase I enzyme activity may account for some of the anticancer properties attributed to garlic.

Changes in bioactivation resulting from a block in cyclooxygenase and lipoxygenase may also partially account for the reduction in tumors following treatment with some carcinogens [93–97]. Ajoene has also been shown to interfere with the COX-2 pathway by using lipopolysaccharide (LPS)-activated RAW 264.7 cells in vitro [98]. Ajoene treatment lead to a dose-dependent inhibition

**Table 3**  Carcinogens Influenced by Garlic and/or Associated Allyl Sulfur Compounds[a]

| Compound | Site | Host |
| --- | --- | --- |
| 1,2-Dimethylhydrazine | Colon | Rat |
| 3-Methylcholanthrene | Cervix | Mouse |
| 4-(Methylnitrosamino)-1-(3-pyridyl)-1-butanone | Nasal | Rat |
| 7,12-Dimethylbenz(a)anthracene | Mammary | Rat |
| 7,12-Dimethylbenz(a)anthracene | Skin | Mouse |
| 7,12-Dimethylbenz(a)anthracene | Forestomach | Hamster |
| 7,12-Dimethylbenz(a)anthracene | Buccal pouch | Hamster |
| Aflatoxin $B_1$ | Liver | Toad |
| Aflatoxin $B_1$ | Liver | Rat |
| Azoxymethane | Colon | Rat |
| Benzo(a)pyrene | Forestomach | Mouse |
| Benzo(a)pyrene | Lung | Mouse |
| Benzo(a)pyrene | Skin | Mouse |
| Benzo(a)pyrene | Bone marrow | Mouse |
| Methylnitronitrosoguanidine | Gastric | Rat |
| N-Methyl-N-nitrosourea | Mammary | Rat |
| N-Nitrosodiethylamine | Colon | Rat |
| N-Nitrosodiethylamine | Nasal | Rat |
| N-Nitrosodimethylamine | Liver | Rat |
| N-Nitrosodimethylamine | Nasal | Rat |
| N-Nitrosodimethylamine | Skin | Mouse |
| N-Nitrosomethylbenzylamine | Esophagus | Rat |
| Vinyl carbamate | Skin | Mouse |

[a] The overall response to garlic and/or specific allyl sulfur components depends on the quantity provided and the amount of carcinogen administered.

in release of LPS (1 μg/mL)-induced prostaglandin $E_2$ in RAW 264.7 macrophages ($IC_{50}$ value: 2.4 μM). This effect was found to be due to an inhibition of COX-2 enzyme activity by ajoene ($IC_{50}$ value: 3.4 μM). Ajoene did not reduce the expression of COX-2, but rather increased LPS-induced COX-2 protein and mRNA expression compared to LPS-stimulated cells. In the absence of LPS, however, ajoene was unable to induce COX-2; the nonsteroidal anti-inflammatory drug indomethacin behaved similarly. These data suggest that ajoene works by a mechanism of action similar to that of nonsteroidal anti-inflammatory drugs.

There is some limited evidence that garlic and associated sulfur components can inhibit lipoxygenase activity [99]. Similarly, evidence for the involvement of lipoxygenase in the bioactivation of carcinogens such as dimethylbenzanthracene (DMBA) comes from the work of Song [100], which demonstrated the feeding of the lipoxygenase inhibitor nordihydroguaiaretic acid to be accompanied by a marked reduction in DMBA-induced DNA adducts in rat mammary tissue. Collectively, these studies pose interesting questions about the role of both cyclooxygenase and lipoxygenase not only in forming prostaglandins and, therefore, modulating tumor cell proliferation and immunocompetence, but also in the bioactivation of carcinogens. Clearly, additional attention is needed to clarify what role, if any, these bioactivation enzymes have in determining the biological response to dietary garlic and its associated allyl sulfur compounds.

## 3. Detoxification and Allyl Sulfur Specificity

Detoxification enzymes involved in the removal of carcinogenic metabolites may also be instrumental to in the observed protection offered by garlic. Singh et al. [101] provided evidence that suppression of NAD(P)H:quinone oxidoreductase activity correlated with the ability of garlic preparations to suppress benzo(a)pyrene tumorigenesis. Similarly, changes in glutathione concentration and the activity of specific glutathione-$S$-transferase, both factors involved in phase II detoxification, may be important in the protection provided by garlic. Feeding garlic powder to rats was accompanied by increased liver and mammary GST activity [90,102,103]; however, not all GST isozymes appear to be influenced equally. Hu et al. [104] found evidence that the induction of glutathione (GSH) $S$-transferase pi (mGSTP1-1) may be particularly important in the anticarcinogenic properties associated with garlic and its allyl sulfur components. Bose et al. [105] demonstrated that mGSTP1 mRNA expression was either unaltered in liver or moderately increased in the forestomach following treatment with dipropyl disulfide, indicating that the allyl group is crucial for the mGSTP1-inducing activity of DADS. Munday and Munday [106] showed that both diallyl disulfide and diallyl trisulfide were important in the anticancer action of garlic, while dipropenyl sulfide maybe involved in the anticancer action of another *Allium*, the onion.

Dietary garlic supplementation has also been found to reduce the incidence of tumors resulting from the treatment with methylnitrosurea (MNU), a direct-acting carcinogen [81] in an animal model. Water-soluble $S$-allylcysteine (57 µmol/kg diet) and lipid-soluble DADS caused comparable reductions in MNU-induced $O^6$-methylguanine adducts bound to mammary cell DNA [107]. In contrast, Cohen et al. [108] did not observe $S$-allylcysteine to inhibit MNU-induced mammary tumors when the compound was added to the diet. The reason for this discrepancy is unclear, but may relate to the quantity of lipid in the diet or to the dose of carcinogen provided. If there truly are effects of DADS and SAC on MNU-induced carcinogenesis, the mechanism(s) by which it brings about this effect are unknown. While it is possible that garlic may influence mammary gland terminal end bud formation and/or change rates of DNA repair, more research is needed to determine the quantity and duration of the supplementation required to produce these effects.

Rarely has a comparison of water- and oil-soluble compounds been conducted within the same study. Nevertheless, what evidence is available suggests that major differences in efficacy among extracts are not of paramount importance [31,62,107,109–111]. While subtle differences between garlic preparations are likely to occur, *quantity* rather than *source* appears to be the primary factor influencing the degree of protection [68]. Differences that do occur between preparations probably relate to the content and effectiveness of individual sulfur-containing compounds. The number of sulfur atoms present in the molecule seems to influence the degree of protection with diallyl trisulfide > diallyl disulfide > diallyl sulfide [59,112]. In a similar manner, the presence of the allyl group generally enhances protection over that provided by the propyl moiety [59,104].

## C. Epigenetic Events and Garlic

Cancer progression is probably also highly dependent on epigenetic changes. Two extensively examined mechanisms for epigenetic gene regulation are patterns of DNA methylation and histone acetylation/deacetylation. Several studies indicate that DNA hypermethylation is an important mechanism for inactivation of key regulatory genes such as E-cadherin, pi-class glutathione *S*-transferase, the tumor suppressor cyclin-dependent kinases (CDKN2), the phosphatase gene (PTEN), and insulin-like growth factor (IGF-II). Targeted histone acetylation/deacetylation results in remodeling of chromatin structure and correlates with activation/repression of transcription (e.g., of IGFBP-2 and p21).

Both DNA methylation and histone acetylation appear to be modified by garlic and/or related allyl sulfur compounds. Studies by Ludeke et al. [113] found that DAS inhibited the formation of $O^6$-methyldeoxyguanosine arising from treatment with *N*-nitrosomethylbenzylamine in esophagus ($-26\%$), nasal mucosa ($-51\%$), trachea ($-68\%$), and lung ($-78\%$). Similarly, other studies [81,107] revealed DADS, SAC, and deodorized garlic to be effective in retarding DNA methylation caused by MNU. Lea et al. [114] reported that at least part of the ability of DADS to induce differentiation in DS19 mouse erythroleukemic cells might relate to its ability to increase histone acetylation. Diallyl disulfide caused a marked increase in the acetylation of H4 and H3 histones in DS19 and K562 human leukemic cells. Similar results were also obtained with rat hepatoma- and human breast cancer cells. In 2001 Lea and Randolph [115] presented evidence that DADS administered to rats could also increase histone acetylation in both the liver and a transplanted hepatoma cell line. The data suggested an increase in the acetylation of core mucosomal histones and enhanced differentiation. Allyl mercaptan was found to be a more potent inhibitor of histone deacetylase than diallyl disulfide. Interestingly, the latter has been also been reported to inhibit the growth of H-ras oncogene-transformed tumors in nude mice [116]; this inhibition correlated with the inhibition of p21H-ras membrane association in the tumor tissue.

## D.  Cell Proliferation and Apoptosis

Several of the lipid- and water-soluble organosulfur compounds have been examined for their antiproliferative efficacy [72,117–128]. Table 4 provides a list of some of the human tumor cells that have been found to be inhibited by garlic and/or its allyl sulfur compounds; some of the more commonly used lipid-soluble allyl sulfur compounds in tumorigenesis research are ajoene, diallyl sulfide (DAS), diallyl disulfide (DADS), and diallyl trisulfide (DATS). A breakdown of allicin appears to be necessary for achieving maximum tumor inhibition. Studies by Scharfenberg et al. [117] found that the $ED_{50}$ for lymphoma cells was two times lower for ajoene than for allicin.

Previous studies reported that lipid-soluble DAS, DADS, and DATS (100 $\mu$M) were more effective in suppressing canine tumor cell proliferation than isomolar water-soluble SAC, $S$-ethylcysteine, and $S$-propylcysteine [119]. $S$-Allylmercaptocysteine, one of the more effective water-soluble allyl sulfur compounds, did not reduce the viability of human erythroleukemina cells until concentrations were about 100 $\mu$M [123]. While treatment of human colon tumor cells (HCT-15) with 100 $\mu$M DADS resulted in complete cessation of growth, approximately 200 $\mu$M SAMC was required to achieve a similar depression [72,128]. No changes in growth were observed with concentrations of SAC up to 500 $\mu$M. Although some studies [119,122] reported that SAC was ineffective in altering the proliferation of canine mammary or human colon cells, other researchers have

**Table 4**  Human Tumor Cells Inhibited by Garlic and/or Associated Allyl Sulfur Compounds

| Site | Cell type |
| --- | --- |
| Colon | HCT-15, HT-29 |
| Endometrial | Ishikawa |
| Erythroleukemia | HEL, OCIM-1 |
| Leukemic | HL-60 |
| Lung—non–small-cell | H460, H1299 |
| Lung | A 549 |
| Mammary | MCF-7 |
| Neuroblastoma | LA-N-5 |
| Prostate | LNCAP, CRL-1740 |
| Skin | SKMEL-2 |

demonstrated that it inhibits growth of melanoma, prostate, and neuroblastoma cells; about 4–16 mM SAC was needed to suppress cell proliferation in these studies [118,120,126]. Such inconsistencies may relate to the concentration of SAC added to the cell culture and to differences in the sensitivity among cells. Thus, not all allyl sulfur compounds from garlic are equally effective in retarding tumor proliferation [117,119,123,127].

In an analogous manner to their effects on chemical carcinogenesis, the antitumorigenic effects of organosulfur compounds are not restricted to a specific tissue or a particular cell type, since cultured human colon, skin, and lung tumor cell proliferation can all be inhibited [59]. Therefore, these compounds probably act by modifying common pathway(s) controlling cell proliferation. Nevertheless, evidence suggests that these allyl sulfur compounds preferentially suppress neoplastic over nonneoplastic cells [117,127]. In vitro addition of DATS (10 $\mu$M) to cultures of A549 lung tumor cells inhibited their proliferation by 47%, whereas it was without effect on non-neoplastic MRC-5 lung cells [127]. The antiproliferative effects of allyl sulfides are generally reversible, assuming that apoptosis is not extensive [127,128]; for instance, 24-hour treatment with 50 $\mu$M DADS resulted in a 43% reduction in HCT-15 tumor cell growth, while refeeding with complete medium resulted in a return toward normal rates of growth [128]. Refeeding with complete medium of A549 tumor cells previously exposed to 10 $\mu$M DATS caused normal intracellular calcium values to occur within an hour [127].

There is evidence that cell number influences the antiproliferative effects of allyl sulfur compounds. The inhibitory effects for both ajoene and SAMC were inversely proportional to cell density [117,123], while Sigounas and associates [123] provided data suggesting that confluent cells are less sensitive to SAMC than rapidly growing cultures. There is evidence from studies with ajoene in lymphoma cells that allyl sulfides are absorbed and metabolized to bring about their effects [117]. Overall, the antiproliferative effect of these compounds depends most likely not only on supply but also their rate of metabolism.

DADS exposure has been reported to lead to a proportional increase in the percentage of cells arrested in the $G_2/M$ phase of the cell cycle [128]; this arrest was evident within 4 hours after treatment with 50 $\mu$M DADS. SAMC and DAS have also been reported to increase the percentage of cells blocked within the $G_2/M$ phase [124,129]. The ability of garlic to block the $G_2/M$ phase is not limited to studies in in vitro systems; Kimura et al. [130] observed an increased number of metaphase arrested tumor cells in MTK-sarcoma III xenographs in rats receiving an aqueous extract of garlic (1–10 mg/100 g b.w.) for 4 days compared to those not receiving the extract. Irregular chromosomal organization was found to accompany the block in metaphase progression.

p34$^{cdc2}$ Kinase is a complex that governs the progression of cells from the $G_2$ into the M phase of the cell cycle; its activation promotes chromosomal condensation and cytoskeletal reorganization through phosphorylation of multiple substrates, including histone H1 [131,132]. Factors that inhibit $p34^{cdc2}$ kinase activity lead to a block in the $G_2/M$ phase. Recent studies provide evidence that the $G_2/M$ phase arrest induced by DADS coincides with the suppression in $p34^{cdc2}$ kinase activity. Exposure of synchronized HCT-15 cells to 50 $\mu$M DADS resulted in a 60% suppression in $p34^{cdc2}$ kinase activity.

$p34^{cdc2}$ Kinase complex formation is controlled by the association of the $p34^{cdc2}$ catalytic unit with the cyclin $B_1$ regulatory unit [133]; activation of this complex is governed by both cyclin $B_1$ protein synthesis and degradation and phosphorylation and dephosphorylation of threonine and tyrosine residues on the $p34^{cdc2}$ subunit [131,133]. A twofold increase in cyclin $B_1$ protein expression was observed in cultured HCT-15 cells after 12-hour exposure to DADS [72]. Overall, the ability of DADS to inhibit p34(cdc2) kinase activation appeared to result from decreased p34(cdc2)/cyclin B(1) complex formation and a change in p34(cdc2) hyperphosphorylation [72].

## E.  Dietary Modifiers of Garlic and Allyl Sulfur Efficacy

The influence of garlic on the cancer process cannot be considered in isolation since several dietary components can markedly influence its overall impact. Among the factors recognized to influence the response to garlic are total fat, selenium, methionine, and vitamin A [30,31,134]. Amagase et al. [30] and Ip et al. [135] reported that selenium supplied either as a component of the diet or as a constituent of a garlic supplement enhanced protection against 7,12-dimethyl-benz(a)anthracene (DMBA)–induced mammary carcinogenesis over that provided by garlic alone. Suppression in carcinogen bioactivation, as indicated by a reduction in DNA adduct formation, may partially account for this combined benefit of garlic and selenium [109]. Both selenium and allyl sulfur compounds are recognized to alter cell proliferation and induce apoptosis [59,122,128,136].

Dietary fatty acid supply can also influence the bioactivation of DMBA to metabolites capable of binding to rat mammary cell DNA and a significant portion of the enhancement in mammary DNA adducts caused by increasing dietary corn oil consumption can be attributed to linoleic acid intake [133]. While exaggerated oleic acid consumption also increases DMBA-induced DNA adducts, it is far less effective in promoting adducts compared to linoleic acid. Chen et al. [137] have recently published evidence that garlic oil and fish oil modulated the antioxidant and drug-metabolizing capacity of rats and that their combined effects on drug-metabolizing enzymes was additive. Interestingly, co-administration of garlic with fish oil was well tolerated in humans and had a beneficial effect on concentrations of serum lipids and lipoproteins by providing a combined lowering of total cholesterol, LDL-C, and triacylglycerol concentrations, the ratios of total choles-terol:HDL-C, and LDL-C:HDL-C [138]. The ability of selected fatty acids to alter DMBA bioactivation may provide clues to a plausible mechanism by which garlic and its allyl sulfur compounds might retard chemically induced tumors.

## IV. SUMMARY AND CONCLUSIONS

Garlic is a plant within the genus *Allium* that may well have significance in promoting health, including reducing cancer incidence and altering tumor proliferation. While preclinical studies are intriguing, additional probing and controlled human intervention studies are needed. The dearth of clinical studies is undeniably a major limitation in our current understanding of the true physiological effects of garlic. Again, while the current findings are highly suggestive, more attention is needed to establish the quantity of garlic needed to bring about a response, the time of consumption required to lead to a change in some aspect of the cancer process and who might benefit maximally from enhanced intake. While it is possible that all *Allium* foods possess similar health benefits, few comparative studies have been conduction with model systems or human subjects.

Garlic appears to have relatively few side effects, and thus there are few disadvantages associated with its enhanced use, except its lingering odor and some interactions with specific drugs. It is important to note, however, that odor does not appear to be a prerequisite for many of the health benefits, since preclinical studies indicate that water-soluble *S*-allylcysteine provides comparable benefits to those obtained with compounds that are linked to odor in retarding chemically induced tumors. However, *S*-allylcysteine does not appear to be as effective in suppressing tumor proliferation and inducing apoptosis, suggesting that the intended use of garlic or its preparations can influence the overall response and thus potential health outcomes. It is probable that garlic and its associated water- and lipid-soluble sulfur compounds influence several key molecular targets in cancer prevention. Unfortunately, the sites of action have not been adequately investigated, and information about which is most important in bringing about a phenotypic change remains unclear.

While much remains to be learned about the health benefits that might be attributed to garlic, much is already known, including its ability to alter microbial growth, influence chemical carcinogenesis, suppress tumor proliferation, and promote apoptosis. While intriguing, garlic is not a "magic bullet" and must be considered as part of an arsenal of dietary components that may influence overall health. Likewise, it is important to recognize that all individuals will probably not respond equally to garlic intake since its effects depend on several environmental factors and genetics.

## REFERENCES

1.  Agarwal KC. Therapeutic actions of garlic constituents. Med Res Rev 1996; 16: 111–124.

2. Rivlin RS. Historical perspective on the use of garlic. J Nutr 2001; 131(suppl): 951S–954S.
3. Fenwick GR, Hanley AB. The genus *Allium*—Part 2. Crit Rev Food Sci Nutr 1985; 22:273–377.
4. Fenwick GR, Hanley AB. The genus *Allium*—Part 3. Crit Rev Food Sci Nutr 1985; 23:1–73.
5. Milner JA. Garlic: its anticarcinogenic and antitumorigenic properties. Nutr Rev 1996; 54:S82–S86.
6. Yoshida H, Katsuzaki H, Ohta R, Ishikawa K, Fukuda H, Fujino T, Suzuki A. An organosulfur compound isolated from oil-macerated garlic extract, and its antimicrobial effect. Biosci Biotechnol Biochem 1999; 63:588–590.
7. Orekhov AN, Grunwald J. Effects of garlic on atherosclerosis. Nutrition 1997; 13: 656–663.
8. Ackermann RT, Mulrow CD, Ramirez G, Gardner CD, Morbidoni L, Lawrence VA. Garlic shows promise for improving some cardiovascular risk factors. Arch Intern Med 2001; 161:813–824.
9. Rahman K. Historical perspective on garlic and cardiovascular disease. J Nutr 2001; 131(suppl):977S–979S.
10. Banerjee SK, Mukherjee PK, Maulik SK. Garlic as an antioxidant: the good, the bad and the ugly. Phytother Res 2003; 17:97–106.
11. Thomson M, Ali M. Garlic [*Allium sativum*]: a review of its potential use as an anti-cancer agent. Curr Cancer Drug Targets 2003; 3:67–81.
12. Jeong HG, Lee YW. Protective effects of diallyl sulfide on N-nitrosodimethylamine-induced immunosuppression in mice. Cancer Lett 1998; 134:73–79.
13. Morioka N, Sze LL, Morton DL, Irie RF. A protein fraction from aged garlic extract enhances cytotoxicity and proliferation of human lymphocytes mediated by interleukin-2 and concanavalin A. Cancer Immunol Immunother 1993; 37:316–322.
14. Ghazanfari T, Hassan ZM, Ebrahimi M. Immunomodulatory activity of a protein isolated from garlic extract on delayed type hypersensitivity. Int Immunopharmacol 2002; 2:1541–1549.
15. Nishiyama N, Moriguchi T, Saito H. Beneficial effects of aged garlic extract on learning and memory impairment in the senescence accelerated mouse. Exp Gerontol 1997; 32:149–160.
16. Sookvanichsilp N, Tiangda C, Yuennan P. Effects of raw garlic on physical performance and learning behaviour in rats. Phytother Res 2002; 16:732–736.
17. Steinmetz KA, Kushi LH, Bostick RM, Folsom AR, Potter JD. Vegetables, fruit, and colon cancer in the Iowa Women's Health Study. Am J Epid 1994; 139:1–15.
18. Mei X, Wang ML, Pan XY. Garlic and gastric cancer 1. The influence of garlic on the level of nitrate and nitrite in gastric juice. Acta Nutr Sin 1982; 4:53–56.
19. Hsing AW, Chokkalingam AP, Gao YT, Madigan MP, Deng J, Gridley G, Fraumeni JF. *Allium* vegetables and risk of prostate cancer: a population-based study. J Natl Cancer Inst 2002; 94:1648–1651.
20. Jappe U, Bonnekoh B, Hausen BM, Gollnick H. Garlic-related dermatoses: case report and review of the literature. Am J Contact Dermat 1999; 10:37–39.
21. Nowak-Wegrzyn A, Conover-Walker MK, Wood RA. Food-allergic reactions in schools and preschools. Arch Pediatr Adolesc Med 2001; 155:790–795.

22. Burnham BE. Garlic as a possible risk for postoperative bleeding. Plast Reconstr Surg 1995; 95:213.

23. Wu CC, Sheen LY, Chen HW, Tsai SJ, Lii CK. Effects of organosulfur compounds from garlic oil on the antioxidation system in rat liver and red blood cells. Food Chem Toxicol 2001; 39:563–569.

24. Munday R, Munday JS, Munday CM. Comparative effects of mono-, di-, tri- and tetrasulfides derived from plants of the *Allium* family: redox cycling in vitro and hemolytic activity and Phase 2 enzyme induction in vivo. Free Radic Biol Med 2003; 34:1200–1211.

25. *Fourth Annual Overview of the Nutrition Industry.* Nutrition Business Journal, Nutrition Business International LLC, San Diego, CA, 1998.

26. Lawson LD. Bioactive organosulfur compounds of garlic and garlic products. In: Kinghorn AD, Balandrin MF, Eds. Human Medicinal Agents from Plants. Washington DC: American Chemical Society, 1994:306–330.

27. Arnault I, Christides JP, Mandon N, Haffner T, Kahane R, Auger J. High-performance ion-pair chromatography method for simultaneous analysis of alliin, deoxyalliin, allicin and dipeptide precursors in garlic products using multiple mass spectrometry and UV detection. J Chromatogr A 2003; 991:69–75.

28. Roberfroid M. Functional food concept and its application to prebiotics. Dig Liver Dis 2002; 34(suppl):S105–S110.

29. Foster MW, McMahon TJ, Stamler JS. S-Nitrosylation in health and disease. Trends Mol Med 2003; 9:160–168.

30. Ip C, Lisk DJ, Stoewsand GS. Mammary cancer prevention by regular garlic and selenium-enriched garlic. Nutr Cancer 1992; 7:279–286.

31. Amagase H, Milner JA. Impact of various sources of garlic and their constituents on 7,12-dimethylbenz(a)anthracene binding to mammary cell DNA. Carcinogenesis 1993; 14:1627–1631.

32. Kubec R, Svobodova M, Velisek J. A gas chromatographic determination of S-alk(en)ylcysteine sulfoxides. J Chromatogr 1999; 862:85–94.

33. Sugii M, Suzuki T, Nagasawa S, Kawashima K. Isolation of gamma-Lglutamyl-S-allylmercapto-L-cysteine and S-allylmercapto-L-cysteine from garlic. Chem Pharm Bull 1964; 12:1114–1115.

34. Block E. The chemistry of garlic and onion. Sci Am 1985; 252:114–119.

35. Tamaki T, Sonoki S. Volatile sulfur compounds in human expiration after eating raw or heat-treated garlic. J Nutr Sci Vitaminol (Tokyo) 1999; 45:213–222.

36. Egen-Schwind C, Eckard R, Kemper FH. Metabolism of garlic constituents in the isolated perfused rat liver. Planta Med 1992; 58:301–305.

37. Teyssier C, Guenot L, Suschetet M, Siess MH. Metabolism of diallyl disulfide by human liver microsomal cytochromes P-450 and flavin-containing monooxygenases. Drug Metab Dispos 1999; 27:835–841.

38. Ripp SL, Overby LH, Philpot RM, Elfarra AA. Oxidation of cysteine S-conjugates by rabbit liver microsomes and cDNA-expressed flavin-containing mono-oxygenases: studies with S-(1,2-dichlorovinyl)-L-cysteine, S-(1,2,2 trichlorovinyl)-L-cysteine, S-allyl-L-cysteine, and S-benzyl-L-cysteine. Mol Pharmacol 1997; 51: 507–515.

39. Germain E, Auger J, Ginies C, Siess MH, Teyssier C. In vivo metabolism of diallyl disulphide in the rat: identification of two new metabolites. Xenobiotica 2002; 32: 1127–1138.

40. Clydesdale FM. A proposal for the establishment of scientific criteria for health claims for functional foods. Nutr Rev 1997; 55:413–422.

41. Milner JA. Functional foods and health promotion. J Nutr 1999; 129(suppl): 1395S–1397S.

42. Milner JA, McDonald SS, Anderson DE, Greenwald P. Molecular targets for nutrients involved with cancer prevention. Nutr Cancer 2001; 41:1–16.

43. Kris-Etherton PM, Hecker KD, Bonanome A, Coval SM, Binkoski AE, Hilpert KF, Griel AE, Etherton TD. Bioactive compounds in foods: their role in the prevention of cardiovascular disease and cancer. Am J Med 2002; 113(suppl):71S–88S.

44. Fleischauer AT, Arab L. Garlic and cancer: a critical review of the epidemiologic literature. J Nutr 2003; 131(suppl):1032S–1040S.

45. Adetumbi MA, Lau BH. *Allium sativum* (garlic)—a natural antibiotic. Med Hypotheses 1983; 12:227–237.

46. Harris JC, Cottrell SL, Plummer S, Lloyd D. Antimicrobial properties of *Allium sativum* (garlic). Appl Microbiol Biotechnol 2001; 57:282–286.

47. Adler BB, Beuchat LR. Death of *Salmonella Escherichia coli* O157:H7, and *Listeria monocytogenes* in garlic butter as affected by storage temperature. J Food Prot 2002; 65:1976–1980.

48. Arora DS, Kaur J. Antimicrobial activity of spices. Int J Antimicrob Agents 1999; 12:257–262.

49. Yin MC, Tsao SM. Inhibitory effect of seven *Allium* plants upon three *Aspergillus* species. Int J Food Microbiol 1999; 49:49–56.

50. Dion ME, Agler M, Milner JA. S-Allylcysteine inhibits nitrosomorpholine formation and bioactivation. Nutr Cancer 1997; 28:1–6.

51. Sivam GP, Lampe JW, Ulness B, Swanzy SR, Potter JD. *Helicobacter pylori*—in vitro susceptibility to garlic (*Allium sativum*) extract. Nutr Cancer 1997; 27:118–21.

52. Hooper LV, Wong MH, Thelin A, Hansson L, Falk PG, Gordon JI. Molecular analysis of commensal host-microbial relationships in the intestine. Science 2001; 291:881–884.

53. Cellini L, Di Campli E, Masulli M, Di Bartolomeo S, Allocati N. Inhibition of *Helicobacter pylori* by garlic extract (*Allium sativum*). FEMS Immunol Med Microbiol 1996; 13:273–277.

54. Chung JG, Chen GW, Wu LT, Chang HL, Lin JG, Yeh CC, Wang TF. Effects of garlic compounds diallyl sulfide and diallyl disulfide on arylamine N-acetyltransferase activity in strains of *Helicobacter pylori* from peptic ulcer patients. Am J Chin Med 1998; 26:353–364.

55. Canizares P, Gracia I, Gomez LA, Martin de Argila C, de Rafael L, Garcia A. Optimization of *Allium sativum* solvent extraction for the inhibition of in vitro growth of *Helicobacter pylori*. Biotechnol Prog 2002; 18:1227–1232.

56. Jonkers D, van den Broek E, van Dooren I, Thijs C, Dorant E, Hageman G, Stobberingh E. Antibacterial effect of garlic and omeprazole on *Helicobacter pylori*. J Antimicrob Chemother 1999; 43:837–839.

57. Lemar KM, Turner MP, Lloyd D. Garlic (*Allium sativum*) as an anti-*Candida* agent: a comparison of the efficacy of fresh garlic and freeze-dried extracts. J Appl Microbiol 2002; 93:398–405.

58. Ledezma E, Lopez JC, Marin P, Romero H, Ferrara G, De Sousa L, Jorquera A, Apitz Castro R. Ajoene in the topical short-term treatment of *tinea cruris* and *tinea corporis* in humans. Randomized comparative study with terbinafine. Arzneimittelforschung 1999; 49:544–547.

59. Sundaram SG, Milner JA. Diallyl disulfide inhibits the proliferation of human tumor cells in culture. Biochim Biophys Acta 1996; 1315:15–20.

60. Rabinkov A, Miron T, Konstantinovski L, Wilchek M, Mirelman D, Weiner L. The mode of action of allicin: trapping of radicals and interaction with thiol- containing proteins. Biochim Biophys Acta 1998; 1379:233–244.

61. Dwivedi C, John LM, Schmidt DS, Engineer FN. Effects of oil-soluble organosulfur compounds from garlic on doxorubicin-induced lipid peroxidation. Anticancer Drugs 1998; 9:291–294.

62. Geng Z, Rong Y, Lau BH. S-Allyl cysteine inhibits activation of nuclear factor kappa B in human T cells. Free Radic Biol Med 1997; 23:345–350.

63. Ide N, Lau BH. S-Allylcysteine attenuates oxidative stress in endothelial cells. Drug Dev Ind Pharm 1999; 25:619–624.

64. Dillon SA, Burmi RS, Lowe GM, Billington D, Rahman K. Antioxidant properties of aged garlic extract: an in vitro study incorporating human low density lipoprotein. Life Sci 2003; 72:1583–1594.

65. Wargovich MJ, Woods C, Eng VW, Stephens LC, Gray K. Chemoprevention of N-nitrosomethylbenzylamine-induced esophageal cancer in rats by the naturally-occurring thioether, diallyl sulfide. Cancer Res 1988; 48:6872–6875.

66. Sumiyoshi H, Wargovich MJ. Chemoprevention of 1,2-dimethylhydrazine-induced colon cancer in mice by natural-occurring organosulfur compounds. Cancer Res 1990; 50:5084–5087.

67. Hussain SP, Jannu LN, Rao AR. Chemopreventive action of garlic on methylcholanthrene-induced carcinogenesis in the uterine cervix of mice. Cancer Lett 1990; 49: 175–180.

68. Liu JZ, Lin RI, Milner JA. Inhibition of 7,12-dimethylbenz(a)-anthracene-induced mammary tumors and DNA adducts by garlic powder. Carcinogenesis 1992; 13: 1847–1851.

69. Shukla Y, Singh A, Srivastava B. Inhibition of carcinogen-induced activity of gamma-glutamyl transpeptidase by certain dietary constituents in mouse skin. Biomed Environ Sci 1999; 12:110–115.

70. Song K, Milner JA. Heating garlic inhibits its ability to suppress 7, 12-dimethylbenz(a)anthracene-induced DNA adduct formation in rat mammary tissue. J Nutr 1999; 129:657–661.

71. Shenoy NR, Choughuley AS. Inhibitory effect of diet related sulphydryl compounds on the formation of carcinogenic nitrosamines. Cancer Lett 1992; 65:227–232.

72. Knowles LM, Milner JA. Diallyl disulfide inhibits p34(cdc2) kinase activity through changes in complex formation and phosphorylation. Carcinogenesis 2000; 21: 1129–1134.

73. Atanasova-Goranova VK, Dimova PI, Pevicharova GT. Effect of food products on endogenous generation of N-nitrosamines in rats. Br J Nutr 1997; 78:335–345.

74. Vermeer IT, Moonen EJ, Dallinga JW, Kleinjans JC, van Maanen JC. Effect of ascorbic acid and green tea on endogenous formation of N-nitrosodimethylamine and N-nitrosopiperidine in humans. Mutat Res 1999; 428:353–361.

75. Brown JL. N-Nitrosamines. Occup Med 1999; 12:839–848.

76. Lijinsky W. N-Nitroso compounds in the diet. Mutat Res 1999; 443:129–38.

77. Chung MJ, Lee SH, Sung NJ. Inhibitory effect of whole strawberries, garlic juice or kale juice on endogenous formation of N-nitrosodimethylamine in humans. Cancer Lett 2002; 182:1–10.

78. Williams D H. S-Nitrosation and the reactions of S-ninitroso compounds. Chem Soc Rev 1983; 15:171–196.

79. Mei X, Lin X, Liu J, Lin XY, Song PJ, Hu JF, Liang XJ. The blocking effect of garlic on the formation of N-nitrosoproline in humans. Acta Nutr Sin 1989; 11: 141–145.

80. Ohshima H, Bartsch H. Quantitative estimation of endogenous N-nitrosation in humans by monitoring N-nitrosoproline in urine. Methods Enzymol 1999; 301: 40–49.

81. Lin X-Y, Liu JZ, Milner JA. Dietary garlic suppresses DNA adducts caused by N-nitroso compounds. Carcinogenesis 1994; 15:349–352.

82. Hong JY, Wang ZY, Smith TJ, Zhou S, Shi S, Pan J, Yang CS. Inhibitory effects of diallyl sulfide on the metabolism and tumorigenicity of the tobacco-specific carcinogen 4-(methylnitrosamino)-1-(3-pyridyl)-1-butanone (NNK) in A/J mouse lung. Carcinogenesis 1992; 13:901–904.

83. Haber-Mignard D, Suschetet M, Berges R, Astorg P, Siess MH. Inhibition of aflatoxin B1- and N-nitrosodiethylamine-induced liver preneoplastic *foci* in rats fed naturally occurring allyl sulfides. Nutr Cancer 1996; 25:61–70.

84. Knasmuller S, de Martin R, Domjan G, Szakmary A. Studies on the antimutagenic activities of garlic extract. Environ Mol Mutagen 1989; 13:357–365.

85. Chen L, Lee M, Hong JY, Huang W, Wang E, Yang CS. Relationship between cytochrome P450 2E1 and acetone catabolism in rats as studied with diallyl sulfide as an inhibitor. Biochem Pharmacol 1994; 48:2199–2205.

86. Yang CS, Chhabra SK, Hong JY, Smith TJ. Mechanisms of inhibition of chemical toxicity and carcinogenesis by diallyl sulfide (DAS) and related compounds from garlic. J Nutr 2001; 131(suppl):1041S–1045S.

87. Park KA, Kweon S, Choi H. Anticarcinogenic effect and modification of cytochrome P450 2E1 by dietary garlic powder in diethylnitrosamine-initiated rat hepatocarcinogenesis. J Biochem Mol Biol 2002; 35:615–622.

88. Jin L, Baillie TA. Metabolism of the chemoprotective agent diallyl sulfide to glutathione conjugates in rats. Chem Res Toxicol 1997; 10:318–327.

89. Pan J, Hong JY, Li D, Schuetz EG, Guzelian PS, Huang W, Yang CS. Regulation of cytochrome P450 2B1/2 genes by diallyl sulfone, disulfiram and other organosulfur compounds in primary cultures of rat hepatocytes. Biochem Pharmacol 1993; 45: 2323–2329.

90. Manson MM, Ball HW, Barrett MC, Clark HL, Judah DJ, Williamson G, Neal GE. Mechanism of action of dietary chemoprotective agents in rat liver:induction of

phase I and II drug metabolizing enzymes and aflatoxin B1 metabolism. Carcinogenesis 1997; 18:1729–1738.

91. Wang BH, Zuzel KA, Rahman K, Billington D. Treatment with aged garlic extract protects against bromobenzene toxicity to precision cut rat liver slices. Toxicology 1999; 132:215–225.

92. Wu CC, Sheen LY, Chen HW, Kuo WW, Tsai SJ, Lii CK. Differential effects of garlic oil and its three major organosulfur components on the hepatic detoxification system in rats. J Agric Food Chem 2002; 50:378–383.

93. Hughes MF, Chamulitrat W, Mason RP, Eling TE. Epoxidation of 7,8-dihydroxy-7,8-dihydrobenzo[a]pyrene *via* a hydroperoxide-dependent mechanism catalyzed by lipoxygenases. Carcinogenesis 1989; 10:2075–2080.

94. Joseph P, Srinivasan SN, Byczkowski JZ, Kulkarni AP. Bioactivation of benzo(a)-pyrene-7,8-dihydrodiol catalyzed by lipoxygenase purified from human term placenta and conceptal tissues. Reprod Toxicol 1994; 8:307–313.

95. Rioux N, Castonguay A. Inhibitors of lipoxygenase:a new class of cancer chemo-preventive agents. Carcinogenesis 1998; 19:1393–1400.

96. Ali M. Mechanism by which garlic (*Allium sativum*) inhibits cyclooxygenase activity. Effect of raw *versus* boiled garlic extract on the synthesis of prostanoids. Prostaglandins Leukot Essent Fatty Acids 1995; 53:397–400.

97. Roy P, Kulkarni AP. Co-oxidation of acrylonitrile by soybean lipoxygenase and partially purified human lung lipoxygenase. Xenobiotica 1999; 29:511–531.

98. Dirsch VM, Vollmar AM. Ajoene, a natural product with non-steroidal anti-inflammatory drug (NSAID)-like properties. Biochem Pharmacol 2001; 61:587–593.

99. Belman S, Solomon J, Segal A, Block E, Barany G. Inhibition of soybean lipoxygenase and mouse skin tumor promotion by onion and garlic components. J Biochem Toxicol 1989; 4:151–160.

100. Song K. Factors Influencing on Garlic's Anticancer Properties. Master's thesis, The Pennsylvania State University, American Chemical Society, 1999.

101. Singh SV, Pan SS, Srivastava SK, Xia H, Hu X, Zaren HA, Orchard JL. Differential induction of NAD(P)H:quinone oxidoreductase by anti-carcinogenic organosulfides from garlic. Vol. 244, 1998:917–920.

102. Hatono S, Jimenez A, Wargovich MJ. Chemopreventive effect of S-allylcysteine and its relationship to the detoxification enzyme glutathione S-transferase. Carcinogenesis 1996; 17:1041–1044.

103. Singh A, Singh SP. Modulatory potential of smokeless tobacco on the garlic, mace or black mustard-altered hepatic detoxication system enzymes, sulfhydryl content and lipid peroxidation in murine system. Cancer Lett 1997; 118:109–114.

104. Hu X, Benson PJ, Srivastava SK, Xia H, Bleicher RJ, Zaren HA, Awasthi S, Awasthi YC, Singh SV. Induction of glutathione S-transferase pi as a bioassay for the evaluation of potency of inhibitors of benzo(a)pyrene-induced cancer in a murine model. Int J Cancer 1997; 73:897–902.

105. Bose C, Guo J, Zimniak L, Srivastava SK, Singh SP, Zimniak P, Singh SV. Critical role of allyl groups and disulfide chain in induction of pi class glutathione transferase in mouse tissues in vivo by diallyl disulfide, a naturally occurring chemopreventive agent in garlic. Carcinogenesis 2002; 23:1661–1665.

106. Munday R, Munday CM. Relative activities of organosulfur compounds derived from onions and garlic in increasing tissue activities of quinone reductase and glutathione transferase in rat tissues. Nutr Cancer 2001; 40:205–210.

107. Schaffer EM, Liu JZ, Green J, Dangler CA, Milner JA. Garlic and associated allyl sulfur components inhibit N-methyl-N-nitrosourea induced rat mammary carcinogenesis. Cancer Lett 1996; 102:199–204.

108. Cohen LA, Zhao Z, Pittman B, Lubet R. S-Allylcysteine, a garlic constituent, fails to inhibit N-methylnitrosourea-induced rat mammary tumorigenesis. Nutr Cancer 1999; 35:58–63.

109. Schaffer EM, Liu JZ, Milner JA. Garlic powder and allyl sulfur compounds enhance the ability of dietary selenite to inhibit 7,12-dimethylbenz[a]anthracene-induced mammary DNA adducts. Nutr Cancer 1997; 27:162–168.

110. Singh A, Shukla Y. Antitumor activity of diallyl sulfide in two-stage mouse skin model of carcinogenesis. Biomed Environ Sci 1998; 11:258–263.

111. Balasenthil S, Arivazhagan S, Ramachandran CR, Nagini S. Effects of garlic on 7,12-dimethylbenz[a]anthracene-induced hamster buccal pouch carcinogenesis. Cancer Detect Prev 1999; 23:534–538.

112. Yu FL, Bender W, Fang Q, Ludeke A, Welch B. Prevention of chemical carcinogen DNA binding and inhibition of nuclear RNA polymerase activity by organosulfur compounds as the possible mechanisms for their anticancer initiation and proliferation effects. Cancer Detect Prev 2003; 27:370–379.

113. Ludeke BI, Domine F, Ohgaki H, Kleihues P. Modulation of N-nitrosomethylbenzylamine bioactivation by diallyl sulfide in vivo. Carcinogenesis 1992; 13:2467–2470.

114. Lea MA, Randolph VM, Patel M. Increased acetylation of histones induced by diallyl disulfide and structurally related molecules. Int J Oncol 1999; 15:347–352.

115. Lea MA, Randolph VM. Induction of histone acetylation in rat liver and hepatoma by organosulfur compounds including diallyl disulfide. Anticancer Res 2001; 21: 2841–2845.

116. Singh SV, Mohan RR, Agarwal R, Benson PJ, Hu X, Rudy MA, Xia H, Katoh A, Srivastava SK, Mukhtar H, Gupta V, Zaren HA. Novel anti-carcinogenic activity of an organosulfide from garlic:inhibition of H-RAS oncogene transformed tumor growth in vivo by diallyl disulfide is associated with inhibition of p21H-ras processing. Biochem Biophys Res Commun 1996; 225:660–665.

117. Scharfenberg K, Wagner R, Wagner KG. The cytotoxic effect of ajoene, a natural product from garlic, investigated with different cell lines. Cancer Lett 1990; 53: 103–108.

118. Welch C, Wuarin L, Sidell N. Antiproliferative effect of the garlic compound S-allyl cysteine on human neuroblastoma cells in vitro. Cancer Lett 1992; 63:211–219.

119. Sundaram SG, Milner JA. Impact of organosulfur compounds in garlic on canine mammary tumor cells in culture. Cancer Lett 1993; 74:85–90.

120. Takeyama H, Hoon DS, Saxton RE, Morton DL, Irie RF. Growth inhibition and modulation of cell markers of melanoma by S-allyl cysteine. Oncology 1993; 50: 63–69.

121. Scharfenberg K, Ryll T, Wagner R, Wagner KG. Injuries to cultivated BJA-B cells by ajoene, a garlic-derived natural compound:cell viability, glutathione metabolism, and pools of acidic amino acids. J Cell Physiol 1994; 158:55–60.

122. Sundaram SG, Milner JA. Diallyl disulfide induces apoptosis of human colon tumor cells. Carcinogenesis 1996; 17:669–673.

123. Sigounas G, Hooker J, Anagnostou A, Steiner M. S-Allylmercaptocysteine inhibits cell proliferation and reduces the viability of erythroleukemia, breast, and prostate cancer cell lines. Nutr Cancer 1997; 27:186–191.

124. Sigounas G, Hooker JL, Li W, Anagnostou A, Steiner M. S-Allylmercaptocysteine, a stable thioallyl compound, induces apoptosis in erythroleukemia cell lines. Nutr Cancer 1997; 28:153–159.

125. Sooranna SR, Patel S, Das I. The effect of garlic on cell growth and cell division in cultured trophoblast and endothelial cell lines. Biochem Soc Trans 1997; 25: 456S.

126. Pinto JT, Qiao C, Xing J, Rivlin RS, Protomastro ML, Weissler ML, Tao Y, Thaler H, Heston WD. Effects of garlic thioallyl derivatives on growth, glutathione concentration, and polyamine formation of human prostate carcinoma cells in culture. Am J Clin Nutr 1997; 66:398–405.

127. Sakamoto K, Lawson LD, Milner J. Allyl sulfides from garlic suppress the in vitro proliferation of human A549 lung tumor cells. Nutr Cancer 1997; 29:152–156.

128. Knowles LM, Milner JA. Depressed p34cdc2 kinase activity and G2/M phase arrest induced by diallyl disulfide in HCT-15 cells. Nutr Cancer 1998; 30:169–174.

129. Zheng S, Yang H, Zhang S, Wang X, Yu L, Lu J, Li J. Initial study on naturally occurring products from traditional Chinese herbs and vegetables for chemoprevention. J Cell Biochem Suppl 1997; 27:106–112.

130. Kimura Y, Yamamoto K. Cytological effect of chemicals on tumors. XXIII. Influence of crude extracts from garlic and some related species on MTK-sarcoma III. GANN 1964; 55:325–329.

131. Nurse P. Universal control mechanism regulating onset of M-phase. Nature 1990; 344:503–508.

132. Hartwell LH, Kastan MB. Cell cycle control and cancer. Science 1994; 266: 1821–1828.

133. Morgan DO. Principles of CDK regulation. Nature 1995; 374:131–134.

134. Schaffer EM, Milner JA. Impact of dietary fatty acids on 7,12-dimethylbenz[a]anthracene-induced mammary DNA adducts. Cancer Lett 1996; 106:177–183.

135. Amagase H, Schaffer EM, Milner JA. Dietary components modify garlic's ability to suppress 7,12-dimethylbenz(a)anthracene induced mammary DNA adducts. J Nutr 1996; 126:817–824.

136. Ganther HE. Selenium metabolism, selenoproteins and mechanisms of cancer prevention:complexities with thioredoxin reductase. Carcinogenesis 1999; 20: 1657–1666.

137. Chen HW, Tsai CW, Yang JJ, Liu CT, Kuo WW, Lii CK. The combined effects of garlic oil and fish oil on the hepatic antioxidant and drug-metabolizing enzymes of rats. Br J Nutr 2003; 89:189–200.

138. Adler AJ, Holub BJ. Effect of garlic and fish-oil supplementation on serum lipid. Am J Clin Nutr 1997; 65:445–450.

# 9
# Polymethylated Flavonoids: Cancer Preventive and Therapeutic Potentials Derived from Anti-inflammatory and Drug Metabolism–Modifying Properties

**Akira Murakami and Hajime Ohigashi**
*Kyoto University, Kyoto, Japan*

## I. INTRODUCTION

### A. Chemical Characteristics of Polymethylated Flavonoids

Flavonoids are a huge group of plant secondary metabolites that are widely and ubiquitously distributed throughout the plant kingdom. Chemically, flavonoids consist of $C_3$-$C_6$-$C_3$ units and are classified into several subgroups according to, for example, their oxidation status, number of aromatic rings, carbon skeleton, and presence or absence of positive charge. A common chemical characteristic of flavonoids is the presence of phenolic and hydroxyl groups. Polymethylated flavonoids (PMFs) have particular chemical characteristics—specifically, hydroxyl groups that are mostly methylated, and an oxidation status of the A-ring that is distinct from that of common flavonoids, since the 6- or 8-position is oxidized in addition to the 5- and 7-positions, where common flavonoids have hydroxyl group(s) (Fig. 1). The above-mentioned hydrophobic characteristics provide PMFs with notable biological activities and metabolic properties, as described herein.

### B. Multistage Carcinogenesis

One outstanding characteristic of PMFs is their potential for cancer chemoprevention and chemotherapy. Biochemical or cellular events during carcinogenesis may

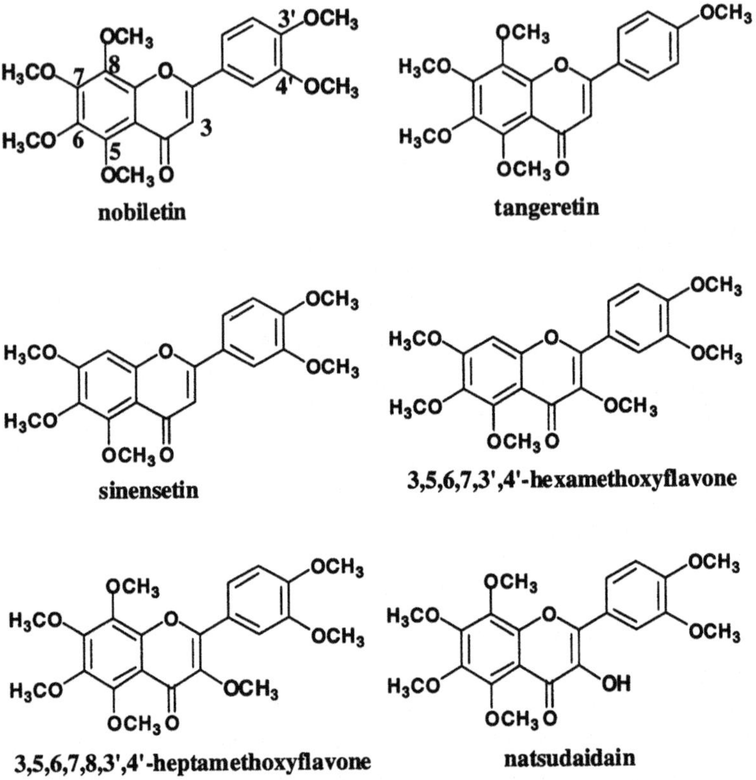

nobiletin

tangeretin

sinensetin

3,5,6,7,3',4'-hexamethoxyflavone

3,5,6,7,8,3',4'-heptamethoxyflavone

natsudaidain

**Figure 1**   Structures of representative PMFs.

be operationally divided into at least three distinct stages [1]: initiation, promotion, and progression (Fig. 2). The conventional multistage carcinogenesis model still remains useful for mechanistic studies, although the development of chemopreventive and chemotherapeutic agents has recently drawn a great deal of attention to additional complexities.

*Initiation*, the first stage of chemical carcinogenesis, is provoked in one or a few tissue cells by the metabolic activation of procarcinogens by enzymes of the cytochrome P450 superfamily [2]. These can convert procarcinogens into ultimate carcinogens, most of which are reactive electrophiles. If not detoxified by phase 2 enzymes, these carcinogenic species can interact with and bind to cellular DNA to trigger gene mutation(s).

*Tumor promotion*, the next stage, is notable for selective and sustained hyperplasia that leads to an expansion of only the initiated cells into benign

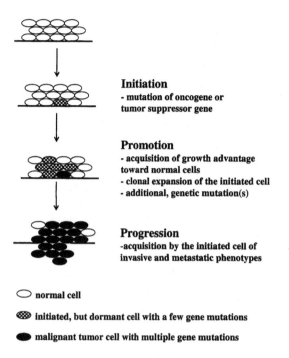

Initiation
- mutation of oncogene or
tumor suppressor gene

Promotion
- acquisition of growth advantage
toward normal cells
- clonal expansion of the initiated cell
- additional, genetic mutation(s)

Progression
-acquisition by the initiated cell of
invasive and metastatic phenotypes

○ normal cell

initiated, but dormant cell with a few gene mutations

malignant tumor cell with multiple gene mutations

**Figure 2** Schematic illustration of multistage carcinogenesis.

tumors. Thus, tumor promoters elicit biochemical changes that confer a selective growth advantage to the initiated population over that of the surrounding noninitiated cells [3]. In addition, the tumor-promotion stage represents the longest and probably the most critical event in tumor development, suggesting the possibility for highly efficient chemopreventative intervention in humans. Tumor promotion itself is a stepwise process and may be regarded as the process of a nongenotoxic event, though it alters or enhances gene expression. Tumor promoters disturb the biochemical pathways of cellular death, differentiation, and proliferation to allow initiated cells to acquire a selective ability for clonal expansion and neoplastic phenotypes.

*Progression*, the final stage in multistage carcinogenesis, is characterized by irreversibility of the process, somatic aneuploidy and karyotypic instability, all of which reflect additional genetic mutation(s) [4]. Acquisition of such genetic and biological alterations by benign tumors results in their conversion to malignant ones, possessing invasiveness and metastatic capability. In a two-stage carcinogenesis experiment in mouse skin, 5% of papillomas were converted to carcinomas spontaneously at approximately 30 weeks [5]. Some tumor progressive agents, most of which are genotoxic, have also recently been identified in mouse skin carcinogenesis experiments.

## C. Anti–Tumor-Initiating Activities

The antimutagenic properties of PMFs have been demonstrated using in vitro models by two independent groups [6,7]. Miyazawa et al. [6] performed activity-guided separation of *Citrus aurantium* and isolated a number of active constituents, including PMFs (tetra-$O$-methylscutellarein, sinensetin, and nobiletin). Further, Iwase and coworkers [8] used in vivo experiments to demonstrate the efficacy of PMFs; nobiletin and 3,3′,4′,5,6,7,8-heptamethoxyflavone exhibited anti–tumor-initiating activity toward two-stage carcinogenesis in mouse skin tumors induced by ($\pm$)-($E$)-methyl-2-[($E$)-hydroxyimino]-5-nitro-6-methoxy-3-hexenamide, a nitric oxide donor, the initiator, and 12-$O$-tetradecanoylphorbol-13-acetate (TPA) as the promoter.

## D. Anti–Tumor-Promoting Activities

Inflammation is closely associated with tumor promotion, and we have published results of in vitro and in vivo experiments showing that nobiletin attenuates certain biological activities relating to inflammation, as described later. During inflammatory processes, enhanced production of eicosanoids such as prostaglandins (PGs) and leukotrienes (LTs) occurs; the former are synthesized by cyclooxygenases (COXs), the latter by lipoxygenases (LOXs). Ito et al. [9] have also reported that nobiletin reduced interleukin (IL)-1–induced production of prostaglandin $E_2$ (PGE$_2$) in synovial cells. In addition, the 15-LOX inhibitory activity of several flavonoids was investigated by Malterud and Rydland, who reported the strongest inhibition being shown by 3,5,6,7,3′,4′-hexamethoxyflavone, while sinensetin, nobiletin, tangeretin, tetramethylscutellarein, and 3,3′,4′,5,6,7,8-heptamethoxyflavone all being less active [10]. Demethylation of PMFs may thus lead to a reduction in activity, as demonstrated by a comparison of the activities of sinensetin with 5-$O$-demethylsinensetin in this study.

It is known that cell-cell communication mediated via gap junctions is suppressed by tumor-promoting treatments [11,12]. Chaumontet et al. [13] have shown two flavones, apigenin and tangeretin, to inhibit a TPA-induced gap junctional intercellular communication (GJIC) blockade in a rat liver epithelial cell line, REL; other flavonoids tested, including naringenin, myricetin, catechin, and chrysin, failed to enhance GJIC or counteract this TPA-induced inhibition [14] suggesting higher anti–tumor promotional potentials of PMFs in comparison to those of general flavonoids.

## E.  Anti–Tumor-Invasive and Apoptosis/Differentiation-Inducing Activities

There is ample evidence that PMFs can suppress or halt tumor progression at the stage that involves tumor invasion and metastasis. One of the earliest findings on the distinct biological activity of PMFs was reported by Kandawaswami et al. in 1991 [15]; while two PMFs, nobiletin and tangeretin, markedly inhibited cancer cell growth, quercetin and taxifolin did not. The authors speculated that the differences in anti–cancer cell proliferation activity were due to the greater membrane uptake of PMFs as compared with quercetin and taxifolin. Yano et al. [16] conducted an extensive study of the anti–cancer cell–proliferating activities of natural flavonoids; in all 27 citrus, flavonoids were examined for their antiproliferative activities against several tumor and normal human cell lines. The order of inhibitory potency was luteolin > natsudaidain > quercetin > tangeretin > eriodictyol > nobiletin > 3,3′,4′,5,6,7,8-heptamethoxyflavone. As regards structure-activity relationships, C-3 hydroxyl and C-8 methoxyl groups were found to be essential for high activity [16]. While quercetin, hesperetin, 7,3′-dimethylhesperetin, and eriodictyol failed to produce an effect on B16F10 or SK-MEL-1 melanoma cell lines, tangeretin was the most effective in inhibiting cell growth of all the flavonoids [17]. The presence of at least three adjacent methoxyl groups was also shown to confer a more potent antiproliferative effect [17].

Hirano et al. [18] have shown an apoptotic cell death–inducing activity of tangeretin using HL-60 cells. PMFs, including nobiletin and tangeretin, had previously been reported to induce differentiation of mouse myeloid leukemia cells [19]. From the results of structure-activity relationship studies, Kawaii et al. [19] concluded that the hydroxyl group at the 3-position and the methoxyl group at the 8-position enhanced the differentiation-inducing activity. Mak and coworkers [20] also isolated nobiletin and tangeretin from *Citrus reticulata* and demonstrated antileukemia properties in studies of their proliferation suppression and differentiation-inducing activity toward a murine myeloid leukemic cell clone, WEHI 3B.

Possibly the best characterized biological activity of PMFs is their anti–tumor-invading property. The super family of matrix metalloproteinases (MMPs) are pivotal in tumor cell invasion as well as angiogenesis and oncogenesis [21–23]. Bracke et al. examined the potential activity of PMFs for suppressing tumor invasion [24–26] and developed an in vitro model in which some flavonoids inhibited the invasion of mouse MO4 cells into embryonic chick heart fragments; potencies were ranked in the following descending order: tangeretin > nobiletin > hesperidin = naringin [26], indicating a higher activity of PMFs than common flavonoids. Unlike ( + )-catechin, tangeretin bound poorly to the extracellular matrix and did not alter the fucosylated surface glycopeptides of MO4 cells.

Ito et al. [9] isolated six flavonoids from *Citrus depressa* (Rutaceae), including tangeretin, 6-demethoxytangeretin, nobiletin, 5-demethylnobiletin, 6-demethoxynobiletin, and sinensetin. While each compound suppressed IL-1–induced production of proMMP-9/progelatinase B in rabbit synovial cells in a dose-dependent manner (<64 μM), nobiletin was most effective in suppressing proMMP-9 production along with a decrease in its mRNA. Taken together, these results suggest nobiletin to be a functional phytochemical that attenuates inflammatory processes and, thus, cancer cell invasion [9,27].

Similarly, Minagawa et al. [28] found that nobiletin inhibited the formation of peritoneal dissemination nodules from a cancer cell line, TMK-1, in a SCID mouse model and observed the total weight of the dissemination nodules to be significantly lower in the treated group than in the vehicle control group, as was the total number disseminated. Moreover, the enzymatic activity of MMP-9 expressed in culture medium obtained from a co-culture of TMK-1 and mouse fibroblastic cells was inhibited by nobiletin in a concentration-dependent manner. Rooprai et al. [29] compared the effects of four anti-invasive agents—nobiletin, tangeretin, swainsonine (a locoweed alkaloid) and captopril (an anti-hypertensive drug)—on various parameters of brain tumor invasion such as MMP secretion, migration, invasion, and adhesion and found nobiletin to be the most effective.

The E-cadherin/catenin complex is a powerful invasion suppressor in epithelial cells; it is expressed in the human MCF-7 breast cancer cell line family, but is functionally defective in the invasive MCF-7/6 variant. Vermeulen et al. showed that tangeretin is capable of upregulating the function of this complex in MCF-7/6 cells [30], thereby implicating PMFs in suppression of tumor invasion.

## F.  Other Biological Activities

An activity-guiding separation of citrus fruit peel also resulted in isolation of nobiletin as a tyrosinase inhibitor with an activity greater than that of kojic acid, the positive control [31]. Sempinska et al. [32] found the anti–platelet adhesive activity of flavonoids to be in the order nobiletin > isorhamnetin 5,7,4′-trimethylether 3-glucoside > isorhamnetin 3-glucoside, and that methylation of the hydroxyl groups of isorhamnetin-3-glucoside markedly increased its antiadhesive activity. In addition, desmethylnobiletin, nobiletin, and tangeretin have been isolated from *Citrus reticulata* peel extracts as antibacterial compounds [33].

## II.  ANTI-INFLAMMATORY ACTIVITY, METABOLISM, AND TOXICITY

### A.  Inflammation as the Critical Etiology of Carcinogenesis

It is well established that certain carcinogenesis processes, including those in the esophagus, stomach, colon, and liver, occur via inflammation, as shown by animal experiments and epidemiological surveys (Fig. 3) [34–37]. In addition, oxidative stress is regarded as an ubiquitous cause of a wide range of types of carcinogenesis as a result of modifying macromolecules such as proteins, lipids and DNA bases [38–42]. It is important to emphasize that inflammatory processes and oxidative stress do not occur independently but rather interact with one another in a highly complex manner.

Nitric oxide (NO) is released at high levels by inducible NO synthase (iNOS) with the concomitant formation of stoichiometric amounts of L-citrulline from L-arginine [43,44]. iNOS-mediated excessive and prolonged NO generation has attracted attention on account of its relevance to epithelial carcinogenesis [45–47] and its reactivity to form carcinogenic $N$-nitrosoamines [48]. With respect to the postinitiation phase, it should be noted that NO also has important

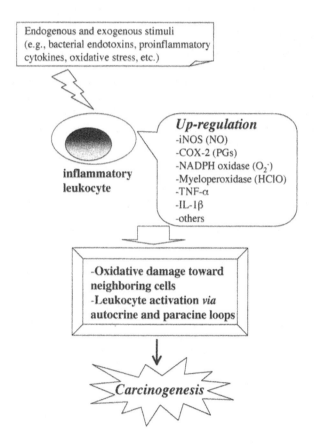

**Figure 3**   Some pivotal roles of activated inflammatory leukocytes in carcinogenesis.

roles in inflammatory responses such as edema formation and hyperplasia, as well as in papilloma development in mouse skin [49,50].

On the other hand, there are extensive data showing COX-2 expression to be involved in the development of certain cancers [51,52]; in marked contrast to COX-1, COX-2 activity is inducible and its elevation enhances the biosynthesis of PGs, including PGE$_2$, one of the physiologically active and stable PGs produced in the pathways downstream of COX enzymes. PGE$_2$ is known to stimulate bcl-2 activity and, thereby, prevent apoptosis [53]. Topical application of TPA to mice led to edema and papilloma formation by enhancing COX-2 protein expression; specific COX-2 inhibitors [54,55] could, therefore, actually counteract these biological events.

Superoxide anion ($O_2^-\cdot$) is a free radical generated through NADPH oxidase, predominantly present in leukocytes, and from xanthine oxidase in epithelial cells; it may be subsequently converted into more reactive intermediates such as the hydroxyl radical responsible for DNA mutations. Wei et al. [56] reported that double applications of TPA to mouse skin led to excessive reactive oxygen species (ROS) production. Ji and Marnett [57] termed such successive application as "priming" (the first stage, illustrated by leukocyte recruitment, maturation, and infiltration of inflammatory leukocytes such as polymorphonuclear leukocytes and macrophages into inflamed lesions) and "activation" (the second stage, illustrated by ROS production from accumulated leukocytes) (Fig. 4). We set out to determine whether nobiletin inhibits the priming and/or activation stages in this double application model [58], as described below, since food phytochemicals able to suppress or inhibit these biological processes are likely to be highly useful in the development of anti-inflammatory and anticarcinogenic strategies.

## B. Suppressive Effects on Induction of iNOS

Stimulation of murine RAW 264.7 macrophages with a combination of a bacterial toxin, lipopolysaccharide (LPS), and a proinflammatory cytokine, interferon (IFN)-$\gamma$, for 18 hours resulted in NO generation and then nitrite ($NO_2^-$) accumulation in the media. Nobiletin, at a concentration range of 25–100 $\mu M$, concentration-dependently suppressed $NO_2^-$ production (Fig. 5B). The inhibitory effect of resveratrol, a red wine polyphenol, was also examined, its potency being similar to that of nobiletin. iNOS protein expression by nobiletin was detected using Western blotting (Fig. 5A); while hardly detectable in the cellular lysate without stimulation, it was remarkably upregulated 18 hours after stimulation. Nobiletin showed a dose-dependent suppression at a concentration range of 25–100 $\mu M$, resveratrol exhibiting a marked inhibition only at 100 $\mu M$.

## C. Suppressive Effects on Induction of COX-2

Stimulation of RAW cells with LPS and IFN-$\gamma$ for 18 hours led to PGE$_2$ production in the media (Fig. 5C). Nobiletin suppressed PGE$_2$ production at concentra-

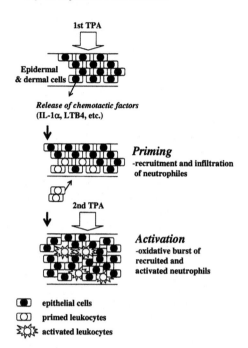

**1st TPA**

Epidermal & dermal cells

*Release of chemotactic factors*
(IL-1α, LTB4, etc.)

*Priming*
-recruitment and infiltration
of neutrophiles

**2nd TPA**

*Activation*
-oxidative burst of
recruited and
activated neutrophils

epithelial cells
primed leukocytes
activated leukocytes

**Figure 4**   Concept of two-stage oxidative stress model in mouse skin.

tions of 50 and 100 μM, whereas resveratrol exhibited higher suppressive rates at 25–100 μM. Stimulation of RAW 264.7 cells also led to a dramatic upregulation of COX-2 protein expression (Fig. 5A), which was hardly detectable in the control. While nobiletin showed a dose-dependent suppression at a concentration range of 25–100 μM, resveratrol only showed marked inhibition at 100 μM.

## D. Suppressive Effects on Generation of $O_2^-$·

We then examined the inhibitory effects of nobiletin on differentiated HL-60 cells, known to generate $O_2^-$· with TPA stimulation and compared them with those of luteolin. As shown in Figure 6, nobiletin exhibited a much higher level of inhibition than luteolin, while the $IC_{50}$ value of nobiletin (1.2 μM) was approximately 8 times lower than that of luteolin (9.8 μM). Because nobiletin did not significantly scavenge $O_2^-$ generated from the xanthine/xanthine oxidase system nor inhibited xanthine oxidase activity up to a concentration of 500 μM (data not shown), it may inhibit the assembly or activity of multicomponent NADPH oxidase system in differentiated HL-60 cells.

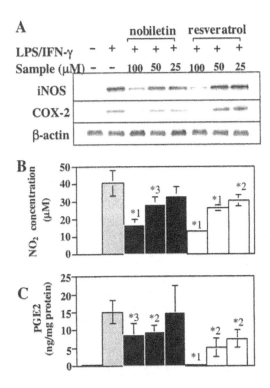

**Figure 5** Inhibitory effects of nobiletin and resveratrol on LPS/IFN-γ–induced iNOS/COX-2 protein expression (A) and on generation of $NO_2^-$ (B) and $PGE_2$ (C) in RAW 264.7 cells. RAW 264.7 cells were treated with LPS (100 ng/mL), $BH_4$ (10 mg/mL), IFN-γ (100 U/mL), L-arginine (2 mM), and the test compound for 18 hours. *1:$p < 0.001$; *2:$p < 0.01$; *3:$p < 0.02$ vs. positive control (LPS/IFN-γ) in Student's $t$-test. Each experiment was done independently in duplicate twice, and data are shown as m ± SD values. One representative picture is shown. (Adapted from Ref. 58.)

## E. Oxidative Stress–Attenuating Effects

We addressed the question as to whether nobiletin inhibits the priming and/or activation stages in mouse skin using the double-TPA-application method, the time schedule of drug applications in each group being shown in Table 1. Groups 1 and 2 represent negative and positive controls, respectively. Nobiletin was applied at the priming stage in group 3, at the activation stage in group 4, and at both stages in group 5.

Double TPA applications, 8.1 nmol each with a 24-hour interval, led to notable edema formation (group 2) as compared with the control (group 1) (Fig.

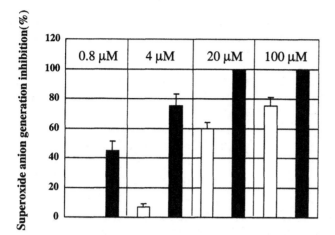

**Figure 6** Inhibitory effects of nobiletin and luteolin on TPA-induced $O_2^-$ generation in differentiated HL-60 cells and their cellular uptake. Differentiated HL-60 cells (1 × $10^6$ cells/mL) were stimulated by 100 nM TPA to induce $O_2^-$, the extracellular concentration of which was measured by a cytochrome $c$ method. Test compounds were added at concentrations of 0.8, 4, 20, and 100 μM. Hatched bars, nobiletin. Open bars, luteolin. Each experiment was done independently in duplicate twice, and data are shown as m ± SD values. (Adapted from Ref. 58.)

**Table 1** Experimental Groups for Examining Suppressive Effects of Nobiletin in Two-Stage Oxidative Stress Model

|  | 0 h | 0.5 h | 24 h | 24.5 h | 25.5 h |
|---|---|---|---|---|---|
| G1 | Ac | Ac | Ac | Ac | Sc |
| G2 | Ac | TPA | Ac | TPA | Sc |
| G3 | NOB | TPA | NOB | TPA | Sc |
| G4 | NOB | TPA | Ac | TPA | Sc |
| G5 | Ac | TPA | NOB | TPA | Sc |

Ac, acetone; Sc, sacrificed; NOB, nobiletin. Five ICR mice were used in each experimental group. Nobiletin (810 nmol in 100 μL acetone) was topically applied to the shaved area of dorsal skin 30 minutes before application of a TPA solution (8.1 nmol/100 μL in acetone). After 24 hours, the same doses of nobiletin or acetone were applied 30 minutes prior to a second TPA application.

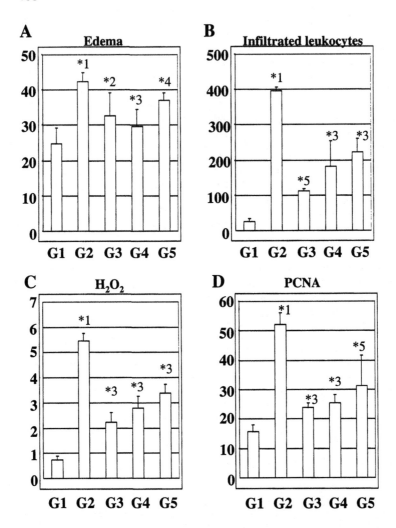

**Figure 7** Inhibitory effects of nobiletin on TPA-induced biological and histological parameters in mouse skin. (A) y-axis, edema formation (mg/punch); (B) y-axis, infiltrated leukocytes (No./mm$^2$); (C) y-axis, $H_2O_2$ (nmol/punch); (D) y-axis, PCNA-labeling index (%). G1–G5, groups 1–5. Statistical analysis was done using a Student's $t$-test. *1:$p <$ 0.001 (vs. group 1); *2:$p < 0.001$; *3:$p < 0.01$; *4:$p < 0.02$; *5:$p < 0.05$ (each vs. group 2). (Adapted from Ref. 58.)

7A). Double pretreatment with nobiletin (810 nmol) (group 3) 30 minutes prior to each TPA application suppressed edema formation. Pretreatment in the priming phase (group 4) was more effective for edema suppression than in the activation phase (group 5) and comparable or higher as a part of double pretreatment. A great number of leukocytes were found to have infiltrated the dermis in the double TPA application mice as compared with the controls (Fig. 7B) (group 1 vs. group 2). Further, double pretreatment with nobiletin markedly inhibited leukocyte infiltration, and a single pretreatment in either the priming or activation phase also significantly suppressed infiltration. Notably, the double TPA application dramatically increased the level of $H_2O_2$ in the mouse epidermis and dermis (Fig. 7C). A higher inhibition by nobiletin was again observed when applied at the priming stage (group 4) rather than at the activation phase (group 5). Double pretreatment with nobiletin afforded the highest inhibition of $H_2O_2$ production among all the three inhibitory groups (group 3). The PCNA-labeling index, a marker for cell proliferation, in the epidermis of the group 2 mice increased 3.3-fold over that of group 1 (Fig. 7C); in contrast, pretreatment (single or double nobiletin application) significantly reduced the PCNA-labeling indices (groups 3–5).

The anti-inflammatory effect of nobiletin was also investigated using a TPA-induced mouse ear edema formation test [59]. In a similar manner to that reported above, topical application of TPA to mouse ears causes acute inflammation, possibly accompanied by free radical generation, enhancement of prostanoid synthesis and increased vascular permeability, thereby resulting in edema formation. Anti-inflammatory activity was evaluated by measuring the increase in ear weights due to TPA-induced inflammation (Fig. 8A). Six hours after TPA application, the weight increase in ear disks (6 mm in diameter) treated only with TPA was $11.2 \pm 2.3$ mg. Pretreatment with nobiletin at a dose 100-fold greater than TPA notably suppressed weight increase, the inhibitory potency being higher than that of a synthetic COX inhibitor, indomethacin.

## F.  In Vivo Chemopreventive Efficacy

The inhibitory effect of a topical application of nobiletin at a dose range of 40–320 nmol on tumor formation was examined in dimethylbenz[a]anthracene (0.19 $\mu$mol)-initiated and TPA (1.6 nmol)-promoted mouse skin. As shown in Fig. 9, the average number of tumors per mouse in the control was $2.06 \pm 1.00$ at 20 weeks after tumor-promoting treatment, and pretreatment with nobiletin at 160 and 320 nmol caused dose-dependent reductions.

The modifying effects of dietary nobiletin on development of azoxymethane (AOM)-induced colonic aberrant crypt foci (ACF) were studied in male F344 rats given subcutaneous injections of AOM (15 mg/kg body weight) once a week for 3 weeks to induce ACF [60]. They also received an experimental diet containing 0.01 or 0.05% nobiletin for 5 weeks, starting 1 week before the initial injection

A

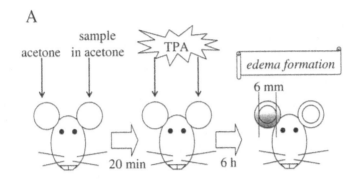

B

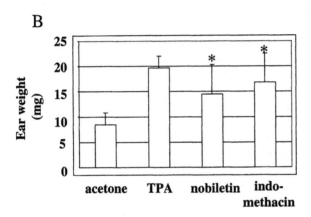

**Figure 8** Illustration of inhibitory test of TPA-induced edema formation in mouse ears (A) and anti-inflammatory activities of nobiletin and indomethacin toward TPA-induced edema formation in ICR mouse skin (B). Five 7-week-old mice were used in each experiment. The test compound (810 nmol/20 μL in acetone) or acetone was applied to an inner part of an ICR mouse ear. After 20 minutes, TPA solutions (8.1 nmol/20 μL in acetone) were applied to the same parts of the ears. After 6 hours, a disk (6 mm in diameter) was obtained from each ear and weighed. *$p < 0.01$ versus TPA using Student's $t$-test. (Adapted from Ref. 59.)

of AOM. AOM exposure produced $139 \pm 35$ ACF per rat at the end of the study (week 5) (Fig. 10A). Dietary administration of nobiletin caused significant reductions: 50% at a dose of 0.01% and 55% at 0.05%. Dietary nobiletin also significantly reduced PGE$_2$ content in colonic mucosa (Fig. 10B). These findings suggest the possible chemopreventive effects of oral feeding of nobiletin through suppression of inflammatory processes.

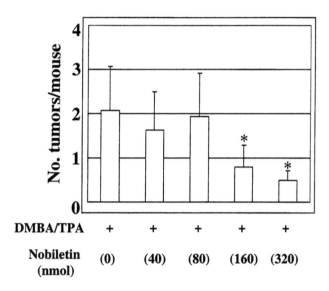

**Figure 9** Anti–tumor promoting activity of nobiletin in DMBA-initiated ICR mouse skin. Each group was composed of 15–17 female ICR mice, and all were given commercial rodent pellets and fresh tap water *ad libitum*, freshly exchanged twice a week. The back of each mouse was shaved with a surgical clipper 2 days before initiation. Mice at 6 weeks old were initiated with DMBA (0.19 μmol/100 μL in acetone). One week after initiation, the mice were promoted with TPA (1.6 nmol/100 μL in acetone) twice a week for 20 weeks. In the other 4 groups, the mice were treated with nobiletin (40, 80, 160, or 320 nmol/100 μL in acetone) 40 minutes before each TPA treatment. *$p < 0.001$ in Student's *t*-test. (Adapted from Ref. 58.)

## G.  Metabolism

We investigated the in vitro absorption and metabolism of nobiletin and compared these with luteolin. Nobiletin preferentially accumulated in a differentiated Caco-2 cell monolayer, used as a model for small intestinal epithelial cells; in contrast, luteolin did not [61]. Treatment of nobiletin with rat liver S-9 mixture led to the formation of 3′-demethylnobiletin, whereas luteolin was unaffected. We concluded, therefore, that PMFs including nobiletin have properties separate and distinct from those of common flavonoids for absorption and metabolism in vitro [61].

In support of our in vitro results, Nielsen et al. [62] have reported the in vivo biotransformation and excretion of tangeretin by analyzing urine and fecal samples from rats after repeated administrations of this flavone at 100 mg/kg body weight/day. Demethylated or hydroxylated metabolites were identified with

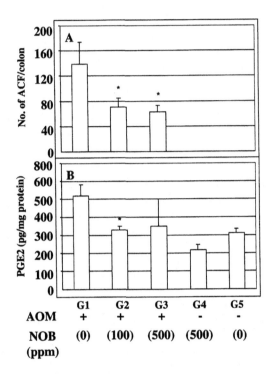

**Figure 10** Effect of nobiletin on AOM-induced colonic ACF (A) and PGE2 (B) formation in F344 rats. A total of 36 male F344 rats were divided into 4 experimental and 1 control groups. Groups 1–3 were initiated with AOM by 3 weekly subcutaneous injections (15 mg/kg body weight). Rats in groups 2 and 3 were fed diets containing nobiletin at 100 and 500 ppm, respectively, for 5 weeks. Group 4 was given the diet containing 500 ppm nobiletin alone. Group 5 was an untreated control. Rats were sacrificed after 5 weeks by an ether overdose. For $PGE_2$ determination, the scraped colonic mucosa of 3 rats from each group was homogenized in 400 mL of phosphate-buffered saline on ice. After centrifugation at 10,000 g for 5 minutes, the supernatants thus obtained were diluted at a ratio of 1:9, and then $PGE_2$ concentration was measured using a commercial experimental kit according to the protocol of the manufacturer. Panel A, *$P < 0.01$ versus group 1 in Welch's t-test. Panel B, *$P < 0.05$ versus group 1 in Welch's t-test. (Adapted from Ref. 60)

metabolic changes primarily occurring in the 4' position of the B-ring. Tangeretin is metabolized in rat and human liver microsomes by an $O$-demethylation reaction involving cytochrome P450 [63].

Datla et al. [64] also investigated the neuroprotective effects of tangeretin in a 6-hydroxydopamine lesion rat model of Parkinson's disease. Following chronic oral administration (10 mg/kg/day for 28 days), significant levels of tangeretin were detected in the hypothalamus, striatum, and hippocampus, indicating transfer across the blood-brain barrier and suggesting a potential for this flavone as a neuroprotective agent.

## III. METABOLISM AND XENOBIOTIC SYSTEMS

Exogenous substances are known to be metabolized by xenobiotic enzymes consisting of phase 1 (e.g., cytochrome P450s that add a hydrophilic functional group) and phase 2 (e.g., GST, which provides glutathione to compounds metabolized by phase 1 enzymes). Most polycyclic aromatic hydrocarbons, procarcinogens, are biologically inactive in their native structure and activated by a phase 1 enzyme, leading to the formation of ultimate carcinogens capable of binding to cellular DNA. Alternatively, these activated carcinogens may be inactivated by phase 2 enzymes and excreted from the body. P-Glycoprotein, known as a plasma membrane glycoprotein, confers multidrug resistance on cells by virtue of its ability to exclude cytotoxic drugs in an ATP-dependent manner. The effects of PMFs on xenobiotic systems and P-glycoprotein are described below.

## A.  Influence on Drug Metabolism

The effects of tangeretin on cytochrome P450 activity were first reported in 1995 by Obermeier et al., who showed 7-ethoxyresorufin-$O$-deethylase (classified as CYP 1A) and nifedipine oxidase (CYP 3A4) in human liver microsomes to be inhibited by tangeretin in a noncompetitive manner [65].

More recently, Lake et al. showed the effects of tangeretin, naringin, and its aglycone, naringenin, on xenobiotic-induced genotoxicity using rat and human precision-cut liver slices [66]. Tangeretin was found to be a potent inhibitor of cooked food mutagen 2-amino-1-methyl-6-phenylimidazo[4,5-b]pyridine (PhIP)–induced unscheduled DNA synthesis (UDS) in human liver slices, whereas naringenin was ineffective and naringin only inhibited genotoxicity at a high concentration. In rat liver slices, tangeretin inhibited 2-acetylaminofluorene-induced UDS, whereas both naringenin and naringin were ineffective [66].

Contradictory data have also been presented by Canivenc-Lavier et al. [67]. Immunoblot analysis of rat liver specimens indicated that flavone and tangeretin increased cytochrome P450 1A1, 1A2, and 2B1 & 2B2 forms, whereas flavanone only enhanced the cytochrome P450 2B isozymes. A Northern blot analysis in the same study showed flavone and tangeretin to increase levels of cytochrome P450 1A2 mRNAs. Flavone and, to a lesser extent, tangeretin increased the activities of ethoxyresorufin O-deethylase, methoxyresorufin O-demethylase, and pentoxyresorufin O-dealkylase, whereas flavanone mainly enhanced only pentoxyresorufin O-dealkylase activity [67]. In addition, acetaminophen oxidation, catalyzed by rat liver P450 3A4, was stimulated by tangeretin, nobiletin, and flavone, but inhibited by 40–60% by myricetin and quercetin [68]. The modifying effects of PMFs on phase 1 enzyme activity may, therefore, depend on the experimental conditions, e.g., the tissues, organs, and species used. It is interesting that, to our knowledge, there is no published work on the effect of PMFs on phase 2 enzyme activity.

As mentioned above, one important mechanism of multidrug resistance involves the multidrug transporter, P-glycoprotein, which confers upon cancer cells the ability to resist lethal doses of certain cytotoxic drugs by pumping the drugs out of the cells and thus reducing their cytotoxicity [69,70]. Sawada et al. showed that nobiletin, tangeretin, and 3,3',4',5,6,7,8-heptamethoxyflavone all enhanced vinblastine uptake by specifically inhibiting drug efflux *via* P-glycoprotein, but did not affect cytochrome P450 isozyme CYP3A4 activity, suggesting that these PMFs may have potential as agents for reversing multidrug resistance or for recovering the bioavailability of certain drugs [71,72]. While it has been established that grapefruit juice has some potential to accelerate the intestinal absorption of certain drugs [73–75], Sawada and coworkers subsequently found that orange juice extract, which includes PMFs, had a higher activity to inhibit P-glycoprotein than grapefruit juice [72].

## B.  Potential Side Effects and Toxicity

Chaumontet et al. conducted an extensive study into the tumor-promoting potentials of four flavonoids (quercetin, tangeretin, flavone, and flavanone) to induce rat liver preneoplastic foci formation and in GJIC activities in vitro and in vivo [76]. Using a GST placental form (GST-P)–positive foci, they compared the effects of flavonoids (1000 ppm in the diet) with those of phenobarbital (PB) on the occurrence of liver preneoplastic lesions and examined the effects of the flavonoids on GJIC in liver samples and in two cell lines. No significant differences in the number and area of GST-P–positive foci were found between any of the flavonoid groups and the control. Further, PB decreased GJIC by 60%, whereas tangeretin alone decreased dye transfer in vivo, suggesting tumor-promoting ability. In contrast, anti–tumor-promotional properties of tangeretin and flavone were observed in REL cells in vitro [76], indicating a discrepancy between the in vivo and in vitro data for tangeretin.

The in vitro effects of several flavonoids, including tangeretin, on nonenzymatic lipid peroxidation in rat brain mitochondria have been studied by Ratty and Das [77]. The flavonoids apigenin, flavone, flavanone, hesperidin, naringin, and tangeretin each promoted ascorbic acid–induced lipid peroxidation, while 18 nonmethoxylated flavonoids also examined were suppressive [77].

Tamoxifen is one of the promising agents for the prevention and cure of breast cancer [78]; Bracke et al. have, however, presented surprising data showing that female nude mice inoculated subcutaneously with human MCF-7/6 mammary adenocarcinoma cells and treated with tangeretin did not show inhibited tumor growth, while the addition of tangeretin to drinking water with tamoxifen completely masked the inhibitory effect of tamoxifen by an unknown mechanism(s) [79]. Therefore, it should be stressed that tangeretin may not have an effect on the potential risks of human cancer when considering its daily intake, and Bracke et al. noted: "Our data certainly do not indicate that citrus flavonoids would be carcinogenic!" [79].

## C. Recent Topics

Molecular mechanisms by which PMFs exert their anti-inflammatory and anti–tumor-promoting, as well as anti–tumor-invading activities remain poorly understood. However, Ito and co-workers [80] recently presented an excellent study indicating that nobiletin suppressed the TPA-induced binding activity of activator protein-1 and that it may interfere in the phosphatidylinositol 3-kinase pathway, which divergently regulates the production of MMP and its inhibitor.

PMFs, including nobiletin, occur exclusively in citrus fruit, with the King orange (*C. nobilis*) being found to contain the highest amount among 34 citrus juices examined; principal component and cluster analyses of readily extractable flavonoids indicated peculiarities of the King and Bergamot varieties. Further, nobiletin and tangeretin occurred at concentrations of 46 and 7 ppm, respectively, in Valencia orange juice [19].

The effect of 6-benzylaminopurine on PMF levels in tangelo Nova fruits and the possible participation of these secondary metabolites in defense mechanisms against *Phytophthora citrophthora* were examined by Ortuno et al. [81]. An in vitro study of the inhibitory effect of these compounds on fungal growth revealed nobiletin to be most active, followed by sinensetin, heptamethoxyflavone, and tangeretin. Treatment with 100 ppm of 6-benzylaminopurine increased the levels of these PMFs in this citrus hybrid and also enhanced in vivo resistance to the fungus by approximately 60%. Such an approach may have value in developing novel functional foods possessing higher physiological functions (anti-inflammatory and anticarcinogenic activity) associated with higher PMF contents.

## IV. CONCLUSION

As mentioned above, most, if not all, reports on the biological activities of PMFs that used in vitro bioassays found that these compounds have significant potential for suppressing inflammatory and carcinogenic processes. A substantial number of characteristics that contribute to this activity are likely to result from their distinct hydrophobicity, as demonstrated previously [58]. In the presence of hydrophobic methyl group(s), PMFs are able to penetrate intracellular compartments in cell cultures and then localize in epithelial lesions in the gastrointestinal tract [82]. In contrast to flavonoids in general, most of whose biochemical and biological properties and activities have been extensively reported, those of PMFs are limited. Although daily intakes of PMFs are much lower than that of general types, they demand greater attention on the basis of data showing their characteristic biological activities. Molecular studies for elucidating the mechanisms of action of PMFs, exhibiting activities, and investigating issues of bioavailability and metabolic fate will provide fertile ground for future study. However, taken together, current data identify PMFs as a promising candidate group of postpolyphenolic flavonoids.

## ACKNOWLEDGMENTS

The authors thank the following collaborators in Japan: K. Koshimizu (Kinki University); Y. Nakamura (Nagoya University); H. Tokuda and H. Nishino (Kyoto Prefectural University of Medicine); K. Ogawa and M. Yano (National Institute of Fruit Tree Science); H. Kohno, S. Yohsitani, Y. Tsukio, and T. Tanaka (Kanazawa Medical University); S. Kuwahara and Y. Takahashi (Wakayama Agricultural Processing Research Corporation); C. Ito and H. Furukawa (Meijo University); M. Ju-ichi (Mukogawa Women's University); Y. Mimaki and Y. Sashida (Tokyo University of Pharmacy and Life Science); and S. Kitanaka (Nihon University). Our work described in this chapter was partly supported by a grant-in-aid from the Program for Promotion of Basic Research Activities for Innovative Biosciences (PROBRAIN).

## REFERENCES

1. Murakami A, Ohigashi H, Koshimizu K. Chemoprevention: insights into biological mechanisms and promising food factors. Food Rev Int 1999; 15:335–395.
2. Nelson DR, Kamataki T, Waxman DJ, Guengerich FP, Estabrook RW, Feyereisen R, Gonzalez FJ, Coon MJ, Gunsalus IC, Gotoh O. The P450 superfamily: update

on new sequences, gene mapping, accession numbers, early trivial names of enzymes, and nomenclature. DNA Cell Biol 1993; 12:1–51.

3. Slaga TJ, DiGiovanni J, Winberg LD, Budunova IV. Skin carcinogenesis: characteristics, mechanisms and prevention. In: McClain PM, Slaga TJ, LeBoeuf R, Pitot H, Eds. Growth Factors and Tumor Promotion: Implications for Risk Assessment. New York: Wiley, 1995:1–20.

4. Pitot HC, Dragan YP. Facts and theories concerning the mechanisms of carcinogenesis. FASEB J 1991; 5:2280–2286.

5. Hennings H, Glick AB, Greenhalgh DA, Morgan DL, Strickland JE, Tennenbaum T, Yuspa SH. Critical aspects of initiation, promotion, and progression in multistage epidermal carcinogenesis. Proc Soc Exp Biol Med 1993; 202:1–8.

6. Miyazawa M, Okuno Y, Fukuyama M, Nakamura S, Kosaka H. Antimutagenic activity of polymethoxyflavonoids from *Citrus aurantium*. J Agric Food Chem 1999; 47:5239–5244.

7. Calomme M, Pieters L, Vlietinck A, Vanden Berghe D. Inhibition of bacterial mutagenesis by citrus flavonoids. Planta Med 1996; 62:222–226.

8. Iwase Y, Takemura Y, Ju-ichi M, Yano M, Ito C, Furukawa H, Mukainaka T, Kuchide M, Tokuda H, Nishino H. Cancer chemopreventive activity of 3,5,6,7,8,3′,4′-heptamethoxyflavone from the peel of citrus plants. Cancer Lett 2001; 163:7–9.

9. Ishiwa J, Sato T, Mimaki Y, Sashida Y, Yano M, Ito A. A citrus flavonoid, nobiletin, suppresses production and gene expression of matrix metalloproteinase 9/gelatinase B in rabbit synovial fibroblasts. J Rheumatol 2000; 27:20–25.

10. Malterud KE, Rydland KM. Inhibitors of 15-lipoxygenase from orange peel. J Agric Food Chem 2000; 48:5576–5580.

11. Rivedal E, Yamasaki H, Sanner T. Inhibition of gap junctional intercellular communication in Syrian hamster embryo cells by TPA, retinoic acid and DDT. Carcinogenesis 1994; 15:689–694.

12. Klann RC, Fitzgerald DJ, Piccoli C, Slaga TJ, Yamasaki H. Gap-junctional intercellular communication in epidermal cell lines from selected stages of SENCAR mouse skin carcinogenesis. Cancer Res 1989; 49:699–705.

13. Chaumontet C, Bex V, Gaillard-Sanchez I, Seillan-Heberden C, Suschetet M, Martel P. Apigenin and tangeretin enhance gap junctional intercellular communication in rat liver epithelial cells. Carcinogenesis 1994; 15:2325–2230.

14. Chaumontet C, Droumaguet C, Bex V, Heberden C, Gaillard-Sanchez I, Martel P. Flavonoids (apigenin, tangeretin) counteract tumor promoter-induced inhibition of intercellular communication of rat liver epithelial cells. Cancer Lett 1997; 114: 207–210.

15. Kandawaswami C, Perkins E, Soloniuk DS, Drzewiecki and Middleton Jr, E. Antiproliferative effects of citrus flavonoids on a human squamous cell carcinoma in vitro. Cancer Lett 1991; 56:147–152.

16. Kawaii S, Tomono Y, Katase E, Ogawa K, Yano M. Antiproliferative activity of flavonoids on several cancer cell lines. Biosci Biotechnol Biochem 1999; 63: 896–899.

17. Rodriguez J, Yanez J, Vicente V, Alcaraz M, Benavente-Garcia O, Castillo J, Lorente J, Lozano JA. Effects of several flavonoids on the growth of B16F10 and SK-MEL-

1 melanoma cell lines: relationship between structure and activity. Melanoma Res 2002; 12:99–107.

18.  Hirano T, Abe K, Gotoh M, Oka K. Citrus flavone tangeretin inhibits leukaemic HL-60 cell growth partially through induction of apoptosis with less cytotoxicity on normal lymphocytes. Br J Cancer 1995; 72:1380–1388.

19.  Kawaii S, Tomono Y, Katase E, Ogawa K, Yano M. HL-60 differentiating activity and flavonoid content of the readily extractable fraction prepared from citrus juices. J Agric Food Chem 1999; 47:128–135.

20.  Mak NK, Wong-Leung YL, Chan SC, Wen J, Leung KN, Fung MC. Isolation of anti-leukemia compounds from *Citrus reticulata*. Life Sci 1996; 58:1269–1276.

21.  Egeblad M, Werb Z. New functions for the matrix metalloproteinases in cancer progression. Nat Rev Cancer 2002; 2:161–174.

22.  Cox G, O'Byrne KJ. Matrix metalloproteinases and cancer. Anticancer Res 2001; 21:4207–4219.

23.  John A, Tuszynski G. The role of matrix metalloproteinases in tumor angiogenesis and tumor metastasis. Pathol Oncol Res 2001; 7:14–23.

24.  Bracke ME, Bruyneel EA, Vermeulen SJ, Vennekens K, Marck VV, Mareel MM. Citrus flavonoid effect on tumor invasion and metastasis. Food Technol 1994; 48: 121–124.

25.  Bracke M, Vyncke B, Opdenakker G, Foidart JM, De Pestel G, Mareel M. Effect of catechins and citrus flavonoids on invasion in vitro. Clin Exp Metastasis 1991; 9:13–25.

26.  Bracke ME, Vyncke BM, Van Larebeke NA, Bruyneel EA, De Bruyne GK, De Pestel GH, De Coster WJ, Espeel MF, Mareel MM. The flavonoid tangeretin inhibits invasion of MO4 mouse cells into embryonic chick heart in vitro. Clin Exp Metastasis 1989; 7:283–300.

27.  Ito A, Ishiwa J, Sato T, Mimaki Y, Sashida Y. The citrus flavonoid nobiletin suppresses the production and gene expression of matrix metalloproteinases-9/gelatinase B in rabbit synovial cells. Ann NY Acad Sci 1999; 878:632–624.

28.  Minagawa A, Otani Y, Kubota T, Wada N, Furukawa T, Kumai K, Kameyama K, Okada Y, Fujii M, Yano M, Sato T, Ito A, Kitajima M. The citrus flavonoid, nobiletin, inhibits peritoneal dissemination of human gastric carcinoma in SCID mice. Jpn J Cancer Res 2001; 92:1322–1328.

29.  Rooprai HK, Kandanearatchi A, Maidment SL, Christidou M, Trillo-Pazos G, Dexter DT, Rucklidge GJ, Widmer W, Pilkington GJ. Evaluation of the effects of swainsonine, captopril, tangeretin and nobiletin on the biological behaviour of brain tumour cells in vitro. Neuropathol Appl Neurobiol 2001; 27:29–39.

30.  Vermeulen S, Van Marck V, Van Hoorde L, Van Roy F, Bracke M, Mareel M. Regulation of the invasion suppressor function of the cadherin/catenin complex. Pathol Res Pract 1996; 192:694–707.

31.  Sasaki K, Yoshizaki F. Nobiletin as a tyrosinase inhibitor from the peel of citrus fruit. Biol Pharm Bull 2002; 25:806–868.

32.  Sempinska E, Kostka B, Krolikowska M, Kalisiak E. Effect of flavonoids on the platelet adhesiveness in repeatedly bred rats. Pol J Pharmacol Pharm 1977; 29:7–10.

33.  Jayaprakasha GK, Negi PS, Sikder S, Rao LJ, Sakariah KK. Antibacterial activity of *Citrus reticulata* peel extracts. Z Naturforsch [C] 2000; 55:1030–1034.

34. Fitzpatrick FA. Inflammation, carcinogenesis and cancer. Int Immunopharmacol 2001; 1:1651–1667.

35. Wong NA, Harrison DJ. Colorectal neoplasia in ulcerative colitis-recent advances. Histopathology 2001; 39:221–234.

36. Chen X, Yang CS. Esophageal adenocarcinoma: a review and perspectives on the mechanism of carcinogenesis and chemoprevention. Carcinogenesis 2001; 22: 1119–1129.

37. Kuper H, Adami HO, Trichopoulos D. Infections as a major preventable cause of human cancer. J Intern Med 2000; 248:171–183.

38. Kovacic P, Jacintho JD. Mechanisms of carcinogenesis: focus on oxidative stress and electron transfer. Curr Med Chem 2001; 8:773–796.

39. Kawanishi S, Hiraku Y, Oikawa S. Mechanism of guanine-specific DNA damage by oxidative stress and its role in carcinogenesis and aging. Mutat Res 2001; 488: 65–76.

40. Bartsch H, Nair J. New DNA-based biomarkers for oxidative stress and cancer chemoprevention studies. Eur J Cancer 2000; 36:1229–1234.

41. Klaunig JE, Xu Y, Isenberg JS, Bachowski S, Kolaja KL, Jiang J, Stevenson DE, Walborg EF. The role of oxidative stress in chemical carcinogenesis. Environ Health Perspect 1998; 106(suppl 1):289–925.

42. Dreher D, Junod AF. Role of oxygen free radicals in cancer development. Eur J Cancer 1996; 32A:30–38.

43. Vanvaskas S, Schmidt HH. Just say NO to cancer?. J Natl Cancer Inst 1997; 89: 406–407.

44. Nathan C, Xie Q. Nitric oxide synthases: roles, tolls and controls. Cell 1994; 78: 915–918.

45. Ohshima H, Bartsch H. Chronic infections and inflammatory processes as cancer risk factors: possible role of nitric oxide in carcinogenesis. Mutat Res 1994; 305: 253–264.

46. Xie K, Huang S, Dong Z, Juang SH, Wang Y, Fidler IJ. Destruction of bystander cells by tumor cells transfected with inducible nitric oxide (NO) synthase gene. J Natl Cancer Inst 1997; 89:421–427.

47. Tsuji S, Kawano S, Tsuji M, Takei Y, Tanaka M, Sawaoka H, Nagano K, Fusamoto H, Kamada T. *Helicobactor pylori* extract stimulates inflammatory nitric oxide production. Cancer Lett 1996; 108:195–200.

48. Miwa M, Stuehr D J, Marletta MA, Wishnok JS, Tannenbaum SR. Nitrosation of amines by stimulated macrophages. Carcinogenesis 1987; 8:955–958.

49. Teixeira MM, Williams TJ, Hellewell PG. Role of prostaglandins and nitric oxide in acute inflammatory reactions in guinea-pig skin. Br J Pharmacol 1993; 110: 1515–1521.

50. Ahmad N, Srivastava RC, Agarwal R, Muktar H. Nitric oxide synthase and skin tumor promotion. Biochem Biphys Res Commun 1997; 232:328–331.

51. Taketo MM. Cyclooxygenase-2 inhibitors in tumorigenesis (part II). J Natl Cancer Inst 1998; 90:1609–1620.

52. Taketo MM. Cyclooxygenase-2 inhibitors in tumorigenesis (part I). J Natl Cancer Inst 1998; 90:1529–1536.

53.  Fosslien E. Molecular pathology of cyclooxygenase-2 in neoplasia. Ann Clin Lab Sci 2000; 30:3–21.

54.  Mueller DK, Kopp SA, Marks F, Seibert K, Fuerstenberger G. Localization of prostaglandin H synthase isoenzymes in murine epidermal tumors: suppression of skin tumor promotion by inhibition of prostagalndin H synthase-2. Mol Carcinog 1998; 23:36–44.

55.  Puignero V, Queralt J. Effect of topically applied cyclooxygenase-2-selective inhibitors on arachidonic acid- and tetradecanoylphorbol acetate-induced dermal inflammation in the mouse. Inflammation 1997; 21:431–442.

56.  Wei H, Wei L, Frenkel K, Bowen R, Barnes S. Inhibition of tumor promoter-induced hydrogen peroxide formation in vitro and in vivo by genistein. Nutr Cancer 1993; 20:1–12.

57.  Ji C, Marnett LJ. Oxygen radical-dependent epoxydation of (7S,8S)-dihydroxy-7,8-dihydrobenzo[a]pyrene in mouse skin in vivo. Stimulation by phorbol esters and inhibition by antiinflammatory steroids. J Biol Chem 1992; 267:17842–17878.

58.  Murakami A, Nakamura Y, Torikai K, Tanaka T, Koshiba T, Koshimizu K, Kuwahara S, Takahashi Y, Ogawa K, Yano M, Tokuda H, Nishino H, Mimaki Y, Sashida Y, Kitanaka S, Ohigashi H. Inhibitory effect of citrus nobiletin on phorbol ester-induced skin inflammation, oxidative stress and tumor promotion in mice. Cancer Res 2000; 60:5059–5966.

59.  Murakami A, Nakamura Y, Ohto Y, Yano M, Koshiba T, Koshimizu K, Tokuda H, Nishino H, Ohigashi H. Suppressive effects of citrus fruits on free radical generation and nobiletin, an anti-inflammatory polymethoxyflavonoid. Biofactors 2000; 12: 187–192.

60.  Kohno H, Yoshitani S, Tsukio Y, Murakami A, Koshimizu K, Yano M, Tokuda H, Nishino H, Ohigashi H, Tanaka T. Dietary administration of citrus nobiletin inhibits azoxymethane-induced colonic aberrant crypt foci in rats. Life Sci 2001; 69:901–913.

61.  Murakami A, Kuwahara S, Takahashi Y, Ito C, Furukawa H, Ju-Ichi M, Koshimizu K. In vitro absorption and metabolism of nobiletin, a chemopreventive polymethoxyflavonoid in citrus fruits. Biosci Biotechnol Biochem 2001; 65:194–197.

62.  Nielsen SE, Breinholt V, Cornett C, Dragsted LO. Biotransformation of the citrus flavone tangeretin in rats. Identification of metabolites with intact flavane nucleus. Food Chem Toxicol 2000; 38:739–746.

63.  Canivenc-Lavier MC, Brunold C, Siess MH, Suschetet M. Evidence for tangeretin O-demethylation by rat and human liver microsomes. Xenobiotica 1993; 23:259–266.

64.  Datla KP, Christidou M, Widmer WW, Rooprai HK, Dexter DT. Tissue distribution and neuroprotective effects of citrus flavonoid tangeretin in a rat model of Parkinson's disease. Neuroreport 2001; 12:3871–3875.

65.  Obermeier MT, White RE, Yang CS. Effects of bioflavonoids on hepatic P450 activities. Xenobiotica 1995; 25:575–584.

66.  Lake BG, Beamand JA, Tredger JM, Barton PT, Renwick AB, Price RJ. Inhibition of xenobiotic-induced genotoxicity in cultured precision-cut human and rat liver slices. Mutat Res 1999; 440:91–100.

67.  Canivenc-Lavier MC, Bentejac M, Miller ML, Leclerc J, Siess MH, Latruffe N, Suschetet M. Differential effects of nonhydroxylated flavonoids as inducers of cyto-

chrome P450 1A and 2B isozymes in rat liver. Toxicol Appl Pharmacol 1996; 136: 348–353.

68. Li Y, Wang E, Patten CJ, Chen L, Yang CS. Effects of flavonoids on cytochrome P450-dependent acetaminophen metabolism in rats and human liver microsomes. Drug Metab Dispos 1994; 22:566–571.

69. Lehne G. P-glycoprotein as a drug target in the treatment of multidrug resistant cancer. Curr Drug Targets 2000; 1:85–99.

70. Ueda K, Yoshida A, Amachi T. Recent progress in P-glycoprotein research. Anticancer Drug Des 1999; 14:115–121.

71. Ikegawa T, Ushigome F, Koyabu N, Morimoto S, Shoyama Y, Naito M, Tsuruo T, Ohtani H, Sawada Y. Inhibition of P-glycoprotein by orange juice components, polymethoxyflavones in adriamycin-resistant human myelogenous leukemia (K562/ADM) cells. Cancer Lett 2000; 160:21–28.

72. Takanaga H, Ohnishi A, Yamada S, Matsuo H, Morimoto S, Shoyama Y, Ohtani H, Sawada Y. Polymethoxylated flavones in orange juice are inhibitors of P-glycoprotein but not cytochrome P450 3A4. J Pharmacol Exp Ther 2000; 293:230–236.

73. Bailey DG, Malcolm J, Arnold O, Spence JD. Grapefruit juice-drug interactions. Br J Clin Pharmacol 1998; 46:101–110.

74. Fuhr U. Drug interactions with grapefruit juice. Extent, probable mechanism and clinical relevance. Drug Saf 1998; 18:251–272.

75. Feldman EB. How grapefruit juice potentiates drug bioavailability. Nutr Rev 1997; 55:398–400.

76. Chaumontet C, Suschetet M, Honikman-Leban E, Krutovskikh VA, Berges R, Le Bon AM, Heberden C, Shahin MM, Yamasaki H, Martel P. Lack of tumor-promoting effects of flavonoids: studies on rat liver preneoplastic foci and on in vivo and in vitro gap junctional intercellular communication. Nutr Cancer 1996; 26:251–63.

77. Ratty AK, Das NP. Effects of flavonoids on nonenzymatic lipid peroxidation: structure-activity relationship. Biochem Med Metab Biol 1988; 39:69–79.

78. Bentrem DJ, Craig Jordan V. Tamoxifen, raloxifene and the prevention of breast cancer. Minerva Endocrinol 2002; 27:127–139.

79. Bracke ME, Depypere HT, Boterberg T, Van Marck VL, Vennekens KM, Vanluchene E, Nuytinck M, Serreyn R, Mareel MM. Influence of tangeretin on tamoxifen's therapeutic benefit in mammary cancer. J Natl Cancer Inst 1999; 91:354–359.

80. Sato T, Koike L, Miyata Y, Hirata M, Mimaki Y, Sashida Y, Yano M, Ito A. Inhibition of activator protein-1 binding activity and phosphatidylinositol 3-kinase pathway by nobiletin, a polymethoxy flavonoid, results in augmentation of tissue inhibitor of metalloproteinases-1 production and suppression of production of matrix metalloproteinases-1 and -9 in human fibrosarcoma HT-1080 cells. Cancer Res 2002; 62:1025–1029.

81. Ortuno A, Arcas MC, Botia JM, Fuster MD, Del Rio JA. Increasing resistance against *Phytophthora citrophthora* in tangelo Nova fruits by modulating polymethoxyflavones levels. J Agric Food Chem 2002; 50:2836–2839.

82. Murakami A, Koshimizu K, Ohigashi H, Kuwahara S, Kuki W, Takahashi Y, Hosotani K, Kawahara K, Matsuoka Y. Characteristic rat tissue accumulation of nobiletin, a chemopreventive polymethoxyflavonoid, in comparison with luteolin. Biofactors 2002; 16:73–82.

# 10

# Biochemical and Molecular Mechanisms of Cancer Chemoprevention by Tea and Tea Polyphenols

**Jen-Kun Lin**
*National Taiwan University, Taipei, Taiwan*

## I. INTRODUCTION

Tea (*Camellia sinensis*) originated in southern China and is consumed by over two thirds of the world population. Tea has an attractive aroma, good taste, and health-promoting effects, and these benefits combine to make it one of the most popular drinks in the world. Tea was used by the Chinese as a medicinal drink as early as 3000 B.C., and as a beverage by the end of the sixth century. Because tea is an excellent beverage and has an effective pharmaceutical activity, tea plants are now widely cultivated in Southeast Asia including mainland China, India, Japan, Taiwan, Sri Lanka, and Indonesia and in a number of countries in central Africa. Hundreds of teas are now produced and are generally classified into three categories: nonfermented green tea, partially fermented oolong and paochong teas, and fully fermented black and puerh tea. The composition of tea varies in a complex manner, according to species, season, age of the leaf, climate, and horticultural practices [1].

Most commercial teas are derived from the young leaf buds of the tea plant [2], which are plucked and treated in one of several ways to convert it into the appropriate form for the tea markets. The bulk of the leaf is processed by one of three distinctly different methods, depending upon the characteristics of the end product. Most tea leaf is used for the manufacture of green or black teas, but a smaller, though still significant, quantity is processed to yield oolong tea [3].

In the case of green tea, it is necessary to avoid oxidative enzymatic activity during the operations between plucking and a subsequent heating step, a particular effort being made to avoid exposure to the sun. Black tea is manufactured through a series of fermentation processes that include withering, bruising, rolling, rerolling (several times) and firing (which stops oxidative processes by heat-inactivating relevant enzymes). Oolong teas represent a middle ground between black and green teas; they are considered semi-fermented, and, therefore, their processing closely resembles that of black tea. The processes involved in oolong tea manufacturing are complicated and must be carefully controlled in order to produce the characteristic, and prized, aroma and taste. There are a number of varieties of oolong teas, their sensory characteristics depending upon the practices of individual factories [4]; these can cause great difficulty when using chemical analysis of tea polyphenols to characterize oolong teas from different sources.

## II.  HEALTH EFFECTS OF TEAS: TEA POLYPHENOLS

The majority of the biological functions of teas may be attributed to their polyphenolic components (see Fig. 1); the chemical structures of the main polyphenols isolated from green, black, oolong, and puerh teas are structurally related but not identical. The monomeric catechins from green tea may be considered as the precursors of the more complex polyphenols found in other teas as a result of their fermentation conditions.

   The major polyphenols in green tea are catechins including catechin itself, gallocatechin, ( − )-epicatechin, ( − )-epigallocatechin, ( − )-epicatechin-3-gallate, and ( − )-epigallocatechin-3-gallate (EGCG) (Fig. 1), the latter being the most abundant. In northern China, most green teas are blended with jasmine flower to promote their aroma and flavor. Many intermediate oxidation products of tea polyphenols have been found in oolong teas and their isolation, characterization, and, particularly, analysis have been a challenge for chemists over many years. Theasinensin A (Fig. 1), a new type of tea polyphenol, has been isolated from oolong tea, its structure being confirmed by synthesis involving free radical oxidation of EGCG. In Taiwan, it is estimated that approximately 80% of tea produced is consumed as oolong tea, although paochong tea, which may be considered as a lightly fermented oolong tea, is quite popular in the northern regions. Partially fermented oolong or paochong teas contain catechins, theaflavins, and, possibly, thearubigins; other components, such as proanthocyanidins, are less well characterized but may also have a role in disease prevention.

   Puerh tea, highly fermented, possesses a rich flavor and was long recognized as a ''tribute'' tea in the imperial court of China. Its major polyphenols are not well characterized, because of their insolubility in most extracting solvents, but are considered to be polymerized catechins of high molecular weight. The aqueous

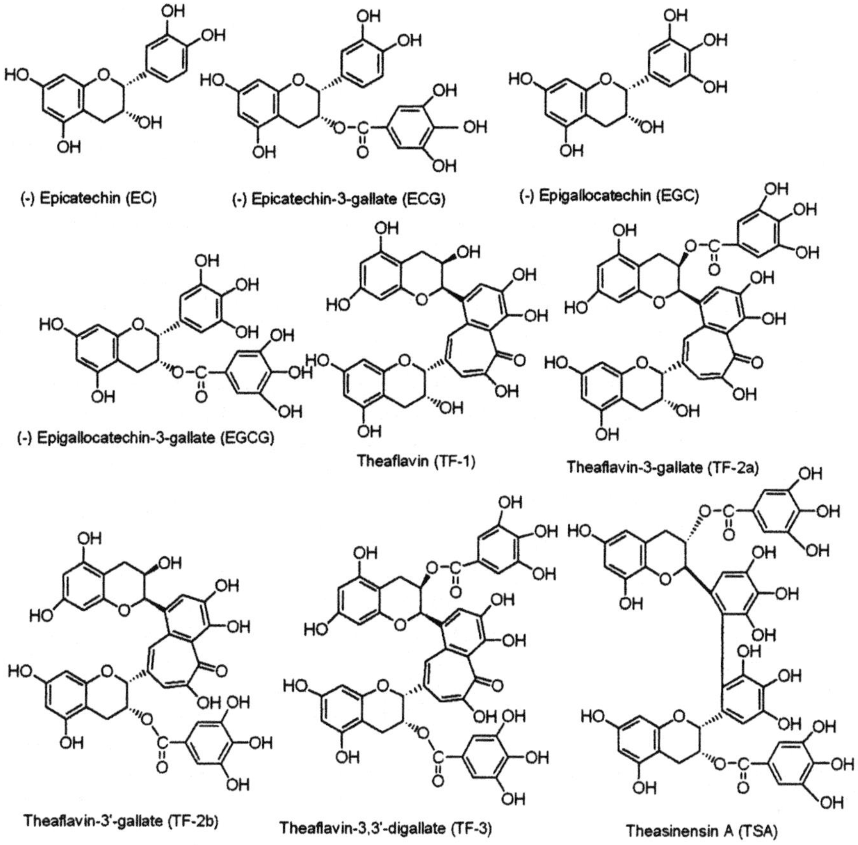

**Figure 1** Chemical structures of tea polyphenols.

extract of puerh tea is active in inhibiting lipopolysaccharide (LPS)-induced inducible nitric oxide synthase activity [5].

## III. BIOCHEMICAL AND MOLECULAR MECHANISMS

Tea polyphenols exhibit a variety of biological properties, including antioxidative effects [6], inhibition of extracellular mitotic signals [7], inhibition of cell cycle at the G1 phase [8], suppression of inducible nitric oxide synthase (iNOS) [7,9], and induction of apoptosis in cancer cells [10]. The natural history of carcinogenesis and cancer provides a strong rationale for a preventive approach to the control of this disease and leads to considerations of the possibility of active pharmacological intervention, or chemoprevention, to arrest or reverse the carcinogenesis prior to invasion and metastasis [11,12].

The inhibitory effects of tea against carcinogenesis have been attributed to the biological activities of its polyphenolics; however, the molecular mechanisms of cancer chemoprevention by tea extract are not fully elucidated. The results of some recent studies that throw light on this important area are discussed below.

## A.   Antioxidative Effects and Scavenging of Reactive Oxygen Species

Tea polyphenols show profound antioxidative effects in various systems; they are strong scavengers of superoxide, hydrogen peroxide, hydroxyl radicals, nitric oxide, and peroxynitrite, the products of various chemical reactions and biological systems.

Theaflavins and EGCG inhibit xanthine oxidase (XO) to produce uric acid and also act as scavengers of superoxide [6]. Theaflavin-3,3′-digallate (TF-3) (Fig. 1) acts as a competitive inhibitor and is the most potent inhibitor of XO among the compounds examined; it also inhibited superoxide production in HL-60 cells. It may be concluded that the antioxidative activity of tea polyphenols is due not only to their ability to scavenge superoxides, but also to their ability to block XO and associated oxidative signal transducers [6].

## B.   Suppression of Tumor Proliferation and Mitogenic Signal Transduction

Tea polyphenols are known to inhibit a broad range of enzymes associated with cell proliferation and tumor progression. Liang et al. [13] have investigated the effects of EGCG on the proliferation of human epidermoid cell line A431. In experiments using a tritiated thymidine incorporation assay, EGCG was found to significantly inhibit the DNA synthesis of A431 cells, to strongly inhibit the protein tyrosine kinase activities of EGF-R, PDGF-R, and FGF-R in vitro, and to exhibit an $IC_{50}$ value of 0.5–1 μg/mL. In contrast, EGCG produced only a minor inhibition of protein kinase activities in pp60[src], PKC, and PKA.

## C.   Inhibition of MAPK Signaling

Overexpression of the transcription factors AP-1 and NF-κB has been identified as a key feature in some carcinogenic pathways, including UVB-induced skin tumorigenesis; this enhanced activity can result from activation of one or more mitogen-activated protein kinase (MAPK) pathways [14]. EGCG and theaflavins have both been shown to inhibit TPA- and EGF-induced transformation of JB6 mouse epidermal cells in a dose-dependent manner. This inhibition was shown to correlate with decreased JNK activation, leading to an inhibition of AP-1 binding to its recognition site.

EGCG and TF-3 affect numerous events in the Ras-MAP kinase signaling pathway [15]. Treatment of 30.7b Ras 12 (Ras-transformed mouse epidermal cells) with 20 μM of either of these polyphenols resulted in decreased levels of phosphorylated Erk1/2 and MEK 1/2. EGCG inhibits the association between Raf-1 and MEK-1, while TF-3, but not EGCG, enhances degradation of Raf-1. In addition to inhibiting the phosphorylation of Erk1/2, both TF-3 and EGCG can also directly inhibit the kinase activity of this protein by competing for access the enzyme active site with its substrate, ELK-1 [15].

Exposure of normal human epidermal keratinocytes (NHEK) to UVB radiation induces intracellular release of hydrogen peroxide (oxidative stress) and phosphorylation of MAPK cell-signaling pathways. Pretreatment of NHEK with EGCG inhibits UVB-induced hydrogen peroxide production and its mediated phosphorylation of MAPK signaling pathways [16]. It has been demonstrated that tea polyphenols inhibited PKC, MAPK, and AP-1 activities in NIH 3T3 cells [17].

## D.  Inhibition of PI3K Pathway

Among the signal transduction pathways that control the high proliferation of cancer cells, the PI3K pathway is one of the most important; the overexpression of EGFR/Ras/PI3K/AKT/GS3K or SP-1 signaling pathway has been demonstrated in several malignant cancers and transformed cell lines. The inhibitory effects of tea polyphenols on UVB-induced phosphatidylinositol-3-kinase (PI3K) activation has been demonstrated in mouse epidermal JB6 C1 41 cells [18]. Pretreatment of cells with EGCG and TF-3 inhibited UVB-induced PI3K activation. Furthermore, UVB-induced activation of the PI3K downstream effectors, Akt and ribosomal p70S6 kinase ($p^{70}$S6-K), were also attenuated by these tea polyphenols. In addition to LY294002, a PI3K inhibitor, pretreatment with MAP-ERK kinase 1 inhibitor, U0126, or a specific p38 kinase inhibitor, SB202190, blocked UVB-induced activation of both Akt and $p^{70}$S6-K. It is noteworthy that UVB-induced $p^{70}$S6-K activation was directly blocked by the addition of EGCG or TF-3, whereas these polyphenols showed only a weak inhibition of UVB-induced Akt activation [18].

## E.  Inhibition of Cell Cycle Progression

The effects of EGCG and other catechins on cell cycle progression have been reported [19]. Studies using DNA flow cytometric analysis indicated that EGCG was able to block cell cycle progression at the G1 phase in asynchronous MCF-

7 cells; over a 24-hour exposure to EGCG, the Rb protein changed from its hyper- to its hypophosphorylated form, and G1 arrest developed. Under the same conditions, protein expression of cyclins D1 and E were both reduced slightly. Immunocomplex kinase experiments showed that EGCG inhibited the activities of cyclin-dependent kinase 2 (Cdk-2) and 4 (Cdk-4) in the cell-free system in a dose-dependent manner; when the cells were exposed to EGCG (30 μM) over 24 hours, a gradual loss of both Cdk2 and Cdk4 kinase activities was observed. EGCG also induced expression of the Cdk inhibitor p21, an effect that correlated with an increase in p53 levels; the level of p21 mRNA also increased under the same conditions. Within 6 hours of EGCG treatment, the expression of the Cdk inhibitor p27 protein was increased. These results suggest that EGCG exerts its growth-inhibitory effects either through modulation of the activities of several key G1 regulatory proteins such as Cdk2 and Cdk4 or via induction of Cdk inhibitors p21 and p27.

## F. Antitumor and Anti-inflammation Effects through iNOS Suppression

The effects of tea polyphenols on iNOS in thioglycolate-elicited and LPS-activated peritoneal macrophages have been studied [7,9]. Gallic acid, EGC, EGCG, TF-1, TF-2, and TF-3 were found to inhibit nitrite production, iNOS protein, and mRNA in activated macrophages. Reverse transcription polymerase chain reaction (RT-PCR) and Western and Northern blot analyses demonstrated that, compared with controls, significantly reduced 130 kDa protein and 4.5 kb mRNA levels of iNOS were expressed in LPS-activated macrophages treated with EGCG or theaflavins. Electrophoretic mobility shift assay indicated that EGCG blocked the activation of NF-κB, a transcription factor necessary for iNOS induction; this flavonoid and the theaflavins also blocked the disappearance of inhibitor IκB from the cytosolic fraction. These results suggest a mechanism of action involving the reduction in expression of iNOS mRNA, possibly occurring via prevention of NF-κB binding to the iNOS promoter, thereby inhibiting the induction of iNOS transcription [9].

## G. Suppression of VCAM and ICAM through Inhibiting IκB Kinase

Expression of vascular cell adhesion molecule-1 (VCAM-1) and intercellular adhesion molecule-1 (ICAM-1) is known to be elevated at sites of inflammation. Studies have been conducted into the effects of EGCG and TF-3 on the expression of these adhesion molecules induced by interleukin-1β (IL-1β) in cultured human umbilical vein endothelial cells (HUVECs) [20]. Both compounds significantly inhibited IL-1β-induced protein expression of VCAM and ICAM in dose-dependent manners and were associated with reduced adhesion of leukocytes to HU-VECs. The mRNA level of VCAM-1 was also inhibited by these tea polyphenolics, as was the NF-κB–dependent transcriptional activity induced by IL-1β. It is concluded that these molecules exhibited anti-inflammatory and anti-invasion properties, probably via a route involving blockage of IκB kinase.

## H.  Suppression of NFκB Activation through Downregulating IκB Kinase

The inhibition of IκB kinase (IKK) activity in LPS-activated murine macrophages (RAW 264.7 cell line) by various polyphenols, including EGCG and theaflavins, has been described by Pan et al. [21]. TF-3 inhibited IKK activity more strongly than the other polyphenols; it strongly inhibited both IKK1 and IKK2 activities and prevented degradation of IκBα and IκBβ in activated macrophage cells. These data suggest that the inhibition of IKK activity by TF-3 and other tea polyphenols could occur by direct effect on IKKs or on upstream events in the signal transduction pathway. Furthermore, TF-3 blocked phosphorylation of IκB from the cytosolic fraction and inhibited both NF-κB activity and increases in iNOS levels in activated macrophages. TF-3 and other polyphenols in tea may, therefore, exert their anti-inflammatory and cancer chemoprevention actions by suppressing the activation of NF-κB through inhibition of IKK activity [21].

## I.  Inhibition of Topoisomerase I and Proteasome Activity

EGCG selectively inhibits the activity of topoisomerase I but not topoisomerase II in human colon cancer cell lines [22]; the doses necessary for this inhibition (10–17 μM) are lower than those required for inhibition of cell growth (IC$_{50}$ = 10–90 μM). EGCG has been shown to inhibit the chymotryptic activity of the 20s proteasome in leukemic, breast cancer, and prostate cancer cell lines [23], leading to an accumulation of p27$^{kip1}$ and IκB, and subsequent cell cycle arrest and inhibition of NF-κB activity, respectively.

## J.  Inhibition of Matrix Metalloproteinase

It has been demonstrated that EGCG can inhibit the matrix metalloproteinase (MMP) activity at a relatively low doses [24]. EGCG inhibits the activity of secreted MMP-2 and MMP-9 at concentrations of 8–13 μM. Meanwhile, 1 μM EGCG was found to increase the expression of TIMP-1 and TIMP-2, proteins that inhibit the activity of activated MMPs. Because such concentrations of tea polyphenols are physiologically significant, this is an attractive mechanism for the observed anti-invasive and antiangiogenic activities of EGCG [25] and other tea polyphenols in vivo. Recently, similar observations on the inhibition of tea polyphenols on the activities of MMPs have been made; TF-3 appears to be the most active of the compounds examined in this respect.

## K. Induction of Apoptosis in Cancer Cells by Tea Polyphenols

Theasinensin A, theaflavin, and theaflavin-3-gallate (TF-1 and TF-2a,b) displayed strong growth inhibitory effects against human histolytic lymphoma U937 ($IC_{50} = 12$ $\mu$M) but were less effective against human acute T-cell leukemia Jurkat, whereas TF-3 and EGCG had lower activities [21]. The mechanisms by which tea polyphenols induce apoptosis were further studied by annexin V apoptosis assay, DNA fragmentation, and caspase activation. Treatment with tea polyphenols caused rapid induction in activity of caspase-3, but not caspase-1, and enhanced proteolytic cleavage of poly(ADP-ribose)-polymerase (PARP). Pretreatment with a potent caspase-3 inhibitor, Z-Asp-Glu-Dal-Asp-fluoromethylketone, inhibited theasinensin A–induced DNA fragmemtation. Experiments using flow cytometry showed that theasinensin A induced a loss of mitochondrial potential, elevation of ROS production, release of mitochondrial cytochrome $c$ into the cytosol, and subsequently induction of caspase-9 activity. Further data indicate that theasinensin A is effective in inducing the degradation of DFF-45 (an inhibitor binding to Dnase), which allows caspase-activated Dnase to enter the nucleus and degrade chromosomal DNA. As a consequence, it has been suggested that induction of apoptosis by theasinensin A and other tea polyphenols affords a pivotal mechanism for their cancer-chemopreventive function [10]; a commentary on the cancer chemoprevention by tea polyphenols through a blockage of mitotic signal transduction has been critically elaborated by Lin et al. [7] and Lin and Liang [26].

## L. Suppression of Fatty Acid Synthase through Downregulating EGFR and the PI3k-AKt Signal Transduction Pathway

Results have recently been published of a long-term feeding trial examining the effect of green tea leaves on levels of cholesterol, lipid, antioxidant, and phase II enzymes in Wistar rats [27]. These indicate that such feeding can reduce total cholesterol, triglyceride, and LDL cholesterol and enhance activities of superoxide dismutase (SOD) in serum and glutathione S-transferase (GST) and catalase in the rat liver [27]. At the 15th week, the average body weights of experimental and control groups were 449 and 510 g, respectively, indicating that oral feeding of green tea leaves resulted in a significant (12%) decrease in body weight ($p < 0.05$). The dose of green tea leaves used in the study did not reduce diet or water consumption throughout the feeding regimen and survival ratios of both groups were 100% (12/12) [27].

The key enzyme for lipogenesis fatty acid synthase (FAS) has been shown to be significantly suppressed by tea extracts and tea polyphenols, including EGCG and TF-3 in human breast carcinoma cells; expression of FAS is enhanced by EGF and inhibited by the presence of tea polyphenols. These experimental data indicate that suppression of FAS may result from downregulation of EGF receptor/PI3k/Akt signal transduction pathway.

## IV. GENERAL DISCUSSION OF THE MECHANISMS OF CANCER CHEMOPREVENTION

It has been recognized that deregulated proliferation and inhibition of apoptosis lie at the heart of all tumor development, and these loci provide two obvious targets for general therapeutic and chemopreventive interventions. Clearly there are numerous mechanisms through which these two defects can occur, and the success of targeted therapy and chemoprevention will depend to a very considerable degree on the molecular fingerprinting of individual tumors [28].

Most receptor tyrosine kinases (RTKs) transduce key intracellular signals that trigger cellular events, such as mitosis and cytoskeletal rearrangement, and serve to orchestrate physiological processes, such as development, wound repair, and oncogenesis. Ligand binding of RTKs mediates these responses by activating a variety of intracellular signaling pathways through an intrinsic kinase activity. It appears that these diverse signaling pathways, activated by growth factor receptors, induce broadly overlapping, rather than independent, sets of genes [29]. RTKs are important regulators of intracellular signal transduction pathways mediating development and multicellular communication in metazoans. Their activity is normally tightly controlled and regulated, and perturbation of RTK signaling by mutation and other genetic alterations leads to deregulated kinase activity and malignant transformation [30].

Several lines of evidence have demonstrated that signal transduction events leading to the activation of the MAPK pathways including ERK, JNK, and P38, NF-κB, and JAK-STAT pathways can result in cell proliferative, survival, differentiating, and apoptotic responses. Figure 2 illustrates the key mechanisms that lead to inhibition of survival gene expression (*c-jun*, *c-fos*, *c-myc*, etc.) and activation of apoptotic signal pathways (caspase 8 and caspase 9 cascades). Two important signaling events, namely the MAPK, NIK (NF-κB–inducing kinase), and caspase cascades (ICE/ced 3 family proteases) pathways have been highlighted. Most tea polyphenols suppress the MAPK and NIK pathways [21,31]

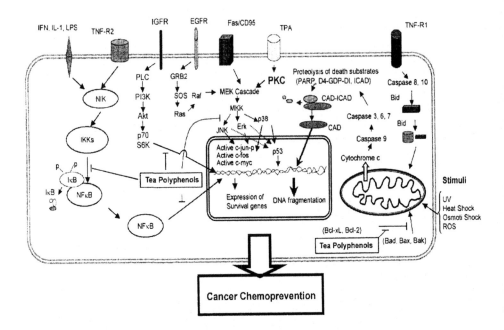

**Figure 2** Proposed biochemical and molecular mechanisms of cancer chemoprevention by tea polyphenols. Several lines of results indicate that antioxidant effects and signal blockades have played important roles in the modulation of cancer chemoprevention by tea and tea polyphenolics. Cell proliferation and differentiation are strictly regulated by programmed networks of intracellular signal transduction pathways through various transducers, including intrinsic factors such as receptor proteins, RTK, PKA, PKB, PKC, MAPK, NFκB, AP-1, *c-jun*, *c-fos*, *c-myc*, iNOS, and ROS and extrinsic factors such as cytokines, growth factors, tumor promoters, TNF, and LPS. The illegitimate regulation or hyperfunction of such signaling processes may lead to induction of carcinogenesis or inflammation. Tea polyphenols were found to suppress the hyperfunction of signaling in various systems that may block carcinogenesis and afford cancer chemoprevention. (From Ref. 38).

but activate the caspase cascade pathways [9]; such a combination of effects will potentiately lead to apoptotic response in the target cells.

Most tea polyphenols with cancer chemopreventive activities possess antioxidant activity [6,32]; it should be emphasized that in addition to acting as ROS scavengers, these compounds can act through multiple mechanisms to modulate the functions of receptors, effectors, protein kinases, protein phosphatases, and protein substrates in the mitogenic and differentiating signaling involved in tumor promotion.

Although most existing cancer drugs are antimitotic, they act by crudely interfering with the basic machinery of DNA synthesis and cell division rather than by targeting the specific lesions responsible for regulated tumor growth. It is now also clear that the surprising selectivity of such crude agents is mainly a result of increased sensitivity to apoptosis afforded to tumor cells by their oncogenic lesions [33,34]. Drugs designed to specifically inhibit growth-deregulating lesions are currently being tested in clinical trials and include inhibitors of RTKs, Ras, downstream signaling kinases such as the MAPK and Akt pathway, and CDKs [35].

## V. CONCLUSION

Hundreds of teas are now produced and are generally classified into nonfermented green tea, partially fermented oolong or paochong tea, and fully fermented black and puerh teas. It has been demonstrated that both green and black teas are equally effective in cancer chemoprevention [36], and the biological functions of both teas have been attributed to their polyphenols, including catechins and theaflavins. Tea polyphenols exhibit a wide variety of biological properties, including antioxidative effects, inhibition of extracellular mitotic signals through blocking the growth receptor signalings, inhibition of cell cycle at G1 phase through cyclin-dependent kinase suppression, suppression of iNOS through inhibiting the activation of IκB kinase (IKK) and NF-κB and induction of apoptosis in cancer cells through releasing cyctochrome $c$ and activating caspase cascades. Recently, tea polyphenols including catechins (ECG, EGCG) and theaflavins (TF-3) have been shown to inhibit IL-1β–induced IKK activity and thus the nuclear translocation of NF-κB leading to the suppression of VCAM-1 and ICAM-1. These findings have provided additional mechanisms for the anti-inflammatory effect of tea polyphenols. It is significant that tea polyphenols suppress the fatty acid synthase (FAS) in human breast carcinoma MCF-7 cells, since several FAS inhibitors have been shown to be potent antitumor agents and FAS is the key enzyme for controlling lipogenesis and body weight gain. Recent results indicate that tea and tea polyphenols may induce hypolipidemic and antiobesity effects through suppressing FAS expression.

In summary, cancer-chemopreventive agents can inhibit tumor growth through arresting cell cycle and inducing cellular apoptosis. During the past few years, experimental results from our laboratory and elsewhere have demonstrated that cancer chemoprevention by tea polyphenols can be achieved by signal transduction blockade [37,38]. Discovering novel therapeutics and chemopreventive agents with clinical utility continues to be the focus of biochemical and pharmacological scientists working in the signal transduction therapy. Developing compounds designed to manipulate kinase pathways and signal events through both inhibitory and stimulatory methods for treating cancer and other diseases offer promising trends for biomedical research.

## ACKNOWLEDGMENTS

Parts of this paper have been presented at the Joint Shizuoka Symposium on Tea and Health Science, Shizuoka, Japan, February 14–15, 2003, and at the 4th International Conference on Phytochemicals, Panoma, CA, October 20–21, 2002. This study was supported by the National Science Council NSC 91-23-20-B-002-068 and NSC91-2311-B-002-037, and by the National Health Research Institute NHRI-EX91-8913BL.

## REFERENCES

1. Lin YL, Juan IM, Chen YL, Liang YC, Lin JK. Composition of polyphenols in fresh tea leaves and association of their oxygen radical-absorbing capacity with antiproliferative actions in fibroblast cells. J Agric Food Chem 1996; 44:1387–1394.
2. James IE. Caffeine and Health. San Diego. CA: Academic Press Inc., 1991:6–10.
3. Harler CR. The Culture and Marketing of Tea. London: Oxford University Press, 1958:61–66.
4. Segall S. Comparing coffee and tea. In: Caffeinated Beverages, Health Benefits, Physiological Effects and Chemistry. Washington. DC: American Chemical Society, 2000:20–28.
5. Lin YS, Tsai Y J, Tsay JS, Lin JK. Factors affecting the levels of tea polyphenols and caffeine in tea leaves. J Agric Food Chem 2003; 51:1864–1873.
6. Lin JK, Chen PC, Ho CT, Lin-Shiau SY. Inhibition of xanthine oxidase and suppression of intracellular reactive oxygen species in HL-60 cells by theaflavin-3,3'-digallate, (−)-epigallocatechin-3-gallate and propyl gallate. J Agric Food Chem 2000; 48:2736–2743.
7. Lin JK, Liang YC, Lin YL, Ho CT. Cancer prevention of tea: Biochemical mechanisms. In: Parliament TH Ho CT, Schieberle P, Eds. Caffeinated Beverages, Health Benefits, Physiological Effects and Chemistry. Washington. DC: American Chemical Society, 2000:78–86.
8. Liang YC, Chen C F, Lin YL, Lin-Shiau SY, Ho CT, Lin JK. Suppression of extracellular signals and cell proliferation by black tea polyphenol, theaflavin-3,3'-digallate. Carcinogenesis 1999; 20:733–736.
9. Lin YL, Lin JK. (−)-Epigallocatechin-3-gallate blocks the induction of nitric oxide synthase by down-regulating LPS-induced activity of NF-κB. Mol Pharmacol 1997; 52:465–472.

10. Pan MH, Liang YC, Lin-Shiau SY, Zhu NQ, Ho CT, Lin JK. Induction of apoptosis by the oolong tea polyphenol theasinensin A through cytochrome c release and activation of caspase-9 and caspase-3 in human U-937 cells. J Agri Food Chem 2000; 48:6337–6346.

11. Wattenberg Lee W. Chemoprevention of cancer. Cancer Res 1985; 45:1–8.

12. Morse MA, Stoner GD. Chemoprevention of cancer: principle and prospects. Carcinogenesis 1993; 14:1737–1746.

13. Liang YC, Lin-Shiau SY, Chen CF, Lin JK. Suppression of extracellular signals and cell proliferation through EGF receptor binding by ( − )-epigallocatechin-3-gallate in human A431 epidermoid carcinoma cells. J Cell Biochem 1997; 67:55–65.

14. Dong Z. Effects of food factors on signal transduction pathways. BioFactors 2000; 12:17–28.

15. Chung JY, Park JO, Phyu H, Dong Z, Yang CS. Mechanisms of inhibition of the Ras-MAP kinase signaling pathway in 30.7b Ras12 cells by tea polyphenol ( − )-epigallocatechin-3-gallate and theaflavin-3,3′-digallate. FASEB J 2001; 15: 2022–2024.

16. Katiyar SK, Afaq F, Azizuddin K, Mukhtou A. Inhibition of UVB-induced oxidative stress-mediated phosphorylation of mitogen-activated green tea polyphenol ( − )-epigallocatechin-3-gallate. Toxicol Appl Pharmacol 2001; 176:110–117.

17. Chen YC, Liang YC, Lin-Shiau SY, Ho CT, Lin JK. Inhibition of TPA-induced protein kinase C and transcription activator protein-1 binding activities by theaflavin-3,3′-digallate from black tea in NIH3T3 cells. J Agric Food Chem 1999; 47: 1415–1421.

18. Nomura M, Kaji A, He Z, Ma WY, Miyamoto K, Yang CS, Dong Z. Inhibitory mechanisms of tea polyphenols on the UVB-activated phosphatidyl-inositol 3-kinase dependent pathway. J Biol Chem 2001; 276:46624–46631.

19. Liang YC, Lin-Shiau SY, Chen CF, Lin JK. Inhibition of cyclin-dependent kinases 2 and 4 activities as well as induction of Cdk inhibitors p21 and p27 during growth arrest of human breast carcinoma cells by ( − )-epigallocatechin-3-gallate. J Cell Biochem 1999; 75:1–12.

20. Ho YS, Lin JK, Hung LF, Lin SY, Pan S, Liang YC. Suppression of leukocyte adhesion molecules through inhibition of IκB kinase by tea polyphenols in human vascular endothelial cells. New Taipei J Med 2002; 4:249–254.

21. Pan MH, Lin-Shiau SY, Ho CT, Lin JH, Lin JK. Suppression of LPS-induced NF-κB activity by theaflavin-3,3′-digallate from black tea and other polyphenols through down-regulation of IKK kinase activity in macrophages. Biochem Pharmacol 2000; 59:357–367.

22. Berger SJ, Gupta S, Belfi CA, Gosky DM, Mukhtar H. Green tea constituent ( − )-epigallocatechin-3-3′-gallate inhibits topoisomerase I activity in human colon carcinoma cells. Biochem Biophys Res Commun 2001; 288:101–105.

23. Nam S, Smith DM, Dou QP. Ester bond-containing tea polyphenols potently inhibit proteasome activity in vitro and in vivo. J Biol Chem 2001; 276:13322–13330.

24. Garbisa S, Sartor L, Biggin S, Salvato B, Benelli R, Albini A. Tumor gelatinases and invasion inhibited by the green tea flavanol epigallocatechin-3-gallate. Cancer 2001; 91:822–832.

25. Yang CS, Maliakai P, Meng X. Inhibition of carcinogenesis by tea. Annu Rev Pharmocol Toxicol 2002; 42:25–54.

26. Lin JK, Liang YC. Cancer chemoprevention by tea polyphenols. Proc Natl Sci Council, ROC (B) 2000; 24:1–13.

27. Lin YL, Cheng CY, Lin YP, Lau YN, Juan IM, Lin JK. Hypolipidemic effect of green tea leaves through induction of antioxidant and phase II enzymes including superoxide dismutase, catalase and glutathione S-transferase. J Agric Food Chem 1998; 46:1893–1899.

28. Evan GI, Vousden KH. Proliferation, cell cycle and apoptosis in cancer. Nature 2001; 411:342–348.

29. Fambrough D, McClure K, Kazlauskas A, Lander ES. Diverse signaling pathways activated by growth factor receptors induced broadly overlapping, rather than independent, sets of genes. Cell 1999; 97:727–741.

30. Blume-Jensen P, Hunter T. Oncogenic kinase signaling. Nature 2001; 411:355–365.

31. Chung JY, Park JO, Phyu H, Dong Z, Yang CS. Mechanisms of inhibition of the Ras-MAP kinase signaling pathway in 30.7b Ras 12 cells by ( − )-epigallocatechin 3-gallate and theaflavin-3,3′-digallate. FASEB J 2001; 15:2022–2024.

32. Chen CW, Ho CT. Antioxidant properties of polyphenols extracted from green and black teas. J Food Lipid 2001; 2:35–46.

33. Schmitt CA, Lowe SW. Apoptosis and therapy. J Pathol 1999; 187:127–137.

34. Evan G, Littlewood TA. A matter of life and cell death. Science 1998; 281:1317–1322.

35. Gibbs JB. Mechanism-based target identification and drug discovery in cancer research. Science 2000; 287:1969–1973.

36. Wang ZY, Huang MT, Lou YR, Xie JG, Reuhl KR, Newmark HL, Ho CT, Yang CS, Conney AH. Inhibitory effects of black tea, green tea, decaffeinated black tea, and decaffeinated green tea on ultraviolet B light-induced skin carcinogenesis in 7,12-dimethylbenz[a]anthracene-initiated SKH-1 mice. Cancer Res 1994; 54:3428–3435.

37. Lin JK, Liang YC, Lin-Shiau SY. Cancer chemoprevention by tea polyphenols through mitotic signal transduction blockade (a commentary). Biochem Pharmacol 1999; 58:911–915.

38. Lin JK. Cancer chemoprevention by tea polyphenols through modulating signal transduction pathways. Archiv Pharma Res 2002; 25:561–571.

# 11

# The Role of Flavonoids in Protection Against Cataract

**Julie Sanderson**
*University of East Anglia, Norwich, Norfolk, United Kingdom*

**W. Russell McLauchlan**
*Institute of Food Research, Colney, Norwich, United Kingdom*

## I. INTRODUCTION

It has been recognized for many years that oxidative stress plays a significant role in the etiology of cataract and that dietary antioxidants may play an equally important role in prevention of this common degenerative disease [1,2]. However, in this respect the flavonoids have received surprisingly little interest, and they are notably absent from recent reviews of nutrition and eye disease [2,3]. Epidemiological research has established that diets rich in fruit and vegetables are associated with a reduced risk of cataract [4,5]. However, research spanning epidemiological studies, in vivo and in vitro laboratory investigations, and, latterly, human intervention trials have focused largely on the antioxidant vitamins, with little attention being given to other potentially beneficial micronutrients [2–8]. In this chapter, epidemiological data relating to flavonoids and cataract will be presented and in vivo and in vitro studies investigating the role of flavonoids in prevention of cararact discussed; consideration will also be given to flavonoid metabolism in the lens and the bioavailability of flavonoids to the eye. Initially, a brief overview of the lens and cataract will be presented (a detailed review can be found in Ref. 9).

## II. THE LENS AND CATARACT

The function of the lens of the eye is to focus incoming light onto the retina. The lens is a highly specialized organ with unique features to enable transparency.

It is avascular and noninnervated and composed of just two cell types, epithelial cells and fiber cells, surrounded by a collagenous capsule (Fig. 1). The epithelial cells form a monolayer on the anterior face of the lens; they divide and elongate at the equatorial region of the lens to form the fiber cells. During the terminal differentiation process all intracellular organelles, including nuclei and the mitochondia, are lost. The fiber cells are highly ordered and densely packed elongated cells arranged in concentric shells with minimal extracellular space. The lens continues to grow throughout life, with the newly differentiated fiber cells forming the outermost layers of the lens (lens cortex) displacing the older fiber cells towards the center of the lens. The very oldest fiber cells, present as the embryonic lens, remain at the center (nucleus) of the organ. The fiber cells contain high concentrations of specialized proteins called crystallins, which constitute 90% of the water-soluble protein in the lens. The highly ordered spacial arrangement of the crystallins is a further specialization to enable transparency.

In cataract the transparency of the lens is lost and the condition is the major cause of blindness worldwide. Cataract surgery is among the most commonly performed operation in the United Kingdom, with over 100,000 extractions taking place annually. The incidence of the most common type of cataract, termed senile cataract, is correlated with increased age, and, therefore, the numbers suffering impaired vision as a result this disease is predicted to increase in coming decades. Reduced lens transparency is a result of loss of structure at the cellular and sub-cellular level; loss of cell-cell contact causes the formation of extracellular vacuoles within the lens and aggregation of crystallins to form large insoluble complexes occurs within the fiber cells. Both changes cause scattering of incident light. Any stress that leads to a disruption of normal lens physiology and biochemistry is potentially cataractogenic. A major stress to which the lens is continually exposed is oxidation, and oxidative mechanisms play an important role in senile cataractogenesis [1]. In addition to the generation of free radicals via oxidative

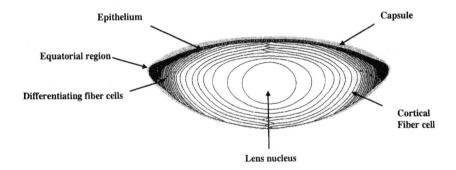

**Figure 1**  Diagram of the adult human lens in cross-section.

metabolism, reactive oxygen species are also produced via photooxidative mechanisms in the lens and aqueous humor. Hydrogen peroxide has been measured at concentrations of 20–30 $\mu$M in normal aqueous humour [10], but may be as high as 660 $\mu$M in the cataractous eye [10].

The lens has considerable protection against oxidative damage; enzymes of the glutathione redox cycle are expressed [11,12] and the lens contains high concentrations of glutathione (6–10 mM) [11]. Superoxide dismutase [12,13], catalase [12], glutathione-S-transferase [14], and thioltransferase [15] activities have all been demonstrated in the human lens. Diurnal animals also have high concentrations of ascorbate in the aqueous humour and lens [16]. Ascorbate is actively transported into the lens: mean concentrations of approximately 1.5 and 3.7 mM are reported for human aqueous humour and lens, respectively, compared with 0.07 mM in plasma [17]. However, analysis of lenses with cataract indicates that during the process of cataract formation, these oxidative defenses have provided insufficient protection and oxidative damage has occurred. In human lenses with cataract, oxidation products of both protein and lipid have been identified [18–20] and loss of glutathione is observed [21]. Oxidative mechanisms of cataractogenesis and the role of antioxidants in prevention of the disease have received a great deal of attention. It has been estimated that delaying cataract formation by 10 years would reduce the need for surgical intervention by approximately 45% [22]. Strategies to achieve this via dietary or pharmacological intervention are hence of considerable interest. Evidence that the flavonoids may have a key role to play will be discussed in the remainder of the chapter.

## III. EPIDEMIOLOGY

There is substantial epidemiological evidence that dietary intakes rich in fruit and vegetables can lower the risk of developing long-term degenerative diseases such as cancer [23,24], cardiovascular disease [25–27], and cataract [4,5]. In an American study [4], a significant increase in risk of cataract was found when average daily consumption of fruit was less than 1.5 servings, daily vegetable consumption was less than 2.0 servings, and daily combined fruit and vegetable consumption was less than 3.5 servings. The risk was greatest for posterior subcapsular cataract across all three of the above intakes; however, there was also an elevated risk of cortical cataract with low combined fruit and vegetable intake. In an Italian case-control study [5], the role played by particular vegetables in protecting against cataract was identified. Odds ratios suggesting a significant reduction in risk were found for cruciferous vegetables, tomatoes, citrus fruits and melon ($p < 0.01$), and also for spinach and peppers ($p < 0.05$). In a Canadian case-control study it was shown that tea, a major source of dietary flavonoids, was protective against cataract. Patients with lenticular opacities were less likely to consume five or more cups of tea per day [28].

A great deal of effort has been directed towards dissecting this association between high intakes of fruit and vegetables and the reduced risk of degenerative disease in order to obtain epidemiological evidence that identifies the individual protective components responsible. For compounds restricted to certain dietary sources, such as the glucosinolates found in cruciferous vegetables, the evidence linking diets rich in these compounds to reduced risk of cancers of the gastrointestinal tract, for example, is strong [29]. This may, in part, be due to the ability of volunteers to reliably recall their intake of these vegetables in food questionnaires, thereby facilitating comparison of large cohorts with low and high intakes. In contrast, for a group of compounds such as flavonoids, which are of ubiquitous occurrence within the plant kingdom [30] and the human diet [31], the gathering of such unequivocal data has been much more difficult with studies, producing contradictory results for cancer [32–34], cardiovascular disease, and stroke [35–43].

In contrast to epidemiological studies of these diseases, there has been just one study that compared flavonoid intake and cataract [44]. In this Finnish study the dietary intakes of over 10,000 men and women were estimated for the year preceding baseline health examination. Health outcomes were determined 28–30 years later using Finnish government health statistics and compared with the highest and lowest intakes. Neither higher total flavonoid intake nor higher intakes of quercetin, kaempferol, myricetin, hesperetin, or naringenin were associated with reduced risk of cataract. However when apple intake—the main source of flavonoids in this study—was adjusted for intake of vegetables and fruit other than apples, a trend towards lower risk was observed but this was not statistically significant.

It is impossible to draw conclusions regarding the role of flavonoids in cataract prevention from a single epidemiological study. The Finnish study did not include tea and red wine, two major sources of dietary flavonoids in most European diets, because their contribution to the Finnish diet was small at the time that the baseline measurements were recorded. In general, northern Europeans eat less fruit and vegetables than, for example, the Greeks or the Spanish. Data on the consumption of vegetables in 10 European countries [45] shows that in 1990 vegetable consumption in Greece was over 2.5 times greater than in Norway, so it would not be unreasonable to extrapolate flavonoid intake in a similar manner. It is vital, therefore, for future studies not only to produce more cataract epidemiology but to include reliable comparisons of flavonoid intake and cataract incidence across Europe.

If it can be shown that flavonoids do play a role in reducing the risk of cataract, they will only account for a proportion of the total effect, as many other plant constituents, such as vitamins and minerals, have also been shown to be effective. It should also be remembered that cataract is a multifactorial disease, with diet but one of many possible risk factors among, for example, poverty, UV radiation, smoking, diabetes, body mass index, age, sex, and hypertension [9], all of which may interact in a highly complex manner to facilitate development of disease.

## IV. LABORATORY STUDIES

There have been a number of laboratory studies investigating the potential protective role of flavonoids against cataract. The earliest studies were focused on diabetic cataract [46–48]. In the 1970s and into the 1980s there was considerable interest in the role of aldose reductase in diabetic cataract [49], and it was proposed that glucose was converted to sorbitol by the enzyme aldose reductase in the lenses of diabetic patients [50,51]. The accumulation of sorbitol, an osmolyte, then resulted in development of an osmotic cataract. It was, therefore, proposed that aldose reductase inhibitors would prevent the development of cataract in diabetic patients. It was found that some flavonoids were effective inhibitors of aldose reductase [46,52]; quercetin, quercitrin (quercetin-3-rhamnoside), and myricitrin (myricetin-3-rhamnoside) were particularly potent, being effective at submicromolar concentrations [46]. Quercetin and quercitrin have been investigated in in vivo animal models of diabetic cataract [47], both being shown to delay onset of cataract when administered either orally and topically. As other more potent and more stable aldose reductase inhibitors were identified, the focus of research moved away from the flavonoids, and it has subsequently become apparent that aldose reductase is unlikely to play a significant role in human diabetic cataract [9]. It has, however, been recognized that oxidative mechanisms are likely to be involved in formation of both diabetic and senile cataract [53], and the explosion of research on the flavonoids in relation to other age-related diseases has contributed towards renewed interest in the flavonoids as potential anticataract compounds.

There have been a number of papers published recently in relation to flavonoids and cataract. Extracts containing flavonoids (tea polyphenols and *Ginkgo biloba*) have been investigated in an experimental model in which cataract is induced in early postnatal rats (days 10–14) by subcutaneous injection of sodium selenite [54,55]. Lens opacification is observed within 48 hours, with dense nuclear cataract occurring at 3–5 days [56]. This selenite model has been used extensively both in the investigation of the mechanisms of cataractogenesis and as an in vivo assay for potential anticataract compounds. Administration of *Ginkgo biloba* extract Egb761 (50 mg) by daily IP injection retarded the development of cataract [55]. Extracts of green and black teas, given by daily IP injection from 2 days prior to selenite exposure, also reduced cataract formation [54,57].

In vitro, green tea polyphenols have also been reported to protect against hydrogen peroxide–, UVA-, UVB-, and x-ray–induced damage in cultured lens epithelial cells [58–60] and in organ cultured rat lenses exposed to selenite [57]. In addition, the flavonoid product venoruton (a mixture of mono-, di-, tri-, and tetrahydroxy-ethylrutosides) has been shown to inhibit loss of lens viability (assessed by LDH leakage) and transparency in an in vitro model of diabetic cataract in which rat lenses were cultured in high-glucose medium [61].

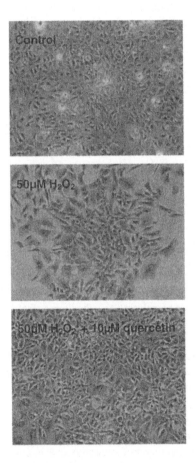

**Figure 2** Protection by quercetin of human lens epithelial cells against hydrogen peroxide–induced cell death. Confluent FHL124 cells, a human lens epithelial cell line, were exposed for 24 hours to hydrogen peroxide (50 μM) in the presence and absence of quercetin (10 μM). Images are presented of fixed cells stained with Coomassie blue.

Our research has focused on the flavonoid quercetin in relation to cataract. It has been shown that quercetin can protect against hydrogen peroxide–induced cell death in the human lens epithelial cell line (FHL124) (Fig. 2). In addition, use of an in vitro oxidative model of cataract, the lens organ culture with hydrogen peroxide (LOCH) model, has enabled direct assessment of both lens function and the particular mechanisms involved in protection of lens transparency [62]. Initial experiments showed that from a series of polyphenolic compounds, quercetin was the most potent in reducing loss of lens transparency. Lenses that have been treated with 500 μM hydrogen peroxide show cortical opacification, which is most pronounced in the equatorial region (Fig. 3). Increased $Ca^{2+}$ influx and loss of glutathione was also observed, both of which are considered to be early events in the aetiology of cataract [1,63]. It was found that quercetin was an inhibitor of $Ca^{2+}$ influx, 10 μM quercetin reducing the $H_2O_2$-induced $Ca^{2+}$ influx by approximately 90% [62]. Interestingly, levels of reduced glutathione were not protected by quercetin, which suggested that quercetin has multiple mechanisms of actions.

Further experiments were carried out comparing the relative protective action of quercetin, epicatechin, and chlorogenic acid. It was concluded that while the antioxidant activity of quercetin had a role to play in protection of the lens against oxidative damage, other mechanisms must also be contributing, since inhibition of opacification and $Ca^{2+}$ influx did not correlate with antioxidant activity. It was likely that protection involved direct inhibition of ion channels that were responsible for the accumulation of intracellular $Ca^{2+}$. Significantly, inhibition of $Ca^{2+}$ influx was recently reported to be one of three distinct mechanisms by which flavonoids protect of neuronal cells from oxidative cell death [64].

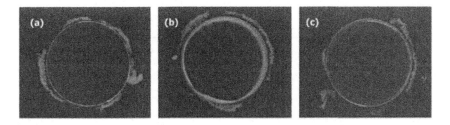

**Figure 3** Dark field images of rat lenses following incubation for 4 hours in (a) control medium, (b) 500 μM $H_2O_2$, or (c) 500 μM $H_2O_2$ + 10 μM quercetin. Note that $H_2O_2$ induces cortical light scatter in the equatorial regions of the lens, which is inhibited in the presence of quercetin.

## V.  FLAVONOID BIOAVAILABILITY AND METABOLISM IN THE LENS

Research on flavonoid metabolism in the lens has focused on quercetin [65]. Metabolism has been investigated in the rat and human lens; incubation of the rat lens in vitro with 10 μM quercetin shows maximal quercetin content between 1 and 2 hours (Fig. 4), declining to approximately 15% of this content between 2 and 6 hours. Over the 6-hour culture period an increase in isorhamnetin (3'-O-methylquercetin) was observed; methylation of quercetin was inhibited by the catechol-O-methyltransferase (COMT) inhibitor 3,5-dinitrocatechol (Fig. 5), and the presence of this enzyme in the rat lens has been confirmed by Western blot techniques. Significantly, over the 6-hour period, there was no evidence of conjugated metabolites such as glucuronides or sulfates. Similar experiments have been carried out using the intact human lens (Sanderson and Cornish, unpublished data); these reveal differences between the rat and the human lens. In the latter, two methylated metabolites are seen and in addition, quercetin and methylquercetin glucuronides are observed at 4 hours. Furthermore, these compounds can also be detected in the bathing medium, suggesting that there is efflux of the glucuronides from the lens following conjugation.

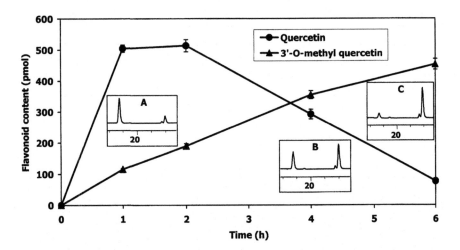

**Figure 4**   The conversion of quercetin to isorhamnetin (3'-O-methyl quercetin) in the whole rat lens. Lenses were incubated in 10 μM quercetin, and flavonoid content was analyzed by HPLC. Inserts show chromatogram peaks for quercetin (first peak, retention time 19 min) and isorhamnetin (second peak, retention time 21 min) at (A) 1 hour, (B) 4 hours, and (C) 6 hours. Peaks were identified by mass spectrometry (quercetin m/z 303; isorhamnetin m/z 317). Data are expressed as mean ± SEM, $n = 4$. (From Ref. 65.)

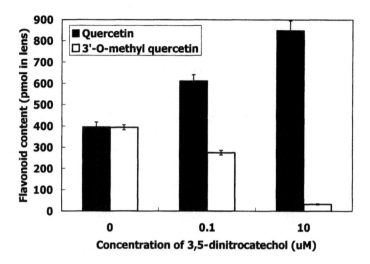

**Figure 5** Dose-dependent inhibition of the conversion of quercetin to isorhamnetin (3′-O-methyl quercetin) by the catechol-O-methyl transferase inhibitor 3,5-dinitrocatechol. Data are expressed as mean ± SEM, $n = 4$. (From Ref. 65.)

In relation to the bioavailability of flavonoids to the eye, evidence is largely circumstantial. Only one report tentatively identifies the presence of epigallocatechin gallate in the mouse lens following the drinking of green tea [60]. The efficacy of tea extracts and quercetin in reducing lens opacification in in vivo models suggests that flavonoids are secreted into the aqueous humour and taken up by the lens [47,54,57]. However, an effect of these compounds in reducing the cataractogenic stress extraocularly cannot be excluded.

There are other indications that flavonoids could be distributed to the lens. It is known, for example, that in humans fluroscein glucuronides can be transported across the ciliary epithelium into the aqueous humor [66]. It would, therefore, seem likely that circulating flavonoid glucuronides would cross the blood–aqueous humor barrier. In human aqueous humour, β-glucuronidase activity has been measured [67], suggesting that flavonoid glucuronides could be deglucuronidated, leaving the aglycones to diffuse into the lens. Uptake would then be dependent upon the lipid solubility of the specific flavonoid. Thus while there are indications that the lens could be exposed to flavonoids following dietary intake, specific experiments specifically addressing the bioavailability to the eye are a priority.

## VI.  CONCLUSIONS

There is increasing evidence from laboratory studies that dietary flavonoids may protect the human lens from cataract. These initial studies provide the basis for future experimental programs to understand the metabolism and cellular effects of this important class of dietary phytochemicals. The recent identification of the major circulating metabolites of quercetin in human plasma following ingestion of a flavonoid-rich meal [68] has facilitated study of their transport and metabolism across the ciliary and lens epithelial barriers. This information will also be important for determining the biological effects of these metabolites at the cellular and molecular level to unravel the mechanism that underlies their ability to protect against cataract.

Future experimental work would greatly benefit from complementary epidemiological studies that looked at the incidence of cataract across Europe and offered correlations with the results of detailed studies of flavonoid intake. In conclusion, while the study of flavonoids in relation to cataract is undoubtedly in its infancy, the data presented here demonstrate that the flavonoids could potentially play a role in reducing the incidence of cataract, adding to the ever-increasing body of data linking dietary flavonoids with benefits to human health.

## REFERENCES

1.  Spector A. Oxidative stress-induced cataract: mechanism of action. FASEB J 1995; 9:1172–1182.
2.  Taylor A. Nutritional and environmental influences on risk for cataract. In: Taylor A, Ed. Nutritional and Environmental Influences on the Eye. Boca Raton. FL: CRC Press, 1999:53–93.
3.  Bunce GE. Animal studies on cataract. In: Taylor A, Ed. Nutritional and Environmental Influences on the Eye. Boca Raton. FL: CRC Press, 1999:105–115.
4.  Jacques PF, Chylack LT. Epidemiologic evidence of a role for the antioxidant vitamins and carotenoids in cataract prevention. Am J Clin Nutr 1991; 53:352S–355S.
5.  Tavani A, Negri E, La Vecchia C. Food and nutrient intake and risk of cataract. Ann Epidemiol 1996; 6:41–46.
6.  Hammond BR, Johnson MA. The age-related eye disease study (AREDS). Nutr Rev 2002; 60:283–288.
7.  Chylack LT, Brown NP, Bron A, Hurst M, Kopcke W, Thien U, Schalch W. The Roche European American Cataract Trial (REACT): a randomized clinical trial to investigate the efficacy of an oral antioxidant micronutrient mixture to slow progression of age-related cataract. Ophthalmic Epidemiol 2002; 9:49–80.
8.  Knekt P, Heliövaara M, Rissanen A, Aromaa A, Aaran R-K. Serum antioxidant vitamins and risk of cataract. Br Med J 1992; 305:1392–1394.
9.  Harding JJ. Cataract. Biochemistry, Epidemiology and Pharmacology. London: Chapman & Hall, 1991.

10. Spector A, Garner WH. Hydrogen peroxide and human cataract. Exp Eye Res 1981; 33:673–681.

11. Reddy VN. Glutathione and its function in the lens—an overview. Exp Eye Res 1990; 50:771–778.

12. Babizhayev MA, Deyev AI, Chernikov AV. Peroxide-metabolizing systems of the crystalline lens. Biochim Biophys Acta 1992; 1138:11–19.

13. Behndig A, Svensson B, Marklund SL, Karlsson K. Superoxide dismutase isoenzymes in the human eye. Invest Ophthalmol Vis Sci 1998; 39:471–475.

14. Singh SV, Srivastava SK, Awasthi YC. Purification and characterization of the 2 forms of glutathione S-transferase present in human lens. Exp Eye Res 1985; 41: 201–207.

15. Qiao F, Xing K, Liu A, Ehlers N, Raghavachari N, Lou MF. Human lens thioltranferase: cloning, purification, and function. Invest Ophthalmol Vis Sci 2001; 42: 743–751.

16. Reiss GR, Werner PG, Zollman PE, Brubaker RF. Ascorbic acid levels in the aqueous humor of nocturnal and diurnal mammals. Arch Ophthalmol 1986; 104:753–755.

17. Taylor A, Jacques PF, Nowell T, Perrone G, Blumberg J, Handelman G, Jozwiak B, Nadle D. Vitamin C in human and guinea pig aqueous, lens and plasma in relation to intake. Curr Eye Res 1997; 16:857–864.

18. Lou MF, Dickerson JE. Protein-thiol mixed disulfides in human lens. Exp Eye Res 1992; 55:889–896.

19. Bhuyan KC, Bhuyan DK, Podos SM. Lipid peroxidation in cataract of the human. Life Sci 1986; 38:1463–1471.

20. Garner MH, Spector A. Selective oxidation of cysteine and methionine in normal and senile cataractous lenses. Proc Natl Acad Sci USA 1980; 77:1274–1277.

21. Harding JJ. Free and protein-bound glutathione in normal and cataractous human lenses. Biochem J 1970; 117:957–960.

22. Kupfer C. The conquest of cataract: a global challenge. Trans Ophthalmol Soc UK 1984; 104:1–10.

23. Block G, Patterson B, Subar A. Fruit, vegetables and cancer prevention. A review of the epidemiological evidence. Nutr Cancer 1992; 18:1–29.

24. Steinmetz KA, Potter JD. Vegetables, fruit and cancer prevention: a review. J Am Diet Assoc 1996; 96:1027–1039.

25. Hu FB, Rimm EB, Stampfer MJ, Ascherio A, Spiegelman D, Willett WC. Prospective study of major dietary patterns and risk of coronary heart disease in men. Am J Clin Nutr 2000; 72:912–921.

26. Bazzano LA, He J, Ogden LG, Loria CM, Vupputuri S, Myers L, Whelton PK. Fruit and vegetable intake and risk of cardiovascular disease in US adults: the first National Health and Nutrition Examination Survey Epidemiologic Follow-up Study. Am J Clin Nutr 2002; 76:93–99.

27. Rissanen TH, Voutilainen S, Virtanen JK, Venho B, Vanharanta M, Mursu J, Salonen JT. Low intake of fruits, berries and vegetables is associated with excess mortality in men: the Kuopio Ischaemic Heart Disease Risk Factor (KIHD) Study. J Nutr 2003; 133:199–204.

28. Robertson JM, Donner AP, Trevithick JR. A possible role for vitamins C and E in cataract prevention. Am J Clin Nutr 1991; 53:346S–351S.

29. Verhoeven DT, Goldbohm RA, van Poppel G, Verhagen H, van den Brandt PA. Epidemiological studies on brassica vegetables and cancer risk. Cancer Epidemiol Biomarkers Prev 1996; 5:733–748.

30. Cook NC, Samman S. Flavonoids—chemistry, metabolism, cardioprotective effects and dietary sources. Nutr Biochem 1996; 7:66–76.

31. Scalbert A, Williamson G. Dietary intake and bioavailability of polyphenols. J Nutr 2000; 130:2073S–2085S.

32. Arts IC, Hollman PC, Bueno De Mesquita HB, Feskens EJ, Kromhout D. Dietary catechins and epithelial cancer incidence: the Zutphen elderly study. Int J Cancer 2001; 92:298–302.

33. Stefani ED, Boffetta P, Deneo-Pellegrini H, Mendilaharsu M, Carzoglio JC, Ronco A, Olivera L. Dietary antioxidants and lung cancer risk: a case-control study in Uruguay. Nutr Cancer 1999; 34:100–110.

34. Garcia-Closas R, Agudo A, Gonzalez CA, Riboli E. Intake of specific carotenoids and flavonoids and the risk of lung cancer in women in Barcelona, Spain. Nutr Cancer 1998; 32:154–158.

35. Geleijnse JM, Launer LJ, van der Kuip DA, Hofman A, Witteman JC. Inverse association of tea and flavonoid intakes with incident myocardial infarction: the Rotterdam Study. Am J Clin Nutr 2002; 75:880–886.

36. Arts IC, Jacobs DR, Harnack LJ, Gross M, Folsom AR. Dietary catechins in relation to coronary heart disease death among postmenopausal women. Epidemiology 2001; 12:668–675.

37. Hirvonen T, Virtamo J, Korhonen P, Albanes D, Pietinen P. Intake of flavonoids, carotenoids, vitamins C and E, and risk of stroke in male smokers. Stroke 2000; 31: 2301–2306.

38. Geleijnse JM, Launer LJ, Hofman A, Pols HA, Witteman JC. Tea flavonoids may protect against atherosclerosis: the Rotterdam Study. Arch Intern Med 1999; 159: 2170–2174.

39. Yochum L, Kushi LH, Meyer K, Folsom AR. Dietary flavonoid intake and risk of cardiovascular disease in postmenopausal women. Am J Epidemiol 1999; 149: 943–949.

40. Rimm E, Katan MB, Ascherio A, Stampfer MJ, Willett WC. Relation between intake of flavonoids and risk for coronary heart disease in male health professionals. Ann Intern Med 1996; 125:384–389.

41. Keli SO, Hertog MG, Feskens EJ, Kromhout D. Dietary flavonoids, antioxidant vitamins and incidence of stroke: the Zutphen study. Arch Intern Med 1996; 156: 637–642.

42. Knekt P, Jarvinen R, Reunanen A, Maatela J. Flavonoid intake and coronary mortality in Finland: a cohort study. Br Med J 1996; 312:478–481.

43. Hertog MG, Kromhout D, Aravanis C, Blackburn H, Buzina R, Fidanza F, Giampaoli S, Jansen A, Menotti A, Nedeljkovic S, Pekkarinen M, Simic BS, Toshima H, Feskens EJM, Hollman PCH, Katan MB. Flavonoid intake and long-term risk of coronary heart disease and cancer in the seven countries study. Arch Intern Med 1995; 155: 381–386.

44. Knekt P, Kumpulainen J, Jarvinen R, Rissanen H, Heliovaara M, Reunanen A, Hakulinen T, Aromaa A. Flavonoid intake and risk of chronic diseases. Am J Clin Nutr 2002; 76:560–568.

45. Trichopoulou A, Vasilopoulou E. Mediterranean diet and longevity. Br J Nutr 2000; 84:S205–S209.

46. Varma SD, Kinoshita JH. Inhibition of lens aldose reductase by flavonoids—their possible role in the prevention of diabetic cataracts. Biochem Pharmacol 1976; 25: 2505–2513.

47. Beyer-Mears A, Farnsworth PN. Diminished sugar cataractogenesis by quercetin. Exp Eye Res 1979; 28:709–716.

48. Varma SD, Mizuno A, Kinoshita JH. Diabetic cataracts and flavonoids. Science 1977; 195:205–206.

49. Kador PF, Kinoshita JH. Diabetic and galactosaemic cataracts. In: Whelan JNP, Ed. Human Cataract Formation. Ciba Foundation Symposium. London: Pitman, 1984: 123–131.

50. van Heyningen R. Formation of polyols by the lens of the rat with ''sugar'' cataract. Nature 1959; 184:194–195.

51. Kinoshita JH. Mechanisms initiating cataract formation. Invest Ophthalmol 1974; 13:713–724.

52. Varma SD, Mikuni I, Kinoshita JH. Flavonoids as inhibitors of lens aldose reductase. Science 1975; 188:1215–1216.

53. Wolff SP, Dean RT. Glucose autooxidation and protein modification. The potential role of ''autoxidative glycosylation'' in diabetes. Biochem J 1987; 245:243–250.

54. Thiagarajan G, Chandani S, Sundari CS, Rao SH, Kulkarni AV, Balasubramanian D. Antioxidant properties of green and black tea, and their potential ability to retard the progression of eye lens cataract. Exp Eye Res 2001; 73:393–401.

55. Thiagarajan G, Chandani S, Rao H, Samuni AM, Chandrasekaran K, Balasubramanian D. Molecular and cellular assessment of Ginkgo biloba extract as a possible ophthalmic drug. Exp Eye Res 2002; 75:421–430.

56. Shearer TR, David LL, Anderson RS. Selenite cataract: a review. Curr Eye Res 1987; 6:289–300.

57. Gupta SK, Halder N, Srivastava SK, Trivedi D, Joshi S, Varma SD. Green tea (Camellia sinensis) protects against selenite-induced oxidative stress in experimental cataractogenesis. Ophthalmic Res 2002; 34:258–263.

58. Lin L-R, Chen SC, Reddy VN, Giblin FJ. Green tea polyphenols protect against H2O2, UVB and X-ray induced damage in cultured dog lens epithelial cells. Exp Eye Res 1998; 67:S141.

59. Ibaraki N, Fan W, Lin L-R, Giblin FJ, Reddy VN. Effects of green tea polyphenols in protecting lens epithelial cells against oxidative damage. Invest Ophthalmol Vis Sci 2000; 41:S1104.

60. Zigman S, Rafferty NS, Rafferty K, McDaniel T, Schultz J, Reddan J. Protection of the lens from oxidative stress by tea polyphenols. Invest Ophthalmol Vis Sci 2000; 41:S1103.

61. Kilic F, Bhardwaj R, Trevithick JR. Modelling cataractogenesis XVIII. In vitro diabetic cataract reduction by venoruton. Acta Ophthalmol Scand 1996; 74:372–378.

62. Sanderson J, McLauchlan WR, Williamson G. Quercetin inhibits hydrogen peroxide-induced oxidation of the rat lens. Free Radic Biol Med 1999; 26:639–645.

63. Duncan G, Williams MR, Riach RA. Calcium, cell signalling and cataract. Prog Retinal Eye Res 1994; 13:623–652.

64. Ishige K, Schubert D, Sagara Y. Flavonoids protect neuronal cells from oxidative stress by three distinct mechanisms. Free Radic Biol Med 2001; 30:433–446.

65. Cornish KM, Williamson G, Sanderson J. Quercetin metabolism in the lens: role in inhibition of hydrogen peroxide induced cataract. Free Radic Biol Med 2002; 33: 63–70.

66. Grotte D, Mattox V, Brubaker R. Fluorescent, physiological and pharmacokinetic properties of fluorescein. Exp Eye Res 1985; 40:23–33.

67. Weinreb RN, Jeng S, Miller AL. Lysosomal enzyme activity in human aqueous humor. Clin Chim Acta 1991; 199:1–6.

68. Day AJ, Mellon F, Barron D, Sarrazin G, Morgan MRA, Williamson G. Human metabolism of dietary flavonoids: identification of plasma metabolites of quercetin. Free Radic Res 2001; 35:941–949.

# 12

# Phytochemicals and the Prevention of Cardiovascular Disease: Potential Roles for Selected Fruits, Herbs, and Spices

**Samir Samman**
*University of Sydney, Sydney, New South Wales, Australia*

## I. INTRODUCTION

Coronary heart disease (CHD) is a multifactorial disease which involves a number of interrelated risk factors (Table 1); however, a major environmental influence is undoubtedly diet [1]. Dietary factors that contribute to disease prevention include the reduction in the intake of saturated fat and an increase in the consumption of a plant-based diet containing fruits, vegetables, and grains. These observations have been translated into guidelines and appear consistently in the dietary goals of many countries and organizations that promote cardiac health, such as the American Heart Association [2]. The inverse association between fruit and vegetable intake and CHD is consistent across different geographical locations and in populations that differ markedly in lifestyle, gender, and age. It has been suggested that the lower rate of CHD is attributed to such factors as the displacement of foods that are high in salt, caloric density, and saturated fat by fruits and vegetables; the increase in intake of dietary fiber, minerals, folate, and vitamins, all of which exhibit antioxidant action; and an increase in intake of plant-derived constituents known as phytochemicals.

The main focus of this brief review is to summarize the involvement of selected phytochemicals in the prevention of CHD; in particular, examples will be sourced from data on fruit, herbs, and spices.

**Table 1**  Risk Factors for Coronary Heart Disease

| Irreversible | Potentially reversible |
| --- | --- |
| Gender | Cigarette smoking |
| Age | Lack of exercise |
| Genetics | Dyslipidemia |
| | Obesity, diabetes |
| | Hypertension |
| | Thrombotic risk |
| | Oxidizability of LDL |
| | Hyperhomocysteinemia |

## II.  Bioactive phytochemicals

Phytochemicals are a broad array of compounds that coexist with nutrients in plant foods; while not considered as essential nutrients per se, some are considered to play a role in the prevention of chronic diseases by acting either independently, synergistically with other phytochemicals or with nutrients that coexist in the plant. Ralph and Provan [3] have conveniently grouped phytochemicals into categories that include:

> *Products of the phenylpropanoid pathway*: These compounds, commonly referred to as polyphenols, are derived from cinnamic acid and include the xanthones and flavonoids. The basic structural unit of the flavonoid family (Fig. 1) comprises two benzene rings linked through a heterocyclic pyran or pyrone ring (C ring); variations in the C ring and the extent of hydroxylation define the major classes [4].

> *Products of the isoprenoid pathway*: These include the carotenoids, steroids, terpenes, and phytosterols [5]. Phytosterols are present in a broad range of plant foods, are not absorbed effectively from the intestine, but can bind cholesterol and prevent it from being absorbed [3].

> *Organosulfur compounds*: These include glucosinolates and allicins, the latter being found in *Allium* vegetables such as onions, chives, and garlic. Although the published literature predominantly addresses their role in cancer prevention, compounds derived from allicin (including sulfides and disulfides) are reported to inhibit endogenous cholesterol synthesis.

## III.  Phytochemicals and CHD

## A.  Epidemiological Studies

Studies of phytochemical intake and its relationship to CHD in humans have been considerably hindered by the lack of availability within food composition

**Generic structure**

**Isoflavones**

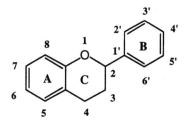

| | 5 | 6 | 7 | 4' |
|---|---|---|---|---|
| Genistein | OH | - | OH | OH |
| Daidzein | - | - | OH | OH |
| Glycitein | - | O-Me | OH | OH |

**Flavonols**

| | 5 | 7 | 3' | 4' | 5' |
|---|---|---|---|---|---|
| Myricetin | OH | OH | OH | OH | OH |
| Quercetin | OH | OH | OH | OH | - |
| Kaempferol | OH | OH | - | OH | - |

**Flavones**

| | 5 | 7 | 3' | 4' |
|---|---|---|---|---|
| Luteolin | OH | OH | OH | OH |
| Apigenin | OH | OH | - | OH |

**Flavanols (Catechins)**

| | 5 | 7 | 3' | 4' | 5' |
|---|---|---|---|---|---|
| (+) Catechin | OH | OH | OH | OH | - |
| (-) Epicatechin *(cis)* | OH | OH | OH | OH | - |
| (-) Epigallocatechin | OH | OH | OH | OH | OH |

**Flavanones**

| | 5 | 7 | 3' | 4' |
|---|---|---|---|---|
| Naringenin | OH | OH | - | OH |
| Hesperetin | OH | OH | OH | O-Me |

**Figure 1**   Structures of the major classes of food flavonoids. Positions of hydroxyl groups on A and B rings are listed for selected examples within each class.

**Table 2**   Risk of Cardiovascular Disease for High Versus Low Flavonoid Intake[a]

| Gender | Men | Men | Men/women | Men |
|---|---|---|---|---|
| Age (y) | 65–84 | 50–69 | 30–69 | 40–75 |
| Outcome | CHD mortality MI incidence | Stroke Incidence | CHD mortality | CHD mortality MI incidence |
| Follow-up (y) | 5 | 15 | 20–25 | 6 |
| N in cohort | 805 | 552 | 5133 | 34,789 |
| RR | 0.32[b] 0.52 | 0.27[b] | 0.67 (men) 0.73 (women) | 0.63 1.09 |
| Ref. | 6 | 7 | 8 | 9 |

RR: Relative risk; CHD: coronary heart disease; MI: myocardial infarction.
[a] Σ quercetin, kaempferol, myricetin, luteolin, apigenin.
[b] Statistically significant.

databases of reliable information on the content of phytochemicals and, in particular, of flavonoids. A second and equally important problem has been the inability of dietary questionnaires to detect differences in the intakes of phytochemicals. Despite such methodological constraints, the balance of evidence supports the view that flavonoids have a protective effect on CHD (Table 2) in populations where the underlying risk of disease is not excessively high [6–9]. The exceptions are described in reports from Wales and Finland, where the intake of fat and, in particular, of saturated fat is known to be the predominant risk factor for CHD.

Further support for the cardioprotective effect of flavonoids has been obtained from a reexamination of food records from 16 cohorts within the Seven Countries Study [10]. During a 25-year follow-up period, an inverse association was observed between CHD mortality and flavonoid intake; this explained a small but significant portion (8%) of the variance in CHD deaths, independently of intakes of alcohol and antioxidant vitamins. CHD mortality was observed to be lowest in Japan with an estimated average flavonoid intake of 61 mg/d, mainly derived from green tea.

## B.   Dietary Sources of Phytochemicals

Based on epidemiological studies [10], the major source of bioactive flavonoids for the Dutch population was black tea, followed by onions and apples. Other studies [7] have shown that tea was a major contributor to flavonoid intake in Netherlands; men who consumed 5 or more cups of tea daily had a lower incidence of stroke than those drinking fewer than 3 cups. The intake of other unmeasured antioxidant polyphenols, such as the catechins in green tea or isoflavones in the traditional Japanese soy staple, together with a diet low in saturated fat may have also contributed to the lowered incidence rates in Japan, despite the high percentage of smokers in that country. Overall, the intake of saturated fat remained the most significant dietary constituent in relation to cardiovascular disease, bring responsible for 73% of the total variance.

As mentioned above, current dietary surveys are hindered by the absence of a supportive comprehensive food compositional database. Nevertheless, if fruit and vegetables are accepted as surrogate markers of a wide range of phytochemicals, the existing data support the view that these compounds make a significant contribution to disease prevention; however, this effect is less powerful than would result from a reduction in intake of saturated fat.

The role of phytochemicals, mainly from spices and herbs, in the prevention of CHD is summarized below.

## 1. Sauces, Seasonings, and Condiments

Considerable research has been devoted to the cardioprotective effects of tea and red wine, while, in comparison, herbs and spices have received much less attention despite their central importance to many dietary practices around the world. Mustard, onions, ginger, vinegar, pepper, and many herbs and spices are key ingredients of sauces, seasonings, and condiments that have the potential to introduce a plethora of phytochemicals into the human diet [11] (Table 3). They are added to food to impart flavor as well as to promote an enhanced appreciation of the dishes presented [12].

## 2. Umbelliferae Family

Parsley contains large quantities of the flavone apigenin and low levels of other flavonoids. The herb has been used to determine the effect of apigenin on oxidation status in humans; in a short-term study, parsley/apigenin consumption pro-

**Table 3**  Selected Phytochemicals Found in Herbs

| Herbal source | Example | Selected phytochemical |
|---|---|---|
| Allium species | Garlic, onions, Chives | Diallyl sulfite, di- and tri-sulfides |
| Labiatae | Basil, marjoram, mint, thyme, rosemary, dill, oregano, sage | Mono-, diterpenes, flavonoids, rosmarol, urosolic acid, phenolic derivatives |
| Umbellifferae | Anise, cumin, caraway, coriander, parsley | Coumarins, phthalides, terpenoids, polyacetylenes |
| Zingiberaceae | Tumeric, ginger | Curcumin, gingerols, diarylheptanoids |

duced an increase in the activities of endogenous phase 2 antioxidant enzymes: glutathione reductase and superoxide dismutase [13].

Similarly, the phenolic antioxidants p-coumaric acid, ferulic acid, curcumin, and caffeic acid, which are found in coriander (as well as members of the Labiatae family, see below), inhibit the formation of 3-nitrotyrosine in vitro and may prevent lipid peroxidation in vivo [14,15]. Caffeic acid and other hydroxycinnamic acids have also been found to exert an inhibitory effect on LDL oxidation, a known risk factor for CHD [16].

## 3. Ginger

Ginger is one of the best known spices to contain mono- and sesquiterpenes; its main antioxidants are gingerols, shaogaols, and related derivatives such as zingerone. In vitro studies have shown extracts of ginger to inhibit cyclooxygenase activity and reduce platelet aggregation. However, clinical trials in humans using raw ginger root (15 g daily) or cooked stem ginger (40 g per day) failed to demonstrate any significant effect on thrombotic tendency [17]. A single high dose (10 g) of ginger powder reduced platelet aggregation but was without effect on blood lipids [18]. In contrast, in apo-E–deficient mice, animals predisposed to heart disease, an ethanolic solution of ginger (5 or 50 mg/L, by mouth) resulted in dose-dependent decreases in atherosclerosis and inhibition of LDL oxidation [19]. This result reinforces previous findings in mice which demonstrated an inhibition of cholesterol biosynthesis [20].

## 4. Garlic

The effects of garlic on cardiovascular disease have been studied extensively. Garlic has been shown to lower the severity of atherosclerosis in animal models by a number of mechanisms [21], including downregulation of cholesterol biosynthesis [22] and inhibition of LDL oxidation [23]. A recent meta-analysis of 13 clinical trials in humans [24] calculated a significant reduction in plasma cholesterol (0.4 mmol/L) following supplementation with doses of garlic ranging from 10 mg of steam-distilled oil to 900 mg of standardized powder over periods of 8–24 weeks. The authors concluded that garlic is superior to placebo in lowering cholesterol concentrations but that the effect was modest.

## 5. Labiatae Family

The use of extracts from rosemary as food preservatives is well established [25]; the phenolic compounds obtained from this source have been shown to act as antioxidants in vitro and reduce the oxidation of dietary lipids in a dose-dependant manner [26]. The constituents of rosemary considered responsible for the majority of this antioxidant activity are rosmarinic acid, carnosol, and carnosic acid [14].

Extracts derived from most members of the Labiatae family (Table 3) have the potential to reduce the oxidation of food products and extend shelf life. This antioxidant action is dependent on the ability of the constituent phenolics to scavenge free radicals and chelate metals [27]. This dual effect has implications for CHD since it has been proposed in some studies, mainly from Europe, that a high status of iron increases the risk of CHD [28]. A number of the active compounds found in rosemary have also been found in sage and other herbs. In oregano, four flavonoids were identified among the active fraction [29], while in thyme, dimers of thymol and flavonoids have been isolated and characterized.

## 6. Red Pepper

Capsaicin, the pungent principle of red pepper, increases energy expenditure and, under some circumstances, decreases energy intake [30]. In a study with both red pepper and caffeine, the magnitude of energy deficit resulting from increased expenditure and reduced intake was equivalent to 4000 kJ/d [31].

## 7. Vinegar

Vinegar is often found as an ingredient of sauces, condiments, and salad dressings, and trials in humans have shown that a limited experimental dose of vinegar (equivalent to 1 g acetic acid; single dose) is sufficient to lower the postprandial glycemic response to a mixed meal. The mechanism of action may involve the delay in the rate of gastric emptying [32].

## 8. Phytosterols—Phytochemicals in Purified Form

Phytosterols, products of the isoprenoid pathway, are able to bind cholesterol and prevent it from being absorbed. These compounds are present in the diet through items such as fortified margarines and mayonnaise. Daily doses of at least 1 g are required to afford a good clinical response, with 2–3 g being considered optimal. The extent of cholesterol lowering is approximately 10% [33].

## C. Contribution of Tropical and Subtropical Fruit to Phytochemical Intake and Prevention of CHD

Generally, there are insufficient data on the phytochemical content of foods, but a selection of phytochemicals to be found in tropical and subtropical fruit is shown in Table 4. Despite the limitations of the data, it is established that these fruits contain active compounds derived from the aforementioned categories of

**Table 4** Examples of Phytochemicals Found in Tropical and Subtropical Fruit

| Common and scientific names | Phytochemical ingredients |
| --- | --- |
| Avocado (*Persea americana*) | Caffeic acid, chlorogenic acid, coumaric acid, cycloartenol, dopamine, phytosterols |
| Durian (*Durio zibethinus*) | Dimethyl trithiolane, methylbutanoate |
| Guava (*Psidium guajava*) | Ascorbigen, bisabolene, cadinene, cinnamyl-acetate, ellagic acid, humulene, leucocyanidin, limonene, mecocyanin, phytin, selinine |
| Jackfruit (*Artocarpus heterophyllus*) | Dihydromorin, cycloartenone, heterophylol, flavonones |
| Lime (*Citrus aurantiifolia*) | Borneol, bergamotene, bisabolene, camphene, cineole, citronellal, coumarin, limonene, methoxypsoralen, naringin, phellandrene, phlobotannin, terpineol, thujene |
| Litchi (*Litchi chinensis*) | Cyanidin, damascenone, guaiacol, malvidin, quercitin |
| Longan (*Dimocarpus longan*) | Acetonylgeraniin |
| Mango (*Mangifera indica*) | Mangiferic acid, mangiferine, neoxanthophyll, phytin, xanthophyll |
| Mangosteen (*Garcinia mangostana*) | Catechins, deoxygartanin, gartanin, mangostin, phytin |
| Papaya (*Carica papaya*) | Germacrene, isocaryophyllene, kryptoflavin, myrcene, ocimene, phellandrene, terpinene |
| Passionfruit (*Passiflora edulis*) | Alkaloids, flavonoids, harman, xanthophylls |
| Pineapple (*Ananas comosus*) | Ananasic acid, chaviol, ellagic acid, phytosterols, serotonin, vanilin |
| Rambutan (*Nephelium lappaceum*) | Damascenone |
| Starfruit (*Averrhoa carambola*) | Cryptochrome, damascenone, phytofluene |

phytochemicals and are capable of impacting favorably on heart disease prevention.

The phytochemical composition of avocado (Table 4), combined with its main fatty acid, oleic acid (18:1), represents a potential for lowering plasma cholesterol [34] and increasing antioxidant status. An added benefit may result from its content of phytosterols. Moreover, the presence of chlorogenic and caffeic acids in avocado increases the antioxidant capacity of plasma. In the same way, jackfruit (*Artocarpus heterophyllus*) contains antioxidants and phytosterols (Table 4) and its flavonoids, particularly the flavonones, have antioxidant activity [4].

Citrus fruits provide a diverse range of flavonoids [35]; naringenin, a flavo-none, has been shown to posses cholesterol-lowering and anticancer properties [36]. The naringin content of lime juice has been estimated to be ~100 μg/g [37], and the lime is also a source of limonene, a biologically active terpene (see above). Other bioactive substances in lime (Table 4) include coumarins that possess anticoagulatory and anti-inflammatory properties.

Carotenoids are the dominant products of the isoprenoid pathway in mango. Phytochemicals in mango have been implicated in protection from cancer [38]. Mango impacts favorably on a number of metabolic functions such as the production of a low plasma glucose concentration relative to other tropical and subtropical fruits commonly consumed in Thailand [39], and also to an increase in plasma vitamin C [40].

Although few data are available on the phytochemical composition of rambutan, the cholesterol-lowering effect [41] of this fruit may be associated with its phytosterol content.

## IV. MOLECULAR TARGETS

The mode of action of phytochemicals in the prevention of CVD is multifaceted and is dependent on the chemical natures of the individual phytochemicals in question and on the consequences of their complex interactions. The most extensively studied group is undoubtedly the flavonoids.

## A. In Vitro Inhibition of LDL Oxidation by Flavonoids

Flavonoids inhibit the formation of lipid peroxides (LPO) at the initiation stage by acting as scavengers of superoxide anions and hydroxyl radicals [42]. It has been proposed that flavonoids terminate chain radical reactions by donating hydrogen atoms to the peroxy radical, forming a flavonoid radical [43] that, in turn, reacts with free radicals, thereby terminating the propagating chain [43]. In addition to these antioxidative properties, some flavonoids can act as metal-chelating agents and inhibit the superoxide-driven Fenton reaction, an important source of active oxygen radicals [42]. The structure-function activity of flavonoids has been reported previously [4].

A number of aglycone flavonoids are potent inhibitors of in vitro oxidative modification of LDL [44]. Phenolic compounds isolated from red wine inhibit the copper-catalyzed oxidation of LDL in vitro (10 mmol/L), significantly more than α-tocopherol [45], possibly by regenerating α-tocopherol [44]. Alternatively, chelation of divalent metal ions by flavonoids may reduce formation of free radicals induced by Fenton reactions [42]. Hydroxylation of the flavone nucleus appears to be advantageous because flavone itself is a poor inhibitor of LDL oxidation, whereas polyhydroxylated flavonoids such as quercetin, morin, hypoleatin, fisetin, gossypetin, and galangin are potent inhibitors of LDL oxidation [44].

The ability of the constituents of tea, particularly ( + )-catechin, to inhibit LDL oxidation has been investigated [46]; as expected, LDL modified by cells or copper-induced oxidation was endocytosed and degraded by macrophages more quickly than native LDL. However, in the presence of ( + )-catechin, the rates of endocytosis and degradation were similar to those of native LDL [46]. In addition to the inhibition of LDL oxidation, flavonoids such as catechin, rutin, and quercetin, at levels of 10–20 mmol/L, minimize the cytotoxicity of oxidized LDL [47]. Moreover, cells preincubated with these flavonoids were observed to be resistant to the cytotoxic effects of previously oxidized LDL (47,48). The postulated mechanisms by which flavonoids protect against the cytotoxicity of oxidized LDL are consistent with their antioxidant and free radical–scavenging properties [4].

Coffee has been reported to inhibit LDL oxidation in vitro. One class of phenolic substances, the hydroxycinnamic acids, is ubiquitous in its occurrence; the most common member of this class is chlorogenic acid, which has been shown to inhibit the oxidative modification of LDL in vitro [49]. Coffee, along with apples and berries, is a major source of chlorogenic acid in the human diet.

## B. Antithrombotic and Vasoprotective Effects of Flavonoids

The antioxidant actions of flavonoids appear to be involved in their observed anti-thrombotic action [50–52]. The antithrombotic and vasoprotective actions of quercetin, rutin, and other flavonoids have been attributed to their ability to bind to platelet membranes and scavenge free radicals [50]. In this manner, flavonoids restore the biosynthesis and action of endothelial prostacyclin and endothelial-derived relaxing factor (EDRF), both of which are known to be inhibited by free radicals [50,53,54]. However, some flavonoids may inhibit arachidonic acid metabolism and platelet function by flavonoid-enzyme interactions rather than by antioxidant effects [55]. In addition to their antiaggregatory effects, flavonoids appear to increase vasodilation by inducing vascular smooth muscle relaxation, which may be mediated by inhibition of protein kinase C, PDEs, or by decreased cellular uptake of calcium [56].

Flavonoids inhibit platelet aggregation and adhesion [50–53,55,57–59]; it has been shown that flavonoids influence several pathways involved in platelet function [58,60,61], such as inhibition of cyclooxygenase and lipoxygenase and antagonism of thromboxane formation and thromboxane receptor function [58]. One of the most potent mechanisms by which flavonoids appear to inhibit platelet aggregation is by mediating increases in platelet cyclic AMP (cAMP) levels, either by stimulating adenylate cyclase or by inhibiting cAMP phosphodiesterase (PDE) activity [52,53,56,57,61–64].

The consumption of red wine is linked to decreased platelet aggregation [65]. It is postulated that the antioxidant properties of phenolic compounds in red wine reduce platelet aggregation and inhibit lipid peroxidation in vitro [66]. If such mechanisms are established in humans, the protective effects of red wine against platelet aggregation may partly explain the long-term advantages of consuming moderate amounts of red wine over other alcoholic beverages [67].

## V. CONCLUSION

Epidemiological, clinical, and in vitro evidence support the hypothesis that phytochemicals benefit health. Herbs, spices, and tropical and subtropical fruits contain a broad range of such compounds, some of which are reported to protect against cardiovascular disease. The presence of a large number of minor dietary factors that protect against disease reinforces the recommendation to increase the intake of plant foods rather than any specific supplement.

## REFERENCES

1. Truswell AS. ABC of Nutrition. 3rd ed.. London: BMJ Books, 1999.
2. American Heart Association. Heart and Stroke Statistical Update. Dallas: American Heart Association, 2000.
3. Ralph A, Provan GJ. Phytoprotectants. In: Garrow JS, James WPT, Ralph A, Eds. Human Nutrition and Dietetics. Edinburgh: Churchill Livingstone, 2000:417–426.
4. Cook NC, Samman S. Flavonoids: chemistry, metabolism, cardioprotective effects and dietary sources. J Nutr Biochem 1996; 7:66–76.
5. Samman S. Lipid metabolism. In: Kuchel PW, Ralston GB, Eds. Schaum's Outlines of Theory and Problems of Biochemistry. New York: McGraw-Hill Book Company, 1998:362–401.
6. Hertog MGL, Feskens EJM, Hollman PCH, Katan MB, Kromhout D. Dietary antioxidant flavonoids and risk of coronary heart disease: The Zutphen Elderly Study. Lancet 1993; 342:1007–1011.
7. Keli SO, Hertog MGL, Feskens EJM, Kromhout S. Dietary flavonoids, antioxidant vitamins and incidence of stroke: The Zutphen Elderly Study. Arch Intern Med 1996; 154:637–642.
8. Knekt P, Jarvinen R, Reunanen A, Maatela J. Flavonoid intake and coronary mortality in Finland: a cohort study. Br Med J 1996; 312:478–481.
9. Rimm EB, Katan MB, Ascherio A, Stampfer MJ, Willett WC. Relation between intake of flavonoids and risk of coronary heart disease in male health professionals. Ann Int Med 1996; 125:384–389.

10. Hertog MG, Kromhout D, Aravanis C, Blackburn H, Buzina R, Fidanza F, Giampaoli S, Jansen A, Menotti A, Nedeljkovic S. Flavonoid intake and long-term risk of coronary heart disease and cancer in the seven countries study. Arch Int Med 1995; 155:381–386.

11. Craig WJ. Health-promoting properties of common herbs. Am J Clin Nutr 1999; 70(suppl):491S–499S.

12. Simon AL, Howe R. A Dictionary of Gastronomy. Sydney: Andre Deutsch, 1978.

13. Nielsen SE, Young JF, Daneshvar B, Lauridsen ST, Knuthsen P, Sandstrom B, Dragsted LO. Effect of parsley (*Petroselinum crispum*) intake on urinary apigenin excretion, blood antioxidant enzymes and biomarkers for oxidative stress in human subjects. Br J Nutr 1999; 81:447–455.

14. Aruoma OI, Halliwell B, Aeschbach R, Loligers J. Antioxidant and pro-oxidant properties of active rosemary constituents: carnosol and carnosic acid. Xenobiotica 1992; 22:257–268.

15. Aruoma OI, Spencer JP, Rossi R, Aeschbach R, Khan A, Mahmood N, Munoz A, Murcia A, Butler J, Halliwell B. An evaluation of the antioxidant and antiviral action of extracts of rosemary and Provencal herbs. Food Chem Toxicol 1996; 34:449–456.

16. Abu-Amsha R, Croft KD, Puddey IB, Proudfoot JM, Beilin LJ. Phenolic content of various beverages determines the extent of inhibition of human serum and low-density lipoprotein oxidation in vitro. Clin Sci 1996; 91:449–458.

17. Janssen PL, Meyboom S, van Staveren WA, de Vegt F, Katan MB. Consumption of ginger (*Zingiber officinale* Roscoe) does not affect ex vivo platelet thromboxane production in humans. Eur J Clin Nutr 1996; 50:772–774.

18. Bordia A, Verma SK, Srivastava KC. Effect of ginger (*Zingiber officinale* Rosc.) and fenugreek (*Trigonella foenumgraecum* L.) on blood lipids, blood sugar and platelet aggregation in patients with coronary artery disease. Pros Leukot Essent Fatty Acids 1997; 56:379–384.

19. Fuhrman B, Rosenblat M, Hayek T, Coleman R, Aviram M. Ginger extract consumption reduces plasma cholesterol, inhibits LDL oxidation and attenuates development of atherosclerosis in atherosclerotic, apolipoprotein E-deficient mice. J Nutr 2000; 130:1124–1131.

20. Tanabe M, Chen YD, Saito K, Kano Y. Cholesterol biosynthesis inhibitory component from *Zingiber officinale* Roscoe. Chem Pharm Bull 1993; 41:710–713.

21. Campbell JH, Efendy JL, Smith NJ, Campbell GR. Molecular basis by which garlic suppresses atherosclerosis. J Nutr 2001; 131(3s):1006S–1009S.

22. Gupta N, Porter TD. Garlic and garlic-derived compounds inhibit human squalene monooxygenase. J Nutr 2001; 131:1662–1667.

23. Lau BH. Suppression of LDL oxidation by garlic. J Nutr 2001; 131(3s):985S–988S.

24. Stevinson C, Pittler MH, Ernst E. Garlic for treating hypercholesterolemia. A meta-analysis of randomized clinical trials. Ann Int Med 2000; 133:420–429.

25. Chipault JR, Mizuno GR, Hawkins JM, Lundberg WO. The antioxidant properties of natural spices. Food Res 1952; 17:46–58.

26. Schwarz K, Ternes W, Schmauderer W. Antioxidative constituents of *Rosemarinus officinalis* and *Salvia officinalis*. III. Stability of phenolic diterpenes of rosemary extracts under thermal stress as required for technological processes. Z Lebensm Forschung 1992; 195:104–107.

27. Samman S, Sandström B, Toft MB, Bukhave K, Jensen M, Sørensen SS, Hansen M. Green tea or rosemary extract added to foods reduce non-heme iron absorption. Am J Clin Nutr 2001; 73:607–612.
28. Salonen JT, Nyyssonen K, Korpela H, Tuomilehto J, Seppanen R, Salonen R. High stored iron levels are associated with excess risk of myocardial infarction in eastern Finnish men. Circ 1992; 86:803–811.
29. Lagouri V, Boskou D. Nutrient antioxidants in oregano. Int J Food Sci Nutr 1996; 47:493–497.
30. Yoshioka M, St-Pierre S, Drapeau V, Dionne I, Doucet E, Suzuki M, Tremblay A. Effects of red pepper on appetite and energy intake. Br J Nutr 1999; 82:115–23.
31. Yoshioka M, Doucet E, Drapeau V, Dionne I, Tremblay A. Combined effects of red pepper and caffeine consumption on 24 h energy balance in subjects given free access to foods. Br J Nutr 2001; 85:203–211.
32. Brighenti F, Castellani G, Benini L, Casiraghi MC, Leopardi E, Crovetti R, Testolin G. Effect of neutralized and native vinegar on blood glucose and acetate responses to a mixed meal in healthy subjects. Eur J Clin Nutr 1995; 49:242–247.
33. Hendriks HF, Weststrate JA, van Vliet T, Meijer GW. Spreads enriched with three different levels of vegetable oil sterols and the degree of cholesterol lowering in normocholesterolaemic and mildly hypercholesterolaemic subjects. Eur J Clin Nutr 1999; 53:319–327.
34. Colquhoun DM, Moores D, Somerset SM, Humphries JA. Comparison of the effects on lipoproteins and apolipoproteins of a diet high in monounsaturated fatty acids, enriched with avocado, and a high carbohydrate diet. Am J Clin Nutr 1992; 56: 671–677.
35. Kawaii S, Tomono Y, Katase E, Ogawa K, Yano M. Quantitation of flavonoid constituents in citrus fruits. J Agric Food Chem 1999; 47:3565–3571.
36. Guthrie N, Kurowska EM. Anticancer and cholesterol lowering activities of citrus flavonoids. In: Wildman REC, Ed. Handbook of Nutraceuticals and Functional Foods. Boca Raton, FL: CRC Press, 2001:113–126.
37. Yusof S, Ghazali HM, King GS. Naringin content in local citrus fruits. Food Chem 1990; 37:113–121.
38. Botting KJ, Young MM, Learson AE, Harris PJ, Ferguson LR. Antimutagens in food plants eaten by Polynesians: micronutrients, phytochemicals and protection against bacterial mutagenicity of the heterocyclic amine 2-amino-3-methylimida-zol[4,5-f]quinoline. Food Chem Toxicol 1999; 37:95–103.
39. Roongpisuthipong C, Banphotkasem S, Komindr S, Tanphaichitr V. Post-prandial glucose and insulin responses to various tropical fruits of equivalent carbohydrate content in non-insulin dependent diabetes mellitus. Diabetes Res Clin Prac 1991; 14:123–131.
40. Bates CJ, Prentice AM, Prentice A, Paul AA, Whitehead RG. Seasonal variations in ascorbic acid status and breast milk ascorbic acid levels in rural Gambian women in relation to dietary intake. Trans Royal Soc Trop Med Hyg 1982; 76:341–347.
41. Mongkolsirikieat S, Areegitranusorn P, Limratana N, Limpaiboon T, Lulitanond V. Hypocholesterolemic effect of rambutan (*Nephelium lappaceum*) supplementation in Thai adults. Nutr Rep Intl 1989; 39:797–803.

42. Afanas'ev IB, Dorozhko AI, Brodskii AV, Kostyuk VA, Potapovitch AI. Chelating and free radical scavenging mechanisms of inhibitory action of rutin and quercetin in lipid peroxidation. Biochem Pharmacol 1988; 38:1763–1769.

43. Torel J, Cillard J, Cillard P. Antioxidant activities of flavonoids and reactivity with peroxy radical. Phytochemistry 1986; 25:383–385.

44. De Whalley CV, Rankin SM, Hoult JRS, Jessep W, Leake DS. Flavonoids inhibit oxidative modification of low density lipoproteins. Biochem Pharmacol 1990; 39: 1743–1749.

45. Frankel EN, Kanner J, Parks E, Kinsella JE. Inhibition of oxidation of human low-density lipoprotein by phenolic substances in red wine. Lancet 1993; 341:454–457.

46. Mangiapane H, Thomson J, Salter A, Brown S, Bell GD, White DA. The inhibition of the oxidation of low density lipoprotein by (+)-catechin, a naturally-occurring flavonoid. Biochem Pharmacol 1992; 43:445–450.

47. Negre-Salvayre A, Salvayre R. Quercetin prevents the cytotoxicity of oxidized LDL on lymphoid cell lines. Free Radical Biol Med 1992; 12:101–106.

48. Negre-Salvayre A, Alomar Y, Troly M, Salvayre R. Ultraviolet-treated lipoproteins as a model system for the study of the biological effects of lipid peroxides on cultured cells. III. The protective effect of antioxidants (probucol, catechin, vitamin E) against the cytotoxicity of oxidized LDL occurs in two different ways. Biochim Biophy Acta 1991; 1096:291–300.

49. Nardini M, D'Aquino M, Tomassi G, Gentili V, Di Felice M, Scaccini C. Inhibition of human low-density lipoprotein oxidation by caffeic acid and other hydroxycinnamic acid derivatives. Free Rad Biol Med 1995; 19:541–52.

50. Gryglewski RJ, Korbut R, Robak J, Swies J. On the mechanism of antithrombotic action of flavonoids. Biochem Pharmacol 1987; 36:317–322.

51. Robak J, Korbut R, Shridi F, Swies J, Rzadkowska-Bodalska H. On the mechanism of antiaggregatory effect of myricetin. Pol J Pharmacol Pharmacol 1988; 40:337–340.

52. Beretz A, Cazenave J-P, Anton A. Inhibition of aggregation and secretion of human platelets by quercetin and other flavonoids: structure-activity relationships. Agent Action 1982; 12:382–387.

53. Beretz A, Cazenave J. The effect of flavonoids on blood-vessel wall interactions. In: Plant Flavonoids in Biology and Medicine II: Biochemical, Cellular, and Medicinal Properties. New York: Alan R. Liss, 1988:187–200.

54. Robak J, Gryglewski RJ. Flavonoids are scavengers of superoxide anions. Biochem Pharmacol 1988; 37:837–841.

55. Mora A, Paya M, Rios JL, Alcaraz MJ. Structure-activity relationships of polymethoxyflavones and other flavonoids as inhibitors of non-enzymic lipid peroxidation. Biochem Pharmacol 1990; 40:793–797.

56. Duarte J, Vizcaino FP, Utrilla P, Jimenez J, Tamargo J, Zarzuelo A. Vasodilatory effects of flavonoids in rat aortic smooth muscle. Structure activity relationships. Biochem Pharmacol 1993; 24:857–862.

57. Beretz A, Anton R, Cazenave J. The effect of flavonoids on cyclic nucleotide phosphodiesterase. In: Plant Flavonoids in Biology and Medicine: Biochemical, Pharmacological and Structure-Activity Relationships. New York: Alan R. Liss, 1986: 281–296.

58. Tzeng S-H, Ko W-C, Ko F-N, Teng C-M. Inhibition of platelet aggregation by some flavonoids. Thromb Res 1991; 64:91–100.

59. Swies J, Robak J, Dabrowski L, Duniec Z, Michalska Z, Gryglewski RJ. Antiaggregatory effects of flavonoids in vivo and their influence on lipoxygenase and cyclooxygenase in vitro. Pol J Pharmacol Pharm 1984; 36:455–463.

60. Elliott AJ, Scheiber SA, Thomas C, Pardini RS. Inhibition of glutathione reductase by flavonoids. A structure-activity study. Biochem Pharmacol 1992; 44:1603–1608.

61. Landolfi R, Mower RL, Steiner M. Modification of platelet function and arachidonic acid metabolism by bioflavonoids: structure-activity relations. Biochem Pharmacol 1984; 33:1525–1530.

62. Bourdillat B, Delautier D, Labat J, Benveniste J, Potier P, Brink C. Mechanism of action of hispidulin, a natural flavone, on human platelets. In: Plant Flavonoids in Biology and Medicine II. Biochemical, Cellular and Medicinal Properties. New York: Alan R. Liss, 1988:211–214.

63. Ferrell JE, Chang Sing PDG, Loew G, King R, Mansour JM, Mansour TE. Structure-activity studies of flavonoids as inhibitors of cyclic AMP phosphodiesterase and relationship to quantum indices. Mol Pharmacol 1979; 16:556–568.

64. Kuppusamy UR, Das NP. Effects of flavonoids on cyclic AMP phosphodiesterase and lipid mobilization in rat adipocytes. Biochem Pharmacol 1992; 44:1307–1315.

65. Renaud S, de Longeril M. Wine, alcohol, platelets and the French paradox for coronary heart disease. Lancet 1992; 339:1523–1526.

66. Ruf J-C, Berger J-L, Renaud S. Platelet rebound effect of alcohol withdrawal and wine drinking in rats: relation to tannins and lipid peroxidation. Atherioscler Thromb Vasc Biol 1995; 15:140–144.

67. Criqui MH, Ringel BL. Does diet or alcohol explain the French paradox?. Lancet 1994; 344:1719–1723.

# 13

# Beneficial Effects of Resveratrol

**Ann M. Bode and Zigang Dong**
*University of Minnesota, Austin, Minnesota, U.S.A.*

## I. INTRODUCTION

Research studies indicate that nonnutrient dietary compounds may be effective in reducing the risk for developing certain diseases, especially cancer and heart disease. However, one of the greatest challenges today is to eliminate the misinformation and myth in the media regarding the health benefits of certain foods or food supplements. The general public is constantly bombarded with propaganda and half-truths because of a desire to find the "magic pill" that will make them thinner, stronger, healthier, and free of disease. Recommendations are often made based on anecdotal evidence and conclusions are formulated from nonscientific observations. Although much of the early information regarding the significance of such dietary compounds is circumstantial and often contradictory, solid mechanistic data are rapidly accumulating elucidating their interactions with molecular pathways related to development of disease. Increasing state-of-the-art research is being directed toward isolating and identifying the active components in various food compounds and studying their molecular mechanism of action. Stilbenes and, in particular, resveratrol (3,5,4'-trihydroxy-*trans*-stilbene) (Fig. 1), its glucoside, piceid, and other analogs, have been proposed as having beneficial health effects including antitumor and cardioprotective properties. Resveratrol is the parent molecule of the viniferin family, known to inhibit fungal infection in certain plants [1,2]. It is a polyphenolic phytoalexin, an antibiotic compound produced by plants from *p*-coumaroyl CoA and malonyl CoA in response to environmental stress including injury, ultraviolet (UV) irradiation, or attack by pathogens [2–4]. The primary purpose of this review is to explore the current state of knowledge regarding the beneficial health effects reported to be associated

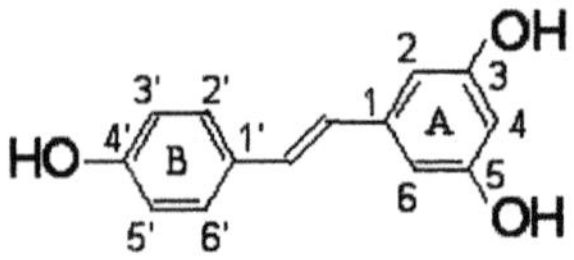

Resveratrol (RSVL)
(3,5,4'-trihydroxy-*trans*-stilbene)

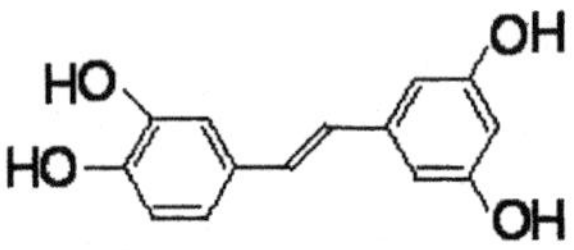

RSVL-1
(3,5,3',4'-tetrahydroxy-*trans*-stilbene)

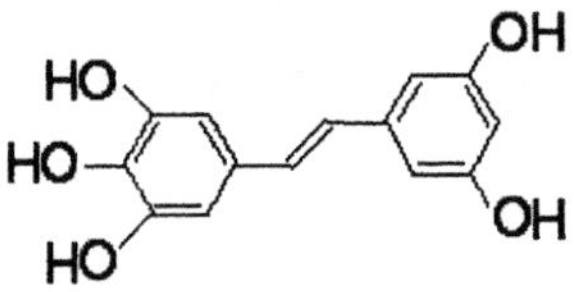

RSVL-2
(3,5,3',4',5'-pentahydroxy-*trans*-stilbene)

**Figure 1**   Structure of resveratrol and two analogues.

with resveratrol and to critically examine the solid mechanistic data available in support of those claims.

## II. SOURCES OF RESVERATROL

Resveratrol is found in both the *cis* and *trans* configuration in more than 70 plant species, many of which are components of the human diet, including mulberries, peanuts, and grapes [5–7]). Relatively high quantities are available in grapes, possibly because of the response of *Vitis vinifera* (Vitaceae) to fungal infection [8–10]). Resveratrol is found constitutively in the roots and stems but is induced in the leaves and fruits in response to fungal attack (reviewed in Ref. 1).

Fresh grape skin contains about 50–100 μg resveratrol/g fresh weight [11]. The biological activity of red wine has been attributed to its polyphenolic constituents, including resveratrol [5,12,13]. In red wine, the concentration of resveratrol is in the range of 1.5–20 mg/L [14–16], and white and rosé wine have lower (0.68–1 mg/L) but detectable amounts [17]. In addition, resveratrol is found in commercial grape juice at a level of about 4 mg/L [12]. Besides wine, peanuts and peanut butter are another significant source of resveratrol [18]. Resveratrol has also been found at substantially high levels in peanut roots, which are usually left in the field as agricultural waste [19]; itadori tea has also been found to contain relatively high levels [20]. Although information regarding the metabolism of resveratrol is becoming more available, the question of its absorption and bioavailability needs to be more fully addressed [2,21].

## III. ABSORPTION, METABOLISM, AND BIOAVAILABILITY OF RESVERATROL

Too many times, in vivo biological activity is extrapolated from in vitro studies with little regard for physiological availability. Because of its potentially important health impact, the bioavailability of resveratrol is of great importance. Some have suggested that this may be too low to achieve tissue concentrations needed to display the effects observed in vitro. Yang et al. very astutely point out that the chemical properties of a compound, conjugation reactions in the intestine, intestinal absorption, and metabolic enzymes are all key factors that must be considered in assessing the bioavailability of a particular food factor [22].

Earlier reports claimed that small doses of resveratrol in red wine resulted in a pharmacological effect on platelet aggregation [23]. Kinetics of *trans*- and *cis*-resveratrol and concentrations were evaluated in plasma, heart, liver, and kidneys of rats following oral administration of red wine. Investigators concluded that tissue concentrations showed a significant cardiac bioavailability and strong affinity for liver and kidneys [24–26]. However, these studies have since been criticized as providing little experimental detail or methodology [27].

More recent data indicate that perfusion of physiologically obtainable amounts of resveratrol stimulated small intestine vascular uptake and a conjugation of most of the absorbed resveratrol to yield resveratrol glucuronide. Smaller amounts of resveratrol, resveratrol sulfate, and resveratrol conjugates were also found in the intestinal tissue, and investigators concluded that the small intestine demonstrated sufficient uptake and metabolic conversion to be physiologically relevant [28]. Others have also studied the absorption and metabolism of resveratrol in a rat small intestine model. The results agreed that resveratrol was most likely in the form of a glucuronide conjugate after crossing the small intestine and entering the blood circulation [29]. Some investigators believe that glucuronation or sulfation may reduce its bioavailability. However, other evidence suggests that additional flavonoids appear to inhibit resveratrol glucuronidation, which, on the other hand, may improve its bioavailability [30–32].

The effective half-life of resveratrol is reported to be relatively short. Recent work by Soleas et al. [27] indicates that, following oral administration, the concentration of *trans*-resveratrol in blood and serum peaked very rapidly in rats. However, its metabolites appear to be more slowly metabolized, and 50–75% of total *trans*-resveratrol seemed to be absorbed in vivo [27]. Another group showed that plasma levels of resveratrol in rabbits, rats, and mice peaked within minutes following intravenous or oral administration (20 mg/kg) [33]. Vascular tissue levels followed plasma levels but were low. Most of the resveratrol measured was found in the *trans* form and was rapidly metabolized by hepatocytes [33]. In contrast, the pharmacokinetics of the aglycone and glucuronide forms of *trans*-resveratrol were studied following intravenous or oral administration to rats. As was observed by others, plasma concentrations of resveratrol peaked and then declined rapidly; however, a second increase in plasma concentration was observed 4–8 hours after administration. This later increase was attributed to enterohepatic recirculation. Investigators concluded that resveratrol is bioavailable and enterohepatic recirculation contributes to the extended tissue exposure [34]. In another study, resveratrol was injected intraperitoneally (30 mg/kg), either during or shortly after occlusion of common carotid arteries, and again at 24 hours after ischemia. A time course study revealed a peak activity of resveratrol in serum, liver, and brain at 1, 4, and 4 hours, respectively. Significantly, resveratrol protected against neuronal cell death following experimentally induced ischemia, demonstrating that it can cross the blood-brain barrier and exert protective effects against cerebral ischemic injury over a prolonged period of time [35]. Thus, most results seem to indicate that resveratrol is absorbed and metabolized, but whether common dietary sources (e.g., red wine and peanuts) can provide physiologically achievable, nontoxic amounts in humans is still not clear.

## IV.  ANTIOXIDANT ACTIVITIES OF RESVERATROL

Many dietary factors are believed to exert their potent antidisease effects because they possess strong antioxidant activities. However, direct compelling evidence for this supposition is lacking [22] and is difficult to obtain. A multitude of descriptive studies correlating the presence of resveratrol and protection against oxidative stress have been performed over the past 1–2 years. Almost all of these reports present indirect, correlative evidence. The data clearly support an antioxidant function for resveratrol but do very little to provide mechanistic evidence for its action in vivo. However, a few of the most recent studies are highlighted.

Resveratrol is a clearly a compound that exerts protective effects against a variety of oxidative stresses, but the effect observed seems to be dependent on the experimental system employed, leading to seemingly conflicting results. In addition, the mechanistic basis explaining its antioxidant effect is not clearly delineated, and some evidence suggests that resveratrol, like any good antioxidant, can be induced to exert pro-oxidant effects. For example, resveratrol has been shown to increase DNA strand breaks induced by $H_2O_2/Cu(II)$, and in the presence of only $Cu(II)$, resveratrol also caused DNA strand breaks in vitro [36]. On the other hand, resveratrol has also not been shown to induce DNA damage in vitro [31].

The interaction of resveratrol and lipoproteins has been studied, and resveratrol has been shown to decrease hydroperoxide accumulation in low-density lipoproteins (LDL) induced by ferrylmyoglobin and also to inhibit LDL apoprotein modification induced by peroxynitrite [37]. However, other in vitro experiments indicate that it can contribute to peroxidase-mediated oxidation of LDL [38]. Resveratrol and several analogs were shown to protect rat liver microsomes against peroxidation induced by 2,2'-azobis(2-amidinopropane hydrochloride) or $Fe(2^+)$/ascorbate [39]. Others have shown that resveratrol and its analogs differentially inhibited lipid peroxidation in rat brain, kidney, and liver homogenates and rat erythrocyte hemolysis [40]. However, a combination of horseradish peroxidase–$H_2O_2$ and resveratrol was observed to inactivate creatine kinase, which was suggested to be an indication of protein damage induced by free radicals [41]. Resveratrol has also been reported to inhibit the production of reactive oxygen intermediates induced by a variety of compounds in macrophages [42].

## A.  Resveratrol and Ischemia/Reperfusion Injury

Considerable attention has been paid to studies of the protective effects of resveratrol on ischemia/reperfusion-induced injury in heart [43] (reviewed in Ref. 44). These data appear to be solid and present compelling evidence that the compound is effective in preventing this type of damage, often associated with coronary artery disease and its repair. Preconditioning the heart with resveratrol has been shown by several groups to provide protection against ischemia/reperfusion injury and the protection appears to be clearly related to the stimulation of inducible nitric oxide synthase (iNOS) [45,46]. Results from perfusing hearts with or without resveratrol plus or minus a variety of nitric oxide inhibitors showed that resveratrol treatment resulted in improved recovery of postischemic ventricular function, smaller myocardial infarct size, and less cardiomyocyte apoptosis [45]. Another group compared the effect of resveratrol in ischemia/reperfusion-injured hearts from iNOS wild-type and knockout mice [46]. Similar results were obtained, but only in wildtype mice—resveratrol had no protective effect on hearts from iNOS knockout mice. The protective effects were associated with an increased iNOS mRNA level [45] and an increased iNOS protein expression induced by resveratrol [46]. Although the protective effect of resveratrol appears be evident, the stimulatory effects on iNOS and nitric oxide production do not seem to occur universally. In somewhat of a contrast to the above-described results, resveratrol was shown to inhibit lipopolysaccharide (LPS)-induced nitric oxide production in macrophage cells (RAW 264.7 and J774) when cells were treated prior to stimulation with LPS [47]. Additionally, iNOS has been suggested to be associated with damage observed in atherosclerosis, and resveratrol has been shown to suppress nitric oxide production by macrophages, mediators of blood vessel damage in atherosclerosis [48]. The observed effects on nitric oxide production and iNOS may be related to cell type or experimental model.

Resveratrol has also been shown to reduce ischemia-reperfusion injury in rat kidney [49,50]. In addition to heart and kidney, resveratrol was shown to protect against damage, including motor impairment and volume of infarct, from focal ischemia induced by middle cerebral artery occlusion in rats [51]. Resveratrol also substantially preserved rat brain mitochondrial function following anoxia-reoxygenation conditions [52]. However, in a rabbit model of ischemia/reperfusion, the compound failed to provide protection [53]. In spite of this, most evidence suggests that resveratrol has a significant protective effect against experimentally induced ischemia/reperfusion injury.

## B.  Coronary Heart Disease and Atherosclerosis

Coronary heart disease and atherosclerosis are major contributors to morbidity and mortality in the United States and other developed countries. Epidemiological studies suggest that decreased mortality from coronary heart disease is associated with moderate consumption of alcohol, and especially of red wine [7]. Numerous studies indicate that resveratrol inhibits platelet aggregation [8–10,12,16,54,55], alters eicosanoid synthesis [15,55], modulates lipoprotein mechanisms [14,56–58], inhibits vascular smooth muscle cell proliferation [59], and acts as an estrogen receptor agonist [60]. On the other hand, at least one earlier study indicated that it promoted atherosclerotic development, rather than protected against it, by a mechanism that appeared to be independent of observed differences in gross animal health, liver function, plasma cholesterol concentrations, or LDL oxidative status [61].

Evidence suggests that resveratrol may protect against atherosclerosis by enhancing endothelium integrity through the inhibition of protein kinase C (PKC)–mediated signaling [62]. However, resveratrol was also shown to inhibit protein kinase D, a member of the PKC family, but did not affect other PKC isoforms [63,64]. Furthermore, effects observed in bovine pulmonary artery endothelial cells subjected to stimulated arterial shear stress were also shown to be unrelated to PKC [65].

Resveratrol may protect against atherosclerosis by reducing the peroxidative deterioration of LDL; it was found to efficiently decrease the accumulation of hydroperoxides in LDL promoted by ferrylmyoglobin [37]. Earlier studies suggested that it may be effective at both the protein and lipid moieties of LDL [57]. Platelet aggregation precedes thrombus formation, which can precipitate occlusion of coronary arteries resulting in myocardial infarction. Very recent studies confirm that platelet aggregation was significantly inhibited by resveratrol in humans and in hypercholesterolemic rabbits [9,10]. In spite of the attention resveratrol has been receiving, the precise molecular and cellular mechanisms explaining its antiatherosclerosis effects are still to be elucidated.

## V. RESVERATROL AND CANCER PREVENTION

Jang et al. [66] were one of the first to report that the strong anticarcinogenesis effects of resveratrol were attributed to the inhibition of one or more stages (e.g., initiation, promotion, or progression) of tumor development. Since that report, a plethora of subsequent research data has confirmed that the compound does, indeed, have potent anticarcinogenesis effects in a variety of cancer types; this has been the subject of a number of recent reviews [11,22,67,68]. Although many studies indicate that a primary mechanism underlying these anticancer effects involves induction of cell cycle arrest and apoptosis, the precise targets and pathways mediating the anticarcinogenesis effects of resveratrol remain largely unknown. The remainder of this chapter will examine the current state of knowledge regarding the signal transduction pathways that appear to provide the mechanistic basis for the anticancer effects of this compound.

### A. Carcinogenesis, Signal Transduction, and Transcription Factors

In order to determine the individual mechanism[s] associated with the anticancer effects of dietary chemicals, the basis of cellular communication must first be understood. Information from an extracellular signal is transmitted from the plasma membrane into the cell and along an intracellular chain of signaling molecules to stimulate a cellular response, a process referred to as signal transduction. The response of an individual cell to a stimulus often involves the activation of gene transcription. Gene transcription is initiated by the binding of one or more proteins, known as transcription factors, to specific sequences of DNA that are located in the promoter region of the target gene. New gene transcription is the most common result of the binding of the transcription factor to the DNA, the process being referred to as transcriptional activation.

When cells are exposed to stress or mitogenic stimuli, a complex response occurs that involves distinct, but interactive, phosphorylation cascades and, in particular, members of the mitogen-activated protein kinase (MAPK) family. Individual MAPKs may be classified into three groups and include p38 kinases, c-Jun N-terminal kinases/stress-activated protein kinases (JNKs/SAPKs) and extracellular signal–regulated protein kinases (ERKs) [69–72]. The latter have been shown to be strongly activated and to play a critical role in transmitting signals initiated by tumor promoters such as TPA and growth factors, including platelet-derived growth factor (PDGF) and epidermal growth factor (EGF) [73,74]. In contrast, JNKs/SAPKs and p38 kinases are strongly activated by various forms of stress such as ultraviolet (UV) irradiation [70]. The activation of these various pathways is, however, not mutually exclusive. For example, EGF also partially activates the JNKs pathway, and heat shock and UV irradiation partially activate the ERKs cascade [71,74]. Evidence strongly suggests that activation of MAPKs by tumor-promoting agents plays a functional role in tumor promoter–induced malignant transformation [75–79]; MAP kinases are activated by translocation to the nucleus, where they influence specific transcription factors [80], including the tumor suppressor, p53, and activator protein-1 [AP-1] and nuclear factor kappa B (NF-κB), all of which are known to be extremely important in tumor promoter–induced malignant transformation and tumor promotion. Thus, MAP kinases appear to have dual, or even opposing, roles depending on the nature of the stimulus and the cell type.

## B.  Experimental Tools Used to Study the Effect of Dietary Factors on Signal Transduction

Genetic susceptibility to neoplastic transformation, promotion, and progression is widely studied using the JB6 mouse epidermal cell system of clonal genetic variants. This model is comprised of a series of cell lines that include tumor promotion–sensitive ($P^+$), tumor promotion–resistant ($P^-$), and transformed (Tx) variants, representing successive stages of preneoplastic to neoplastic progression [81–85]. $P^-$ variants gain $P^+$ phenotype upon transfection with mutated p53 [86,87], whereas $P^+$ cells gain Tx phenotype following treatment with tumor promoters [88,89]. Tumor promoters, such as TPA, EGF, tumor necrosis factor α (TNF-α), and arsenic induce high-frequency formation of large, tumorigenic, anchorage-independent colonies in $P^+$ cells grown in soft agar. In contrast, $P^-$ cells exhibit a response in soft agar that is ~1% that of $P^+$ cells; Moreover, their colonies are considerably smaller [81–85,88–92].

The use of specific cells transfected with a specific transcription factor-luciferase reporter gene and transgenic mouse models represents another important paradigm for studying transcriptional activation in cancer development. Both cells and mice carry a transcription factor sequence linked to a luciferase reporter gene, thereby allowing visualization of the particular transcription factor–linked luciferase activity using a luminometer [75–77]. These models are used extensively to monitor the activity of specific transcription factors in vivo [93,94]. In addition, dominant-negative mutants, knockout models, or specific inhibitors (employed to identify mediators) are beginning to be extensively used to evaluate potential chemopreventive compounds, including resveratrol.

## C.  Chemoprevention, Apoptosis, and Carcinogenesis

The complex multistage process in which normal cellular growth processes and genes become altered is known as carcinogenesis [95]. Precancerous or cancerous cells may result from clonal proliferation when chemical carcinogens induce mutations in genes that control normal growth [95]. The toxicological evaluation of chemicals is, therefore, frequently based on examination and analysis of cell proliferation data [96].

Programmed cell death, or apoptosis, of individual cells is considered to be an important mechanism that affords protection against cancer development in an organism by eliminating genetically damaged, or otherwise compromised, cells [97,98]. This concept is supported by the fact that many commonly used chemotherapeutic drugs are designed to induce apoptosis [99,100]. Accordingly, apoptosis appears to be a common mechanism by which cells containing irreparable genetic lesions are removed from the host organisms; disruption or suppression of apoptosis results in the survival and outgrowth of damaged or initiated cells, thereby causing carcinogenesis. A variety of tumor promoters act by suppressing or disrupting apoptosis, induced by such factors as exposure to ionizing radiation, low-energy β-radiation, acute serum deprivation [101], and chemicals, including colchicine, etoposide, or methylprednisolone [102].

Many cancer types are associated with loss of p53 function, resulting in a deficiency of normal apoptotic response to genotoxic damage. The induction of apoptosis in human and murine cells following DNA damage is critically dependent upon normal p53 function [103–106]; thus, mice deficient in p53 ($p53^{-/-}$) show almost no apoptosis in thymocytes or intestinal crypt cells following irradiation [104–106] and develop tumors spontaneously at a very high and rapid rate [107]. Thus, compelling evidence suggests that deficiency or disruption of normal p53 function allows a population of genetically damaged cells to escape apoptotic deletion and permits them to proliferate into preneoplastic or neoplastic clones [108–110].

Induction of apoptosis by many chemopreventive agents, including retinoic acid, perillyl alcohol, isothiocyanates, aspirin, tea compounds, ginger, and curcumin, is considered to be responsible, at least in part, for their chemopreventive activities [108,111–116]. Although some investigators have proposed that resveratrol acts as an antimutagenic or anticarcinogenic agent by preventing oxidative DNA damage [117] or inhibiting cyclooxygenase 1 and 2 [118–120], considerable evidence indicates that resveratrol acts primarily by inducing apoptosis in a wide variety of cancer cell types.

## D.  Induction of Apoptosis by Resveratrol

Low levels (0.1–1 μg/mL) of resveratrol have been reported to enhance cell proliferation, whereas higher amounts (10–100 μg/mL) cause apoptosis in a vari-

ety of tumor and endothelial cells, including JB6 epidermal cells, human promye-
loctye leukemia HL-60 cells, THP-1 human monocytic leukemia cells, U937
human promonocytic cells, various colon cancer cell lines, human mammary
cancer cell lines, and human prostate cancer cells [121–131]. In somewhat of a
contrast to these studies, a very recent report indicated that resveratrol reduced
paclitaxel-induced apoptosis in a human neuroblastoma cell line (SH-SY5Y) by
blocking paclitaxel-induced phosphorylation of JNKs, Raf-1, and Bcl-2 [132]. Of
particular interest is a recent study in which it was shown to induce apoptosis in
leukemic human lymphocytes but, under similar conditions, had no effect on the
survival of normal peripheral blood mononuclear cells. The induction of apoptosis
was associated with decreased expression of iNOS and Bcl-2, reduced mitochondrial membrane potential, and an activation of caspase 3 in the cancer cells [133].

Although the precise mechanism and targets for the induction of apoptosis
remain to be revealed, accumulating evidence shows that the anticancer properties
of resveratrol are related to its ability to cause cell cycle arrest in various, but
well-specified, stages of the cell cycle [134–139]. These studies suggest that the
compound has a marked effect on cell cycle regulation. Inhibition of growth and
S/G2 cell cycle arrest by resveratrol have been shown to be associated with a
decrease in ornithine decarboxylase (ODC) activity, which is commonly enhanced
in cancers and is crucial for polyamine biosynthesis [139]. Decreased expression
of various proteins linked with cell cycle regulation has also been associated
with resveratrol treatment [129]. In human epidermoid carcinoma (A431) cells,
resveratrol appeared to act by modulating proteins connected with cell cycle
regulation, resulting in growth inhibition, cell cycle arrest, and apoptosis [136].
In particular, treatment was found to induce WAF1/p21, thereby presumably
accounting for the observed decrease in the expression of cyclins D1, D2, and
E, and the decreased expression and activity of cyclin-dependent kinases 2, 4,
and 6 [136]. Ahmad et al. have provided further evidence that cell cycle inhibition
by resveratrol involves downregulation of retinoblastoma (pRb) phosphorylation
and the E2F family of transcription factors, which play a role in cell cycle progression near the G1/S phase transition [135].

Resveratrol has also been shown to have a marked effect on the NF-κB
signaling pathway, inhibiting NF-κB activation and NF-κB–dependent gene
expression as a result of its inhibitory effect on IκB kinase [140]. Other researchers have also reported that it may exert its effect by suppressing NF-κB activation
as a consequence of blocking TNF-induced activation of NF-κB in myeloid (U-
937), lymphoid (Jurkat), and epithelial cells [141]. Both NF-κB–dependent reporter gene transcription and TNF-induced phosphorylation and translocation of
the p65 subunit of NF-κB were inhibited, coupled with suppression of TNF-
induced activation of AP-1, JNKs, and mitogen-activated protein kinase kinase
(MAPKK) [141]. Others have shown that a naturally occurring structural analog
of resveratrol, piceatannol, also effectively suppressed TNF-induced DNA bind-

ing of NF-κB in human myeloid, lymphocyte, and epithelial cells [142]. Furthermore, piceatannol blocked TNF-induced matrix metalloprotease-9, COX-2, and cyclin D1 expression. The mechanism appeared to be related to an inhibition of IκBα and p65 phosphorylation and IκBα kinase activation and p65 translocation [142]. Recent work showed that resveratrol suppressed the phosphorylation and transactivation of the p65 subunit but not the activation or translocation of the NF-κB/Rel proteins [143].

Resveratrol has been shown to inhibit proliferation, induce differentiation, and enhance the expression of adhesion molecules (CD11a, CD11b, CD18, CD54) in a variety of myeloid leukemia cell lines [144]. The inhibitory effect of resveratrol on cell survival and proliferation of human breast cancer cells was shown to be estrogen receptor dependent [145]. Other studies indicated that it was very effective in inhibiting growth of 4T1 breast cancer cells in culture but had little effect on 4T1 tumor cell growth in mice [146]. In stark contrast, resveratrol was tested against mammary tumors induced by 7,12-dimethylbenz(a)anthracene (DMBA) in rats [147]. In this case, reductions in incidence, multiplicity, and latency period of tumor development were observed. Additional observations included reduced DMBA-induced activation of COX-2, matrix metalloprotease, and NF-κB in resveratrol treated rats. Similar results were also obtained in human breast cancer MCF-7 cells treated with resveratrol [147].

In JB6 cells, resveratrol inhibited tumor promoter TPA- and EGF-induced cell transformation in a dose-dependent manner over a range of 2.3–40 μM [148] and also induced apoptosis. The relationship between the chemical structure of resveratrol and its anticancer activity has been investigated. Two derivatives, RSVL-1 (3,5,3',4'-tetrahydroxy-*trans*-stilbene) and RSVL-2 (3,5,3',4',5'-pentahydroxy-*trans*-stilbene), (Fig. 1) were synthesized and their activities compared with that of resveratrol. Results show that RSVL-2 exhibited a more potent inhibitory effect on EGF-induced cell transformation [149], whereas RSVL-1 showed a reduced inhibition of cell transformation. In contrast to resveratrol, RSVL-2 appeared to exert its anticarcinogenic effect by targeting phosphatidylinositol-3 (PI-3) kinase and was significantly less toxic than the parent compound [149]. Lu et al. [150] have also studied structural analogs of resveratrol and found at least one that specifically inhibits the growth of transformed WI38 cells, but has little effect on normal WI38 cells. This growth inhibition was linked to an increased expression of p53, GADD45, and Bax with corresponding suppression of Bcl2 [150], suggesting the possibility that analogs of resveratrol and other chemopreventive compounds that specifically target tumor cells with little or no toxicity can be successfully developed.

## E. The Role of p53 in Resveratrol-Induced Apoptosis

Normal expression of p53 is critical for tumor suppression because loss or mutation of p53 protein or gene expression is observed in more than half of all human cancers. The p53 protein is, therefore, crucial for induction of apoptosis, and lack of p53 expression or function is associated with an enhanced risk of tumor formation [103–106].

Several reports indicate that p53 is required in resveratrol-induced apoptosis in different cell systems. It has been shown to suppress tumor promoter–induced cell transformation and to stimulate p53-dependent transcriptional activation [148]. Moreover, resveratrol-induced apoptosis occurred only in fibroblasts expressing wild-type p53 ($p53^{+/+}$) and not in p53-deficient ($p53^{-/-}$) cells [148]. In cultured human promyelocytic leukemia (HL-60) cells, apoptosis was induced, resulting in reduced viability and DNA synthesis [122]. These changes were linked to a gradual decrease in the expression of the antiapoptotic protein Bcl-2. Bcl-2 and the pro-apoptosis gene, *Bax*, are two of several apoptosis-related transcriptional targets for p53 [151]. Further studies indicated that cells overexpressing Bcl-2 were markedly resistant to resveratrol-induced caspase-3 activation and apoptosis and exhibited less cytochrome *c* release during subsequent apoptosis [152]. Tessitore et al. [153] studied the effect of resveratrol on azoxymethane (AOM)-induced colon carcinogenesis and found the number and multiplicity of aberrant crypt foci (ACF) to be significantly reduced; this was attributed to changes in Bax and p21 expression induced by resveratrol. The *p21* gene is also a transcriptional target of p53 [154,155]. Additional work indicated that resveratrol induced the expression of NAG-1 [nonsteroidal anti-inflammatory (NSAID) drug–activated gene-1], a member of the TGF-β family of proteins, which is associated with apoptosis and the induction of NAG-1 expression was mediated by p53 [156].

In certain cell lines, however, no effect of resveratrol on p53 expression could be observed [157], but in others, such as p53-inactivated HCT116 colon carcinoma and HCT116, resveratrol was found to induce apoptosis independently of p53 [158,159]. Such reports obviously indicate a dependence on cell type.

### F. Phosphorylation of p53 and Kinases Involved in Resveratrol-Induced Apoptosis

On the basis of the studies reported above, a possible mechanism may involve extracellular signal–regulated protein kinases (ERKs) and p38 kinase-mediated p53 activation and induction of apoptosis [148,160] (Fig. 2). In the mouse JB6 epidermal cell line [161,162], resveratrol induced apoptosis to inhibit tumor promoter–induced cell transformation through enhanced transactivation of p53 activity [148]. Resveratrol-induced activation of p53 and apoptosis was shown to be dependent on the activities of ERKs and p38 kinase and on their phosphorylation of p53 at serine 15 [160], which is known to be critical for the stabilization, upregulation, and functional activation of this protein [101,103]. Tredici et al. have also shown that resveratrol can induce phosphorylation of ERKs in human neuro-

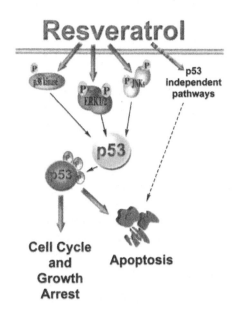

**Figure 2** Resveratrol may induce apoptosis and cell cycle arrest through p53 phosphorylation mediated by MAPKs.

blastoma SH-SY5Y cells [163], and more recently it has also been reported to induce activation of JNKs [164]. The stable expression of a dominant negative mutant of JNK1, or the disruption of the *Jnk1* or *Jnk2* genes, markedly inhibited resveratrol-induced p53-dependent transcription activation and induction of apoptosis [160,164]. Resveratrol-activated JNKs were also shown to phosphorylate p53 in vitro, but this was repressed in cells expressing a dominant negative mutant of JNK1, or in JNK1 or JNK2 knockout (*Jnk1*$^{-/-}$ or *Jnk2*$^{-/}$) cells.

Taken together, these data suggested that ERKs, p38 kinase, and JNKs mediate resveratrol-induced activation of p53 and apoptosis and that this may occur, at least partially, through p53 phosphorylation [160,164] (Fig. 2). Resveratrol treatment can stimulate apoptosis in androgen-insensitive DU 145 prostate cancer cells through activation of ERKs and increased accumulation of total p53 and serine-15 phosphorylated p53, and a p53-stimulated increase in *p21* messenger RNA [165]. Resveratrol also induced apoptosis mediated by activation and nuclear translocation of ERKs in papillary thyroid carcinoma and follicular thyroid carcinoma cell lines, which was associated with accumulation of p53 protein, serine phosphorylation of p53, and abundance of *c-fos*, *c-jun*, and *p21* mRNAs [166]. The effects on the activity of activator protein 1 (AP-1) and MAP kinase pathways have been examined in HeLa cells [167], and inhibition of UVC- or phorbol ester–induced AP-1 transcriptional activation and ERKs, JNK1, and p38 kinase activation was observed. These observations were believed to result from inhibition of both protein tyrosine kinases and protein kinase C [167]. In contrast to this report, treatment of human gastric adenocarcinoma KATO-III and RF-1 cells led to cell cycle arrest, apoptosis, and a significant inhibition of PKCα, but no effect on ERKs was observed [168].

Chemoprevention of carcinogenesis using nontoxic chemical compounds is now accepted as a promising alternative strategy to therapy for the control of human cancers. Convincing scientific data have recently been presented, which show that naturally occurring substances can protect against experimental carcinogenesis. One such compound, resveratrol (3,5,4'-trihydroxy-*trans*-stilbene), found in many dietary plants including grapes and peanuts, has been shown to elicit anticancer effects in a variety of systems that are considered to arise from its inhibition of diverse cellular events associated with tumor initiation, promotion, and progression. These inhibitory effects are strongly linked to an ability to induce cell cycle arrest and apoptosis in cancer cells (summarized in Fig. 3). Although such effects are dependent on cell type and stimulus, they involve activation and phosphorylation of p53 mediated by MAP kinase pathways. While considerable knowledge has been gained in recent years, the nature and complexity of the

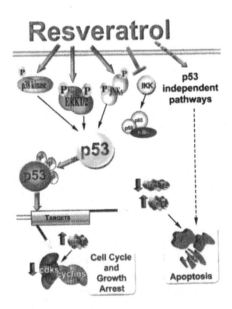

**Figure 3**   Resveratrol inhibits cellular events associated with tumor initiation, promotion, and progression by inducing cell cycle arrest and apoptosis.

interactions occurring between dietary compounds and molecular pathways requires more study to gain an understanding of the molecular mechanistic bases for the chemopreventive effects of resveratrol and its derivatives. Such an understanding offers considerable promise for the designing of a new generation of nontoxic and highly effective chemopreventive agents for control of human cancer.

## VI. CONCLUSIONS AND PERSPECTIVES REGARDING RESVERATROL AND ITS FUTURE IMPACT

Numerous substances derived from foods have been linked to decreased risk of developing cancer and other diseases. However, the available information is often confusing and contradictory due to the complexity of the many interactions that occur between dietary compounds and molecular pathways related to development of disease. Solid mechanistic research data are critically needed to dispel the myths and misinformation surrounding diet and its effect on cancer and other diseases.

Although much of the early information regarding resveratrol is circumstantial in nature and often contradictory, solid mechanistic data are rapidly accumulating elucidating the interactions between resveratrol and molecular pathways related to development of disease. A strong research emphasis on resveratrol as an anticancer agent has occurred during the last 1–2 years. But many questions still need to be addressed. Red wines and peanuts are major sources of resveratrol; unfortunately, many humans exhibit a dangerous allergic reaction to peanuts, and the wisdom of consuming large amounts of red wine is questionable and may cause damage to other organs, in addition to impairing normal motor and mental functions. Although the final metabolic products have not been fully identified, recent experimental results indicate that resveratrol is probably absorbed in the small intestine and mostly metabolized in the liver. Plasma levels of experimentally administered resveratrol peak early and then decline but may show a second peak of activity several hours later. This suggests that the compound could have a good bioavailability; however, whether common dietary sources can provide physiologically relevant levels of resveratrol is not clear.

Consistent research results regarding the effectiveness and mechanisms of the action of resveratrol in the prevention of atherosclerosis and coronary heart disease are still elusive. Resveratrol most likely exerts some of its beneficial effects, which are correlated with a decreased risk of atherosclerosis and coronary heart disease, through nonspecific antioxidant actions. However, much of the evidence applied to the human condition appears to be circumstantial and extrapolated from in vitro nonhuman studies. The strongest substantiation is that obtained regarding the effectiveness of resveratrol in preventing experimentally induced ischemia/reperfusion injury in a variety of animal tissues, including heart, kidney, and brain. Whether these results can be extrapolated to the human condition, e.g., preventing ischemia/reperfusion damage often associated with coronary bypass surgery, remains to be determined.

Probably the greatest impact of resveratrol in the last few years has been its highly touted effectiveness as an anticancer agent. When Jang et al. [66] presented their findings regarding resveratrol in *Science*, an explosion of research interest focused on resveratrol and cancer. Resveratrol was reported to act at all three stages of cancer development by inhibiting free radical formation, inducing phase II enzymes and COX-1 and COX-2 hydroperoxidase activities, inducing differentiation in human promyelocytic leukemia cells, and finally preventing tumor development in the DMBA/TPA two-stage mouse skin model. Since that report, considerable solid experimental evidence indicates that resveratrol acts specifically on cancer cells with little effect on normal cells and acts primarily by causing cell cycle arrest and inducing apoptosis in tumor cells. In many tumor types, resveratrol appears to act through inhibition of the NF-κB signaling pathways and mainly through activation of the p53 tumor suppressor protein mediated by MAPKs. Unfortunately, this may mean that the compound will not be effective in inhibiting tumors with a deficient or mutated p53, a condition observed in almost 50% of all human cancers. However, several analogs of resveratrol have been shown to have greater anticancer effects and less toxicity than the parent compound. This suggests that chemopreventive or chemotherapeutic compounds that specifically target tumor cells with little or no toxicity can be successfully developed.

More long-term studies and clinical trials are needed to assess the bioavailability, toxicity, effectiveness, and other side effects of resveratrol and other dietary factors. Solid mechanistic research data are critically needed to dispel the fabrications and misconceptions surrounding diet and its effect on cancer and other diseases. More of the molecular pathways need to be studied in order to understand why and how dietary components can exert antidisease effects. The use of knockout cells and mice, dominant-negative mutants, small interfering RNA, and specific chemical inhibitors will aid in elucidating the function and role of each of the signaling molecules involved in dietary-induced signaling and disease prevention. These types of studies should improve our understanding of the importance of diet and lead to the identification of new dietary factors and their targets of action.

## ACKNOWLEDGMENTS

This work was supported by The Hormel Foundation and grants funded by the American Institute for Cancer Research (99A062) and National Institutes of Health (CA81064).

## REFERENCES

1. Bavaresco L, Fregoni C, Cantu E, Trevisan M. Stilbene compounds: from the grapevine to wine. Drugs Exp Clin Res 1999; 25:57–63.
2. Soleas GJ, Diamandis EP, Goldberg DM. Resveratrol: a molecule whose time has come? And gone. Clin Biochem 1997; 30:91–113.
3. Dercks W, Creasy LL. The significance of stilbene phytoalexins in the *Plasmopara viticola*-grapevine interaction. Physiol Mol Plant Path 1989; 34:189–202.
4. Dercks W, Creasy LL. The influence of fosetyl-Al on phytoalexin accumulation in the *Plasmopara viticola*-grapevine interaction. Physiol Mol Plant Path 1989; 34: 203–213.
5. Goldberg DM, Yan J, Ng E, Diamandis EP, Karumanchiri A, Soleas GJ, Waterhouse AL. A global survey of trans-resveratrol concentrations in commercial wines. Am J Enol Viti 1995; 46:159–165.
6. Romero-Perez AI, Lamuela-Raventos RM, Waterhouse AL, de la Torre-Boronat MC. Levels of cis- and trans-resveratrol and their glucosides in white and rose vitius vinifera wines from Spain. J Agric Food Chem 1996; 44:2124–2128.
7. Kopp P. Resveratrol, a phytoestrogen found in red wine. A possible explanation for the conundrum of the 'French paradox'. Eur J Endocrinol 1998; 138:619–620.
8. Bertelli AA, Giovannini L, Giannessi D, Migliori M, Bernini W, Fregoni M, Bertelli A. Antiplatelet activity of synthetic and natural resveratrol in red wine. Int J Tissue React 1995; 17:1–3.
9. Wang Z, Zou J, Huang Y, Cao K, Xu Y, Wu JM. Effect of resveratrol on platelet aggregation in vivo and in vitro. Chin Med J (Engl) 2002; 115:378–380.
10. Wang Z, Huang Y, Zou J, Cao K, Xu Y, Wu JM. Effects of red wine and wine polyphenol resveratrol on platelet aggregation in vivo and in vitro. Int J Mol Med 2002; 9:77–79.
11. Gusman J, Malonne H, Atassi G. A reappraisal of the potential chemopreventive and chemotherapeutic properties of resveratrol. Carcinogenesis 2001; 22:1111–1117.
12. Pace-Asciak CR, Rounova O, Hahn SE, Diamandis EP, Goldberg DM. Wines and grape juices as modulators of platelet aggregation in healthy human subjects. Clin Chim Acta 1996; 246:163–182.
13. Fremont L. Biological effects of resveratrol. Life Sci 2000; 66:663–673.
14. Fremont L, Belguendouz L, Delpal S. Antioxidant activity of resveratrol and alcohol-free wine polyphenols related to LDL oxidation and polyunsaturated fatty acids. Life Sci 1999; 64:2511–2521.
15. Pinto MC, Garcia-Barrado JA, Macias P. Resveratrol is a potent inhibitor of the dioxygenase activity of lipoxygenase. J Agric Food Chem 1999; 47:4842–4846.
16. Olas B, Wachowicz B, Stochmal A, Oleszek W. Anti-platelet effects of different phenolic compounds from *Yucca schidigera* Roezl bark Platelets. 2002; 13: 167–173.
17. Sato M, Suzuki Y, Okuda T, Yokotsuka K. Contents of resveratrol, piceid, and their isomers in commercially available wines made from grapes cultivated in Japan. Biosci Biotechnol Biochem 1997; 61:1800–1805.

18.  Ibern-Gomez M, Roig-Perez S, Lamuela-Raventos RM, de la Torre-Boronat MC. Resveratrol and piceid levels in natural and blended peanut butters. J Agric Food Chem 2000; 48:6352–6354.
19.  Chen RS, Wu PL, Chiou RY. Peanut roots as a source of resveratrol. J Agric Food Chem 2002; 50:1665–1667.
20.  Burns J, Yokota T, Ashihara H, Lean ME, Crozier A. Plant foods and herbal sources of resveratrol. J Agric Food Chem 2002; 50:3337–3340.
21.  Soleas GJ, Diamandis EP, Goldberg DM. Wine as a biological fluid: history, production, and role in disease prevention. J Clin Lab Anal 1997; 11:287–313.
22.  Yang CS, Landau JM, Huang MT, Newmark HL. Inhibition of carcinogenesis by dietary polyphenolic compounds. Annu Rev Nutr 2001; 21:381–406.
23.  Bertelli A, Bertelli AA, Gozzini A, Giovannini L. Plasma and tissue resveratrol concentrations and pharmacological activity. Drugs Exp Clin Res 1998; 24:133–138.
24.  Bertelli AA, Giovannini L, Stradi R, Urien S, Tillement JP, Bertelli A. Kinetics of trans- and cis-resveratrol (3,4',5-trihydroxystilbene) after red wine oral administration in rats. Int J Clin Pharmacol Res 1996; 16:77–81.
25.  Bertelli AA, Giovannini L, Stradi R, Bertelli A, Tillement JP. Plasma, urine and tissue levels of trans- and cis-resveratrol (3,4',5-trihydroxystilbene) after short-term or prolonged administration of red wine to rats. Int J Tissue React 1996; 18:67–71.
26.  Bertelli AA, Giovannini L, Stradi R, Urien S, Tillement JP, Bertelli A. Evaluation of kinetic parameters of natural phytoalexin in resveratrol orally administered in wine to rats. Drugs Exp Clin Res 1998; 24:51–55.
27.  Soleas GJ, Angelini M, Grass L, Diamandis EP, Goldberg DM. Absorption of trans-resveratrol in rats. Methods Enzymol 2001; 335:145–154.
28.  Andlauer W, Kolb J, Siebert K, Furst P. Assessment of resveratrol bioavailability in the perfused small intestine of the rat. Drugs Exp Clin Res 2000; 26:47–55.
29.  Kuhnle G, Spencer JP, Chowrimootoo G, Schroeter H, Debnam ES, Srai SK, Rice-Evans C, Hahn U. Resveratrol is absorbed in the small intestine as resveratrol glucuronide. Biochem Biophys Res Commun 2000; 272:212–217.
30.  De Santi C, Pietrabissa A, Spisni R, Mosca F, Pacifici GM. Sulphation of resveratrol, a natural compound present in wine, and its inhibition by natural flavonoids. Xenobiotica 2000; 30:857–866.
31.  De Salvia R, Festa F, Ricordy R, Perticone P, Cozzi R. Resveratrol affects in a different way primary versus fixed DNA damage induced by $H_2O_2$ in mammalian cells in vitro. Toxicol Lett 2002; 135:1–9.
32.  De Santi C, Pietrabissa A, Mosca F, Pacifici GM. Glucuronidation of resveratrol, a natural product present in grape and wine, in the human liver. Xenobiotica 2000; 30:1047–1054.
33.  Asensi M, Medina I, Ortega A, Carretero J, Bano MC, Obrador E, Estrela JM. Inhibition of cancer growth by resveratrol is related to its low bioavailability. Free Radic Biol Med 2002; 33:387–398.
34.  Marier JF, Vachon P, Gritsas A, Zhang J, Moreau JP, Ducharme MP. Metabolism and disposition of resveratrol in rats: extent of absorption, glucuronidation, and enterohepatic recirculation evidenced by a linked-rat model. J Pharmacol Exp Ther 2002; 302:369–373.

35. Wang Q, Xu J, Rottinghaus GE, Simonyi A, Lubahn D, Sun GY, Sun AY. Resveratrol protects against global cerebral ischemic injury in gerbils. Brain Res 2002; 958:439–447.

36. Win W, Cao Z, Peng X, Trush MA, Li Y. Different effects of genistein and resveratrol on oxidative DNA damage in vitro. Mutat Res 2002; 513:113–120.

37. Brito P, Almeida LM, Dinis TC. The interaction of resveratrol with ferrylmyoglobin and peroxynitrite; protection against LDL oxidation. Free Radic Res 2002; 36: 621–631.

38. Pietraforte D, Turco L, Azzini E, Minetti M. On-line EPR study of free radicals induced by peroxidase/$H_2O_2$ in human low-density lipoprotein. Biochim Biophys Acta 2002; 1583:176–184.

39. Cai YJ, Fang JG, Ma LP, Yang L, Liu ZL. Inhibition of free radical-induced peroxidation of rat liver microsomes by resveratrol and its analogues. Biochim Biophys Acta 2003; 1637:31–38.

40. Lu M, Cai YJ, Fang JG, Zhou YL, Liu ZL, Wu LM. Efficiency and structure-activity relationship of the antioxidant action of resveratrol and its analogs. Pharmazie 2002; 57:474–478.

41. Miura T, Muraoka S, Fujimoto Y. Inactivation of creatine kinase induced by stilbene derivatives. Pharmacol Toxicol 2002; 90:66–72.

42. Leiro J, Alvarez E, Garcia D, Orallo F. Resveratrol modulates rat macrophage functions. Int Immunopharmacol 2002; 2:767–774.

43. Hung LM, Su MJ, Chu WK, Chiao CW, Chan WF, Chen JK. The protective effect of resveratrols on ischaemia-reperfusion injuries of rat hearts is correlated with antioxidant efficacy. Br J Pharmacol 2002; 135:1627–1633.

44. Wu JM, Wang ZR, Hsieh TC, Bruder JL, Zou JG, Huang YZ. Mechanism of cardioprotection by resveratrol, a phenolic antioxidant present in red wine (review). Int J Mol Med 2001; 8:3–17.

45. Hattori R, Otani H, Maulik N, Das DK. Pharmacological preconditioning with resveratrol: role of nitric oxide. Am J Physiol Heart Circ Physiol 2002; 282: H1988–1995.

46. Imamura G, Bertelli AA, Bertelli A, Otani H, Maulik N, Das DK. Pharmacological preconditioning with resveratrol: an insight with iNOS knockout mice. Am J Physiol Heart Circ Physiol 2002; 282:H1996–2003.

47. Cho D, Koo N, Chung W, Kim T, Ryu S, Im S, Kim K. Effects of resveratrol-related hydroxystilbenes on the nitric oxide production in macrophage cells: structural requirements and mechanism of action. Life Sci 2002; 71:2071–2082.

48. Chan MM, Mattiacci JA, Hwang HS, Shah A, Fong D. Synergy between ethanol and grape polyphenols, quercetin, and resveratrol, in the inhibition of the inducible nitric oxide synthase pathway. Biochem Pharmacol 2000; 60:1539–1548.

49. Bertelli AA, Migliori M, Panichi V, Origlia N, Filippi C, Das DK, Giovannini L. Resveratrol, a component of wine and grapes, in the prevention of kidney disease. Ann NY Acad Sci 2002; 957:230–238.

50. Giovannini L, Migliori M, Longoni BM, Das DK, Bertelli AA, Panichi V, Filippi C, Bertelli A. Resveratrol, a polyphenol found in wine, reduces ischemia reperfusion injury in rat kidneys. J Cardiovasc Pharmacol 2001; 37:262–270.

51. Sinha K, Chaudhary G, Kumar Gupta Y. Protective effect of resveratrol against oxidative stress in middle cerebral artery occlusion model of stroke in rats. Life Sci 2002; 71:655–665.

52. Zini R, Morin C, Bertelli A, Bertelli AA, Tillement JP. Resveratrol-induced limitation of dysfunction of mitochondria isolated from rat brain in an anoxia-reoxygenation model. Life Sci 2002; 71:3091–3108.

53. Hale SL, Kloner RA. Effects of resveratrol, a flavinoid found in red wine, on infarct size in an experimental model of ischemia/reperfusion. J Stud Alcohol 2001; 62: 730–735.

54. Bertelli AA, Giovannini L, Bernini W, Migliori M, Fregoni M, Bavaresco L, Bertelli A. Antiplatelet activity of cis-resveratrol. Drugs Exp Clin Res 1996; 22:61–63.

55. Pace-Asciak CR, Hahn S, Diamandis EP, Soleas G, Goldberg DM. The red wine phenolics trans-resveratrol and quercetin block human platelet aggregation and eicosanoid synthesis: implications for protection against coronary heart disease. Clin Chim Acta 1995; 235:207–219.

56. Bhavnani BR, Cecutti A, Gerulath A, Woolever AC, Berco M. Comparison of the antioxidant effects of equine estrogens, red wine components, vitamin E, and probucol on low-density lipoprotein oxidation in postmenopausal women. Menopause 2001; 8:408–419.

57. Belguendouz L, Fremont L, Gozzelino MT. Interaction of transresveratrol with plasma lipoproteins. Biochem Pharmacol 1998; 55:811–816.

58. Zou J, Huang Y, Chen Q, Wei E, Cao K, Wu JM. Effects of resveratrol on oxidative modification of human low density lipoprotein. Chin Med J (Engl) 2000; 113: 99–102.

59. Araim O, Ballantyne J, Waterhouse AL, Sumpio BE. Inhibition of vascular smooth muscle cell proliferation with red wine and red wine polyphenols. J Vasc Surg 2002; 35:1226–1232.

60. Gehm BD, McAndrews JM, Chien PY, Jameson JL. Resveratrol, a polyphenolic compound found in grapes and wine, is an agonist for the estrogen receptor. Proc Natl Acad Sci USA 1997; 94:14138–14143.

61. Wilson T, Knight TJ, Beitz DC, Lewis DS, Engen RL. Resveratrol promotes atherosclerosis in hypercholesterolemic rabbits. Life Sci 1996; 59:L15–21.

62. Slater SJ, Seiz JL, Cook AC, Stagliano BA, Buzas CJ. Inhibition of protein kinase C by resveratrol. Biochim Biophys Acta 2003; 1637:59–69.

63. Haworth RS, Avkiran M. Inhibition of protein kinase D by resveratrol. Biochem Pharmacol 2001; 62:1647–1651.

64. Stewart JR, Christman KL, O'Brian CA. Effects of resveratrol on the autophosphorylation of phorbol ester- responsive protein kinases: inhibition of protein kinase D but not protein kinase C isozyme autophosphorylation. Biochem Pharmacol 2000; 60:1355–1359.

65. Bruder JL, Hsieh Tc T, Lerea KM, Olson SC, Wu JM. Induced cytoskeletal changes in bovine pulmonary artery endothelial cells by resveratrol and the accompanying modified responses to arterial shear stress. BMC Cell Biol 2001; 2:1.

66. Jang M, Cai L, Udeani GO, Slowing KV, Thomas CF, Beecher CW, Fong HH, Farnsworth NR, Kinghorn AD, Mehta RG, Moon RC, Pezzuto JM. Cancer chemo-

preventive activity of resveratrol, a natural product derived from grapes. Science 1997; 275:218–220.

67. Bhat KPL, Kosmeder JW, Pezzuto JM. Biological effects of resveratrol. Antioxid Redox Signal 2001; 3:1041–1064.

68. Savouret JF, Quesne M. Resveratrol and cancer: a review. Biomed Pharmacother 2002; 56:84–87.

69. Kyriakis JM, Banerjee P, Nikolakaki E, Dai T, Rubie EA, Ahmad MF, Avruch J, Woodgett JR. The stress-activated protein kinase subfamily of c-Jun kinases. Nature 1994; 369:156–160.

70. Kallunki T, Su B, Tsigelny I, Sluss HK, Derijard B, Moore G, Davis R, Karin M. JNK2 contains a specificity-determining region responsible for efficient c-Jun binding and phosphorylation. Genes Dev 1994; 8:2996–3007.

71. Davis RJ. MAPKs: new JNK expands the group. Trends Biochem Sci 1994; 19: 470–473.

72. Boulton TG, Nye SH, Robbins DJ, Ip NY, Radziejewska E, Morgenbesser SD, DePinho RA, Panayotatos N, Cobb MH, Yancopoulos GD. ERKs: a family of protein-serine/threonine kinases that are activated and tyrosine phosphorylated in response to insulin and NGF. Cell 1991; 65:663–675.

73. Cowley S, Paterson H, Kemp P, Marshall CJ. Activation of MAP kinase kinase is necessary and sufficient for PC12 differentiation and for transformation of NIH 3T3 cells. Cell 1994; 77:841–852.

74. Minden A, Lin A, McMahon M, Lange-Carter C, Derijard B, Davis RJ, Johnson GL, Karin M. Differential activation of ERK and JNK mitogen-activated protein kinases by Raf-1 and MEKK. Science 1994; 266:1719–1723.

75. Dong Z, Huang C, Brown RE, Ma WY. Inhibition of activator protein 1 activity and neoplastic transformation by aspirin. J Biol Chem 1997; 272:9962–9970.

76. Dong Z, Ma W, Huang C, Yang CS. Inhibition of tumor promoter-induced activator protein 1 activation and cell transformation by tea polyphenols, ( − )-epigallocatechin gallate, and theaflavins. Cancer Res 1997; 57:4414–4419.

77. Huang C, Ma WY, Dawson MI, Rincon M, Flavell RA, Dong Z. Blocking activator protein-1 activity, but not activating retinoic acid response element, is required for the antitumor promotion effect of retinoic acid. Proc Natl Acad Sci USA 1997; 94: 5826–5830.

78. Huang C, Ma WY, Young MR, Colburn N, Dong Z. Shortage of mitogen-activated protein kinase is responsible for resistance to AP-1 transactivation and transformation in mouse JB6 cells. Proc Natl Acad Sci USA 1998; 95:156–161.

79. Watts RG, Huang C, Young MR, Li JJ, Dong Z, Pennie WD, Colburn NH. Expression of dominant negative Erk2 inhibits AP-1 transactivation and neoplastic transformation. Oncogene 1998; 17:3493–3498.

80. Kharbanda S, Saxena S, Yoshida K, Pandey P, Kaneki M, Wang Q, Cheng K, Chen YN, Campbell A, Sudha T, Yuan ZM, Narula J, Weichselbaum R, Nalin C, Kufe D. Translocation of SAPK/JNK to mitochondria and interaction with Bcl-x(L) in response to DNA damage. J Biol Chem 2000; 275:322–327.

81. Hsu TC, Young MR, Cmarik J, Colburn NH. Activator protein 1 (AP-1)- and nuclear factor kappaB (NF-kappaB)- dependent transcriptional events in carcinogenesis. Free Radic Biol Med 2000; 28:1338–1348.

82. Dong Z, Cmarik JL, Wendel EJ, Colburn NH. Differential transformation efficiency but not AP-1 induction under anchorage-dependent and -independent conditions. Carcinogenesis 1994; 15:1001–1004.

83. Dong Z, Watts SG, Sun Y, Colburn NH. Progressive elevation of AP-1 activity during preneoplastic-to neoplastic progression as modeled in mouse JB6 cell variants. Int J Oncol 1995; 7:359–364.

84. Dong Z, Birrer MJ, Watts RG, Matrisian LM, Colburn NH. Blocking of tumor promoter-induced AP-1 activity inhibits induced transformation in JB6 mouse epidermal cells. Proc Natl Acad Sci USA 1994; 91:609–613.

85. Huang C, Ma WY, Dong Z. Requirement for phosphatidylinositol 3-kinase in epidermal growth factor-induced AP-1 transactivation and transformation in JB6 P + cells. Mol Cell Biol 1996; 16:6427–6435.

86. Huang C, Schmid PC, Ma WY, Schmid HH, Dong Z. Phosphatidylinositol-3 kinase is necessary for 12-O- tetradecanoylphorbol-13-acetate-induced cell transformation and activated protein 1 activation. J Biol Chem 1997; 272:4187–4194.

87. Sun Y, Nakamura K, Hegamyer G, Dong Z, Colburn N. No point mutation of Ha-ras or p53 genes expressed in preneoplastic-to- neoplastic progression as modeled in mouse JB6 cell variants. Mol Carcinog 1993; 8:49–57.

88. Colburn NH, Former BF, Nelson KA, Yuspa SH. Tumour promoter induces anchorage independence irreversibly. Nature 1979; 281:589–591.

89. Colburn NH, Wendel E, Srinivas L. Responses of preneoplastic epidermal cells to tumor promoters and growth factors: use of promoter-resistant variants for mechanism studies. J Cell Biochem 1982; 18:261–270.

90. Huang C, Li J, Ma WY, Dong Z. JNK activation is required for JB6 cell transformation induced by tumor necrosis factor-alpha but not by 12-O-tetradecanoylphorbol-13-acetate. J Biol Chem 1999; 274:29672–29676.

91. Huang C, Ma WY, Li J, Goranson A, Dong Z. Requirement of Erk, but not JNK, for arsenite-induced cell transformation. J Biol Chem 1999; 274:14595–14601.

92. Colburn NH, Wendel EJ, Abruzzo G. Dissociation of mitogenesis and late-stage promotion of tumor cell phenotype by phorbol esters: mitogen-resistant variants are sensitive to promotion. Proc Natl Acad Sci USA 1981; 78:6912–6916.

93. Zhong S, Quealy JA, Bode AM, Nomura M, Kaji A, Ma WY, Dong Z. Organ-specific activation of activator protein-1 in transgenic mice by 12-o-tetradecanoylphorbol-13-acetate with different administration methods. Cancer Res 2001; 61:4084–4091.

94. Zhong SP, Ma WY, Quealy JA, Zhang Y, Dong Z. Organ-specific distribution of AP-1 in AP-1 luciferase transgenic mice during the maturation process. Am J Physiol Regul Integr Comp Physiol 2001; 280:R376–R381.

95. Pitot HC, Sirica AE. The stages of initiation and promotion in hepatocarcinogenesis. Biochim Biophys Acta 1980; 605:191–215.

96. Cohen SM, Ellwein LB. Cell proliferation in carcinogenesis. Science 1990; 249: 1007–1011.

97. Tang DG, Porter AT. Apoptosis: a current molecular analysis. Pathol Oncol Res 1996; 2:117–131.

98. Thompson CB. Apoptosis in the pathogenesis and treatment of disease. Science 1995; 267:1456–1462.

99. Hickman JA. Apoptosis induced by anticancer drugs. Cancer Metastasis Rev 1992; 11:121–139.

100. Barry MA, Behnke CA, Eastman A. Activation of programmed cell death (apoptosis) by cisplatin, other anticancer drugs, toxins and hyperthermia. Biochem Pharmacol 1990; 40:2353–2362.

101. Tomei LD, Kanter P, Wenner CE. Inhibition of radiation-induced apoptosis in vitro by tumor promoters. Biochem Biophys Res Commun 1988; 155:324–331.

102. Forbes IJ, Zalewski PD, Giannakis C, Cowled PA. Induction of apoptosis in chronic lymphocytic leukemia cells and its prevention by phorbol ester. Exp Cell Res 1992; 198:367–372.

103. Lowe SW, Ruley HE, Jacks T, Housman DE. p53-dependent apoptosis modulates the cytotoxicity of anticancer agents. Cell 1993; 74:957–967.

104. McCarthy SA, Symonds HS, Van Dyke T. Regulation of apoptosis in transgenic mice by simian virus 40 T antigen-mediated inactivation of p53. Proc Natl Acad Sci USA 1994; 91:3979–3983.

105. Merritt AJ, Potten CS, Kemp CJ, Hickman JA, Balmain A, Lane DP, Hall PA. The role of p53 in spontaneous and radiation-induced apoptosis in the gastrointestinal tract of normal and p53-deficient mice. Cancer Res 1994; 54:614–617.

106. Clarke AR, Gledhill S, Hooper ML, Bird CC, Wyllie AH. p53 dependence of early apoptotic and proliferative responses within the mouse intestinal epithelium following gamma-irradiation. Oncogene 1994; 9:1767–1773.

107. Donehower LA, Harvey M, Slagle BL, McArthur MJ, Montgomery CA, Butel JS, Bradley A. Mice deficient for p53 are developmentally normal but susceptible to spontaneous tumours. Nature 1992; 356:215–221.

108. Samaha HS, Kelloff GJ, Steele V, Rao CV, Reddy BS. Modulation of apoptosis by sulindac, curcumin, phenylethyl-3-methylcaffeate, and 6-phenylhexyl isothiocyanate: apoptotic index as a biomarker in colon cancer chemoprevention and promotion. Cancer Res 1997; 57:1301–1305.

109. Wright SC, Zhong J, Zheng H, Larrick JW. Nicotine inhibition of apoptosis suggests a role in tumor promotion. FASEB J 1993; 7:1045–1051.

110. Wright SC, Zhong J, Larrick JW. Inhibition of apoptosis as a mechanism of tumor promotion. FASEB J 1994; 8:654–660.

111. Reddy BS, Wang CX, Samaha H, Lubet R, Steele VE, Kelloff GJ, Rao CV. Chemoprevention of colon carcinogenesis by dietary perillyl alcohol. Cancer Res 1997; 57:420–425.

112. Nomura M, Kaji A, Ma W, Miyamoto K, Dong Z. Suppression of cell transformation and induction of apoptosis by caffeic acid phenethyl ester. Mol Carcinog 2001; 31: 83–89.

113. Kelloff GJ, Boone CW, Steele VE, Crowell JA, Lubet RA, Greenwald P, Hawk ET, Fay JR, Sigman CC. Mechanistic considerations in the evaluation of chemopreventive data. IARC Sci Publ 1996; 139:203–219.

114. Huang C, Ma WY, Li J, Hecht SS, Dong Z. Essential role of p53 in phenethyl isothiocyanate-induced apoptosis. Cancer Res 1998; 58:4102–4106.

115. Fesus L, Szondy Z, Uray I. Probing the molecular program of apoptosis by cancer chemopreventive agents. J Cell Biochem Suppl 1995; 22:151–161.

116. Bode AM, Ma WY, Surh YJ, Dong Z. Inhibition of epidermal growth factor-induced cell transformation and activator protein 1 activation by [6]-gingerol. Cancer Res 2001; 61:850–853.

117. Sgambato A, Ardito R, Faraglia B, Boninsegna A, Wolf FI, Cittadini A. Resveratrol, a natural phenolic compound, inhibits cell proliferation and prevents oxidative DNA damage. Mutat Res 2001; 496:171–180.

118. Subbaramaiah K, Chung WJ, Michaluart P, Telang N, Tanabe T, Inoue H, Jang M, Pezzuto JM, Dannenberg AJ. Resveratrol inhibits cyclooxygenase-2 transcription and activity in phorbol ester-treated human mammary epithelial cells. J Biol Chem 1998; 273:21875–21882.

119. Subbaramaiah K, Michaluart P, Chung WJ, Tanabe T, Telang N, Dannenberg AJ. Resveratrol inhibits cyclooxygenase-2 transcription in human mammary epithelial cells. Ann NY Acad Sci 1999; 889:214–223.

120. Subbaramaiah K, Dannenberg AJ. Resveratrol inhibits the expression of cyclooxygenase-2 in mammary epithelial cells. Adv Exp Med Biol 2001; 492:147–157.

121. Szende B, Tyihak E, Kiraly-Veghely Z. Dose-dependent effect of resveratrol on proliferation and apoptosis in endothelial and tumor cell cultures. Exp Mol Med 2000; 32:88–92.

122. Surh YJ, Hurh YJ, Kang JY, Lee E, Kong G, Lee SJ. Resveratrol, an antioxidant present in red wine, induces apoptosis in human promyelocytic leukemia (HL-60) cells. Cancer Lett 1999; 140:1–10.

123. Tsan MF, White JE, Maheshwari JG, Bremner TA, Sacco J. Resveratrol induces Fas signalling-independent apoptosis in THP-1 human monocytic leukaemia cells. Br J Haematol 2000; 109:405–412.

124. Wolter F, Akoglu B, Clausnitzer A, Stein J. Downregulation of the cyclin D1/Cdk4 complex occurs during resveratrol- induced cell cycle arrest in colon cancer cell lines. J Nutr 2001; 131:2197–2203.

125. Bertelli AA, Ferrara F, Diana G, Fulgenzi A, Corsi M, Ponti W, Ferrero ME, Bertelli A. Resveratrol, a natural stilbene in grapes and wine, enhances intraphagocytosis in human promonocytes: a co-factor in antiinflammatory and anticancer chemopreventive activity. Int J Tissue React 1999; 21:93–104.

126. Clement MV, Hirpara JL, Chawdhury SH, Pervaiz S. Chemopreventive agent resveratrol, a natural product derived from grapes, triggers CD95 signaling-dependent apoptosis in human tumor cells. Blood 1998; 92:996–1002.

127. Dorrie J, Gerauer H, Wachter Y, Zunino SJ. Resveratrol induces extensive apoptosis by depolarizing mitochondrial membranes and activating caspase-9 in acute lymphoblastic leukemia cells. Cancer Res 2001; 61:4731–4739.

128. Ferry-Dumazet H, Garnier O, Mamani-Matsuda M, Vercauteren J, Belloc F, Billiard C, Dupouy M, Thiolat D, Kolb JP, Marit G, Reiffers J, Mossalayi MD. Resveratrol inhibits the growth and induces the apoptosis of both normal and leukemic hematopoietic cells. Carcinogenesis 2002; 23:1327–1333.

129. Joe AK, Liu H, Suzui M, Vural ME, Xiao D, Weinstein IB. Resveratrol induces growth inhibition, S-phase arrest, apoptosis, and changes in biomarker expression in several human cancer cell lines. Clin Cancer Res 2002; 8:893–903.

130. Morris GZ, Williams RL, Elliott MS, Beebe SJ. Resveratrol induces apoptosis in LNCaP cells and requires hydroxyl groups to decrease viability in LNCaP and DU 145 cells. Prostate 2002; 52:319–329.

131. Tsan MF, White JE, Maheshwari JG, Chikkappa G. Anti-leukemia effect of resveratrol. Leuk Lymphoma 2002; 43:983–987.
132. Nicolini G, Rigolio R, Scuteri A, Miloso M, Saccomanno D, Cavaletti G, Tredici G. Effect of trans-resveratrol on signal transduction pathways involved in paclitaxel-induced apoptosis in human neuroblastoma SH-SY5Y cells. Neurochem Int 2003; 42:419–429.
133. Billard C, Izard JC, Roman V, Kern C, Mathiot C, Mentz F, Kolb JP. Comparative antiproliferative and apoptotic effects of resveratrol, epsilon-viniferin and vine-shots derived polyphenols (vineatrols) on chronic B lymphocytic leukemia cells and normal human lymphocytes. Leuk Lymphoma 2002; 43:1991–2002.
134. Bernhard D, Tinhofer I, Tonko M, Hubl H, Ausserlechner MJ, Greil R, Kofler R, Csordas A. Resveratrol causes arrest in the S-phase prior to Fas-independent apoptosis in CEM-C7H2 acute leukemia cells. Cell Death Differ 2000; 7:834–842.
135. Adhami VM, Afaq F, Ahmad N. Involvement of the retinoblastoma (pRb)-E2F/DP pathway during antiproliferative effects of resveratrol in human epidermoid carcinoma (A431) cells. Biochem Biophys Res Commun 2001; 288:579–585.
136. Ahmad N, Adhami VM, Afaq F, Feyes DK, Mukhtar H. Resveratrol causes WAF-1/p21-mediated G(1)-phase arrest of cell cycle and induction of apoptosis in human epidermoid carcinoma A431 cells. Clin Cancer Res 2001; 7:1466–1473.
137. Hsieh TC, Wu JM. Differential effects on growth, cell cycle arrest, and induction of apoptosis by resveratrol in human prostate cancer cell lines. Exp Cell Res 1999; 249:109–115.
138. Ragione FD, Cucciolla V, Borriello A, Pietra VD, Racioppi L, Soldati G, Manna C, Galletti P, Zappia V. Resveratrol arrests the cell division cycle at S/G2 phase transition. Biochem Biophys Res Commun 1998; 250:53–58.
139. Schneider Y, Vincent F, Duranton B, Badolo L, Gosse F, Bergmann C, Seiler N, Raul F. Anti-proliferative effect of resveratrol, a natural component of grapes and wine, on human colonic cancer cells. Cancer Lett 2000; 158:85–91.
140. Holmes-McNary M, Baldwin AS. Chemopreventive properties of trans-resveratrol are associated with inhibition of activation of the IkappaB kinase. Cancer Res 2000; 60:3477–3483.
141. Manna SK, Mukhopadhyay A, Aggarwal BB. Resveratrol suppresses TNF-induced activation of nuclear transcription factors NF-kappa B, activator protein-1, and apoptosis: potential role of reactive oxygen intermediates and lipid peroxidation. J Immunol 2000; 164:6509–6519.
142. Ashikawa K, Majumdar S, Banerjee S, Bharti AC, Shishodia S, Aggarwal BB. Piceatannol inhibits TNF-induced NF-kappaB activation and NF-kappaB-mediated gene expression through suppression of IkappaBalpha kinase and p65 phosphorylation. J Immunol 2002; 169:6490–6497.
143. Pendurthi UR, Meng F, Mackman N, Rao LV. Mechanism of resveratrol-mediated suppression of tissue factor gene expression. Thromb Haemost 2002; 87:155–162.
144. Asou H, Koshizuka K, Kyo T, Takata N, Kamada N, Koeffier HP. Resveratrol, a natural product derived from grapes, is a new inducer of differentiation in human myeloid leukemias. Int J Hematol 2002; 75:528–533.
145. Brownson DM, Azios NG, Fuqua BK, Dharmawardhane SF, Mabry TJ. Flavonoid effects relevant to cancer. J Nutr 2002; 132:3482S–3489S.

146. Bove K, Lincoln DW, Tsan MF. Effect of resveratrol on growth of 4T1 breast cancer cells in vitro and in vivo. Biochem Biophys Res Commun 2002; 291:1001–1005.

147. Banerjee S, Bueso-Ramos C, Aggarwal BB. Suppression of 7,12-Dimethylbenz(a)anthracene-induced mammary carcinogenesis in rats by resveratrol: role of nuclear factor-kappaB, cyclooxygenase 2, and matrix metalloprotease 9. Cancer Res 2002; 62:4945–4954.

148. Huang C, Ma WY, Goranson A, Dong Z. Resveratrol suppresses cell transformation and induces apoptosis through a p53-dependent pathway. Carcinogenesis 1999; 20: 237–242.

149. She QB, Kaji A, Wang MW, Ho CH, Dong Z. Inhibition of cell transformation by resveratrol and its derivatives: structure-activity relationship and mechanisms involved. Oncogene 2003; 22:2143–2150.

150. Lu J, Ho CH, Ghai G, Chen KY. Resveratrol analog, 3,4,5,4'-tetrahydroxystilbene, differentially induces pro-apoptotic p53/Bax gene expression and inhibits the growth of transformed cells but not their normal counterparts. Carcinogenesis 2001; 22:321–328.

151. Basu A, Haldar S. The relationship between BcI2, Bax and p53: consequences for cell cycle progression and cell death. Mol Hum Reprod 1998; 4:1099–1109.

152. Park JW, Choi YJ, Suh SI, Baek WK, Suh MH, Jin IN, Min DS, Woo JH, Chang JS, Passaniti A, Lee YH, Kwon TK. Bcl-2 overexpression attenuates resveratrol-induced apoptosis in U937 cells by inhibition of caspase-3 activity. Carcinogenesis 2001; 22:1633–1639.

153. Tessitore L, Davit A, Sarotto I, Caderni G. Resveratrol depresses the growth of colorectal aberrant crypt foci by affecting bax and p21(CIP) expression. Carcinogenesis 2000; 21:1619–1622.

154. Kokontis JM, Wagner AJ, O'Leary M, Liao S, Hay N. A transcriptional activation function of p53 is dispensable for and inhibitory of its apoptotic function. Oncogene 2001; 20:659–668.

155. Chen X, Ko LJ, Jayaraman L, Prives C. p53 levels, functional domains, and DNA damage determine the extent of the apoptotic response of tumor cells. Genes Dev 1996; 10:2438–2451.

156. Baek SJ, Wilson LC, Eling TE. Resveratrol enhances the expression of non-steroidal anti-inflammatory drug-activated gene (NAG-1) by increasing the expression of p53. Carcinogenesis 2002; 23:425–434.

157. Soleas GJ, Goldberg DM, Grass L, Levesque M, Diamandis EP. Do wine polyphenols modulate p53 gene expression in human cancer cell lines. Clin Biochem 2001; 34:415–420.

158. Mahyar-Roemer M, Katsen A, Mestres P, Roemer K. Resveratrol induces colon tumor cell apoptosis independently of p53 and precede by epithelial differentiation, mitochondrial proliferation and membrane potential collapse. Int J Cancer 2001; 94:615–622.

159. Mahyar-Roemer M, Roemer K. p21 Waf1/Cip1 can protect human colon carcinoma cells against p53- dependent and p53-independent apoptosis induced by natural chemopreventive and therapeutic agents. Oncogene 2001; 20:3387–3398.

160. She QB, Bode AM, Ma WY, Chen NY, Dong Z. Resveratrol-induced activation of p53 and apoptosis is mediated by extracellular-signal-regulated protein kinases and p38 kinase. Cancer Res 2001; 61:1604–1610.

161. Bode AM, Dong Z. Signal transduction pathways: targets for chemoprevention of skin cancer. Lancet Oncol 2000; 1:181–188.
162. Dong Z. Effects of food factors on signal transduction pathways. Biofactors 2000; 12:17–28.
163. Tredici G, Miloso M, Nicolini G, Galbiati S, Cavaletti G, Bertelli A. Resveratrol, map kinases and neuronal cells: might wine be a neuroprotectant. Drugs Exp Clin Res 1999; 25:99–103.
164. She QB, Huang C, Zhang Y, Dong Z. Involvement of c-jun NH(2)-terminal kinases in resveratrol-induced activation of p53 and apoptosis. Mol Carcinog 2002; 33: 244–250.
165. Lin HY, Shih A, Davis FB, Tang HY, Martino LJ, Bennett JA, Davis PJ. Resveratrol induced serine phosphorylation of p53 causes apoptosis in a mutant p53 prostate cancer cell line. J Urol 2002; 168:748–755.
166. Shih A, Davis FB, Lin HY, Davis PJ. Resveratrol induces apoptosis in thyroid cancer cell lines via a MAPK- and p53-dependent mechanism. J Clin Endocrinol Metab 2002; 87:1223–1232.
167. Yu R, Hebbar V, Kim DW, Mandlekar S, Pezzuto JM, Kong AN. Resveratrol inhibits phorbol ester and UV-induced activator protein 1 activation by interfering with mitogen-activated protein kinase pathways. Mol Pharmacol 2001; 60:217–224.
168. Atten MJ, Attar BM, Milson T, Holian O. Resveratrol-induced inactivation of human gastric adenocarcinoma cells through a protein kinase C-mediated mechanism. Biochem Pharmacol 2001; 62:1423–1432.

# 14

# The Role of Lycopene in Human Health

**Regina Goralczyk and Ulrich Siler**
*DSM Nutritional Products, Basel, Switzerland*

## I. INTRODUCTION

The carotenoids form a class of highly colored, unsaturated, lipophilic plant pigments. Their great abundance in fruits and vegetables with colors ranging from yellow through red is responsible for the attractive appearance of peaches, oranges, carrots, peppers, and tomatoes. Carotenoids share a C40 carbon skeleton comprising 8 isoprene units, joined in such a manner that their arrangement is reversed at the center of the molecule. From a total of more than 600 carotenoids found in nature, around 40 occur in the human food chain; of these, about 25 are found in human blood or tissues together with 9 metabolites and oxidation products [1]. Dietary carotenoids are considered to provide health benefits by decreasing the risks of chronic diseases, particularly cancer, cardiovascular disease, and age-related eye diseases, such as cataract and macular degeneration. The most studied carotenoids in this respect are β-carotene, lycopene, lutein, and zeaxanthin. Lycopene, present in tomatoes and their products, has attracted particular attention since the significant reduction in risk for prostate cancer from the Health Professional Follow-Up Study was reported [2]. The interest of the research community was further stimulated by the work of Giovannucci and Clinton [3], who reviewed 72 epidemiological studies, 57 of which showed a significant inverse relationship between lycopene intake or serum levels and risk of cancers at various sites, including those of the prostate, lung, digestive tract, and breast. More recent evidence also points to a role for lycopene in the prevention of heart disease [4,5]. This chapter offers an overview on the properties and health effects of this important phytochemical and discusses several possible mechanisms for its physiological function in the human body.

## II.  PROPERTIES AND PREVALENCE OF LYCOPENE IN FOOD AND THE HUMAN BODY

### A.  Lycopene in Food

Lycopene ($\Psi,\Psi$-carotene) is a red, acyclic carotenoid found in high concentrations in the tomato (Fig. 1), its high microcrystalline deposition in chromoplasts of the peel and flesh being responsible for their characteristic red color. Lycopene contributes about 80% to the total carotenoid content of tomatoes; the remainder includes carotenoid precursors, such as phytoene (5–10%) and phytofluene (2–3%), as well as β-carotene (3–5%) and minor constituents, including γ-carotene, ζ-carotene, neurosporene, and lutein [6]. Raw tomatoes and processed tomato products are the richest source of lycopene in western diets, containing up to 100 μg lycopene/g fresh or processed weight, depending on variety, season, and type of product [6–8]. Other significant dietary sources are watermelon (23–72 μg/g fresh weight), pink guava (~54 μg/g), pink grapefruit (up to ~33 μg/g), and papaya (20–53 μg/g). The Gac fruit (*Momordica cochinchinensis* Spreng.) has recently been identified as a particularly rich source of lycopene [9]; seed membrane concentrations of up to 380 μg/g, i.e., 10-fold greater than found in lycopene-rich fruit and vegetables, have been reported. Lycopene concentrations are enriched in processed tomato products, including pastes (up to 1500 μg/g), juices (50–116 μg/g), and ketchups (100–130 μg/g) [6,8].

   In whole tomatoes and derived concentrates or juices, all-E(*trans*)-lycopene predominates, whereas processed tomato products, such as ketchups and spaghetti sauces, also contain 5-Z(*cis*), 9-Z, 13-Z, and 15-Z-isomers (Fig. 1; Table 1) [8]). Acid and especially heat treatments induce isomerization, converting all-E-lycopene to its Z-isomeric forms; concentrations of the latter also increase with temperature and length of processing [10]. An increase of Z-isomers of up to 16% with a corresponding decrease in all-E-lycopene was reported in tomatoes processed under differing dehydration conditions [6]. The yellow or tangerine tomato is a particularly rich source of prolycopene (7Z,9Z,7'Z,9'Z-tetra-*cis*-lycopene), this isomer replacing 90% of the usual all-E-lycopene as a result of loss in function of the CrtISO gene, which encodes for a carotenoid isomerase required for desaturation [11–13].

   In purified form, carotenoids in general, and lycopene in particular, are sensitive to oxidation and degradation, leading to bleaching and losses from food as a result of processing [14] or improper storage. In contrast, controlled thermal processing can increase the antioxidative activity of tomato products because the lycopene becomes more bioaccessible. Other tomato ingredients, such as polyphenols and vitamin C, may help to stabilize lycopene [15].

### B.  Bioavailability

Various investigators have consistently shown that the bioavailability of lycopene from processed tomatoes, e.g., pastes, soups, or sauces, is higher than that from

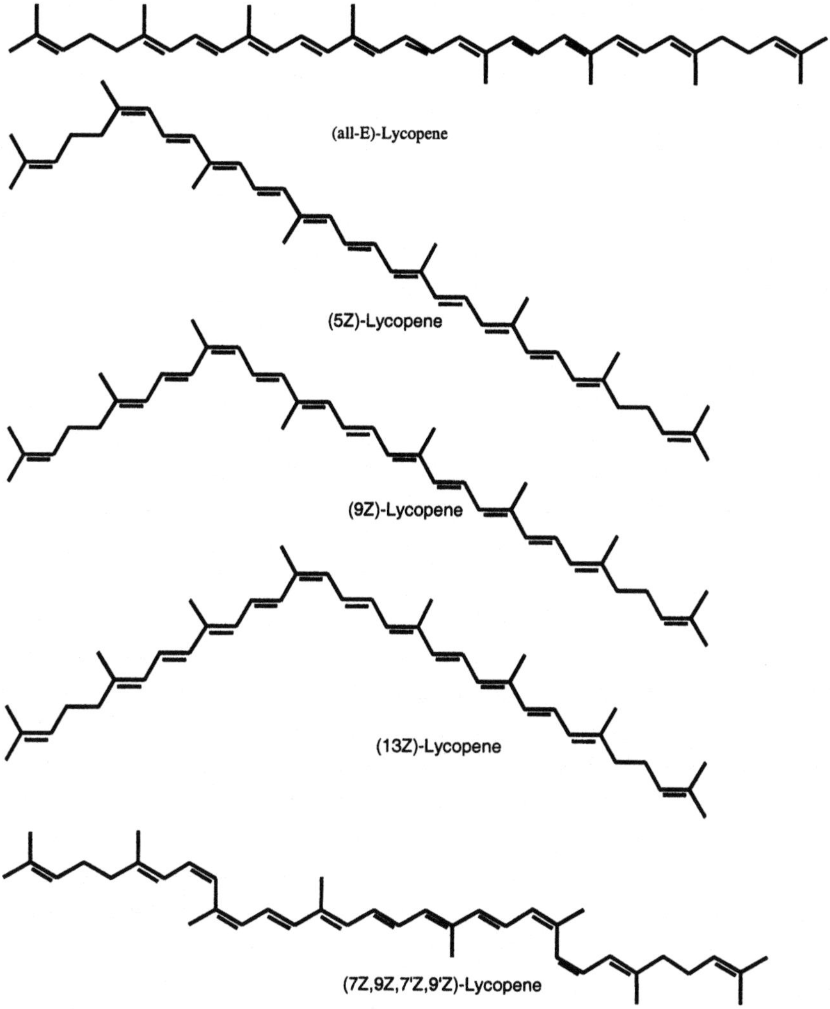

**Figure 1** Chemical structures of geometric isomers of lycopene commonly detected in processed tomato products, human plasma, and tissues

**Table 1**  Contents of Lycopene and Its (E/Z) Isomers in Foodstuffs and Meals

| Product | Total lycopene (mg/g) | All-E (%) | 5-Z (%) | 9-Z (%) | 13-Z + 15-Z (%) | x-Z (%) |
|---|---|---|---|---|---|---|
| Tomato paste | 502 | 96 | 4 | <1 | <1 | <1 |
| Tomato ketchup | 30 | 77 | 11 | 5 | 7 | 1 |
| Bolognese sauce | 92 | 67 | 14 | 6 | 5 | 8 |
| Bolognese sauce (cooked 10 min) | 93 | 60 | 14 | 8 | 7 | 11 |
| Canned tomatoes | 71 | 84 | 5 | 3 | 5 | 3 |
| Canned tomatoes (prepared as spaghetti sauce I) | 34 | 35 | 27 | 14 | 3 | 22 |
| Canned tomatoes (prepared as spaghetti sauce II) | 62 | 62 | 16 | 9 | 7 | 13 |
| Red palm oil (Malaysia) | 450 | 16 | 19 | 6 | 30 | 29 |

*Source:* Modified from Ref. 8.

fresh, raw tomato [10,16–18]. The presence of oil or other lipids in the diet can significantly improve the efficiency of lycopene absorption, as is also found with other carotenoids and lipid-soluble vitamins. Food processing may improve the bioavailability of lycopene by disrupting cell wall and tissue matrices, thus increasing its accessibility. The composition and structure of the food may also impact on bioavailability, and thus affect release of lycopene crystals from the tissue matrix [10]. Z-isomers are reported to be slightly better absorbed than the all-E-form [16]; hence, higher proportions of Z-isomers have been found in human serum and tissues as compared to those in common tomato products. Such preferential tissue accumulation of Z-lycopene could be a consequence of better solubility in mixed micelles, increased stability of Z-isomers, preferential absorption at the brush border level, and/or transport into tissues or increased degradation and/ or metabolism of the all-E form in tissues. Boileau et al. [18,19] demonstrated an enhanced solubility of Z-lycopene in bile acid–containing mixed micelles in vitro. Using the lymph-cannulated ferret as model system, results showed that the Z-isomer fraction of the mucosa and mesenteric lymph was significantly increased when compared to the fraction in the applied lycopene dose, or to the Z-fraction recovered from the stomach and intestine. It was concluded that Z-isomers, in addition to being more soluble, might also be more efficiently absorbed by mucosal cells and preferentially incorporated into chylomicrons. Discrimination at later stages of the transport process is considered less likely, since the Z-isomer fraction was found to be similar in human chylomicrons, very low-density lipoprotein (VLDL), low-density lipoprotein (LDL), high-density lipoprotein (HDL), and plasma [20]. Processing of tomatoes resulted in systemic lycopene availability comparable to that from synthetic lycopene formulated in tablets [21]; in a randomized, parallel group study, relative plasma responses to lycopene intake from tomato juice, soup, or tablets were investigated. Volunteers ($n = 6$ per group) restricted in dietary lycopene intake for 2 weeks before and during the experimental period were dosed with lycopene (20 mg daily for 7 days) in tablets, tomato juice, or soup. Lycopene kinetics was monitored for a further 3 weeks using HPLC. Irrespective of lycopene treatment, the all-E-isomer predominated in plasma, while the most abundant Z-isomer was the 5-Z-compound. The half-life of lycopene was 5–7 days, irrespective of dietary form (tablet or soup). Consumption of tomato juice did not lead to significant increase in plasma concentrations of lycopene. These data again confirm the importance of food processing for improving carotenoid availability and demonstrate that synthetic, adequately formulated lycopene possesses bioavailability characteristics similar to those of individual processed tomato products.

## C.  Lycopene in Human Blood and Tissues

The lycopene concentrations in blood samples vary considerably between countries (Table 2); this largely reflects the habit of tomato product consumption rather than the more general intake of fruit and vegetables [22]. In some western populations, (i.e., United States, southern Europe, and UK) lycopene and β-carotene are the predominant carotenoids found in plasma; levels may be quite high, for example, up to 0.8 or even 1.7 μmol/L in U.S. children, a result of their frequent intake of tomato juice, ketchup, pizza, and spaghetti sauce [23]. Vegetarians usually exhibit higher plasma lycopene concentrations [22,24]. In Asian countries such as Japan or China, where consumption of tomato-based foods is rare, plasma levels are comparatively low, i.e., 0.3 μmol/L. In China, lutein has been found to be the dominant plasma carotenoid (Fig. 2). In humans, Z-isomers contribute up to about 50% of total plasma lycopene content [16,20]. Schierle et al. [8] and others [26] found the 5-Z-isomer to be the next most abundant form after all-E-lycopene (Fig. 3). It should be noted that some analytical methods are unable to resolve the 5-Z- and all-E-lycopene isomers because of congruency in the UV/Vis spectra (e.g., from photo-diode array detection) used for identification (Fig. 3, insert). 13-Z-, 15-Z-, and 9-Z-Lycopene isomers are also detectable in human plasma at low concentrations, typically less than 5% of the all-E compound.

Lycopene, like other carotenoids, is transported in plasma exclusively by lipoproteins, primarily in LDL particles [22,27,28], and the isomeric pattern reflects that of the blood. Li et al. [29] employed surface enhanced Raman spectroscopy to demonstrate that both lycopene and β-carotene were concentrated in the

**Table 2**  Lycopene Plasma Levels[a] in Various Populations

| Population | Range (μmol/L) | Comment | Ref. |
|---|---|---|---|
| Spain, granada | 0.68 | M, F | 24,113 |
| Northern Spain | 0.47 | | |
| Spain (Madrid) | 0.46 (0.46–0.62) | M, F | 114 |
| Spain (Madrid) | 0.35–0.4 | Patients plus controls | 115 |
| France | 0.68 | F | 24 |
| France | 0.66 (0.58–0.77) | M, F, | 114 |
| Italy (Perugia) | 0.75 ± 0.25 | Controls | 116 |
| Italy (Turin, Florence) | 0.95 | M, F, | 24 |
| Italy (Naples, Ragusa) | 1.29 | M, F, | 24 |
| Germany | 0.6 | M, F | 24 |
| UK (Cambridge) | 0.7 | M, F, | 24 |
| UK vegetarians | 0.93 | M, F, | 24 |
| Northern Ireland | 0.64 (0.54–0.76) | M, F | 114 |
| Northern Ireland | 0.319 | F, nonsmoker | 75 |
| Irish Republic | 0.73 (0.59–0.9) | M,F, | 114 |
| Netherlands | 0.5 | M,F, | 24 |
| Netherlands | 0.134 | Rotterdam Study M, F, baseline controls, | 59 |
| Netherlands | 0.54 | M, F | 114 |
| Denmark | 0.55 | M, F | 24 |
| Sweden (Malmö) | 0.49–0.63 | M, F | 24,113 |
| Finland | 0.17 | Kuopio Study, controls | 58 |
| Finland (Ŭmea) | 0.49 | M, F | 24 |
| Greece | 0.8 | M, F | 24 |
| Japan | 0.23–0.28 | M. F | 117 |
| Costa Rica | 0.6 | M,F, adolescents | 118 |
| USA | 0.01–1.79 | Various larger populations | 23,48,119,120 |

[a] Compiled from stated references: when levels from males (M) and females (F) were cited, a mean was calculated. All values were transformed into μmol/L when given in mg/dL.

shell of the LDL particles. However, lycopene exhibited a greater proximity for the LDL-binding region of the apo-B100 protein than did β-carotene, suggesting one reason for differences in tissue uptake of the two compounds. In contrast to β-carotene and xanthophylls, lycopene concentrations in the lipoproteins of menopausal women did not show cyclic fluctuations [28].

Lycopene oxidation products have been extensively studied by Khachik and coworkers [1,30–33]; two oxidative metabolites, 2,6-cyclolycopene-1,5-diols A and B, present in tomatoes at very low concentrations were identified in human

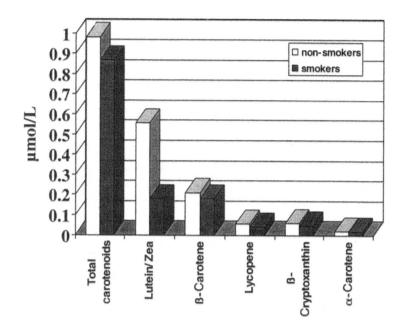

**Figure 2**  Plasma carotenoid concentrations in a population of smokers and nonsmokers in Shanghai. (Adapted from Ref. 25.)

serum, breast milk, and organs (prostate, lung, liver, colon, breast). Recently it has been suggested that the presence of these products may result from the in vivo antioxidation properties of lycopene, and they have been proposed as potential markers of oxidative stress [34].

Lycopene has been reported [35–38] to accumulate predominantly in liver, adrenal, testicular, and prostate tissue (Table 3); in benign or malignant prostate tissue, mean lycopene concentrations of 0.8 nmol/g tissue have been measured, with up to 2.6 nmol/g being found in some individuals [36]. However, no significant correlation was observed between lycopene concentration and that of any other carotenoid. As for serum, the ratio of lycopene Z- to all E-isomers was higher in tissues [36] than in tomato-based foods. While local isomerization cannot be entirely excluded, no responsible enzyme has yet been identified. The Z-isomer fraction in prostate tissue can account for up to nearly 90% of the total lycopene [18], with the 5-Z-compound predominating among the of 14–18 detectable Z-isomeric peaks [26,36]. In summary, the specific accumulation of lycopene and its isomers in the benign and malignant prostate strongly supports a physiological role in this organ.

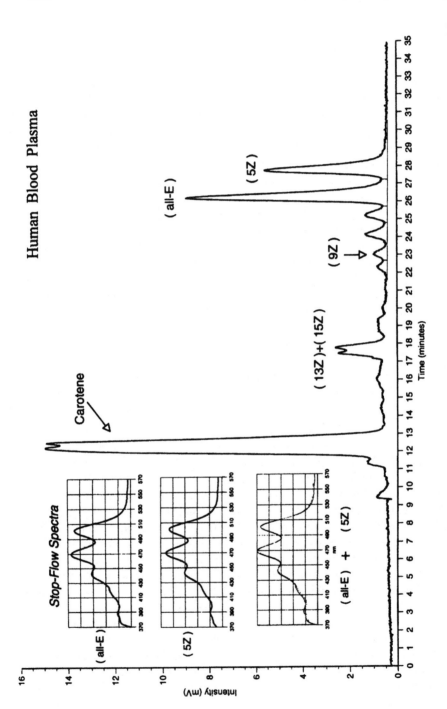

**Figure 3** Pattern of lycopene isomers in a human blood sample. (Insert) Photo-diode array detection spectra of all-E-lycopene and 5-Z-lycopene. (From Ref. 8.)

**Table 3**  Lycopene Concentrations in Human Tissues

| Tissue | Total lycopene (nmol/g tissue) |
|---|---|
| Testes | 4.3–21.4 |
| Adrenal gland | 1.9–21.6 |
| Liver | 1.3–5.7 |
| Prostate | 0.2–0.9 |
| Prostate after treatment with 30 mg lycopene as pasta sauce for 3 weeks [26] | 1.1 |
| Breast | 0.4–0.7 |
| Pancreas | 0.7 |
| Skin | 0.4–1.7 |
| Colon | 0.3–1.0 |
| Ovary | 0.3 |
| Lung | 0.2–0.5 |
| Stomach | 0.2 |
| Kidney | 0.2–0.6 |
| Fat | 0.2 |
| Cervix | 0.2 |

*Source*: Data ranges complied from Refs. 1, 26, 35, 36, 38, 121, and 122.

## III. LYCOPENE IN PREVENTION OF CHRONIC DISEASE

The potential beneficial effects of lycopene in human health have been reviewed extensively in recent years [39–45]; numerous observational studies have consistently shown an inverse relationship between the consumption of lycopene-rich diets (tomato or tomato-based foods) or plasma lycopene levels with the risk of cancers at various sites [46,47]. The strongest inverse relationship was that for prostate cancer [2,46–49], one of the most prevalent male cancers in the western populations and common across the world; other significant inverse relationships were found with lung and stomach cancers.

   The Health Professionals Follow-Up Study, one of the largest male cohort studies conducted in the United States, followed the development of prostate cancer over a 12-year period. Three dietary questionnaires were included in the evaluation [49]; responses showed lycopene consumption to be significantly associated with a decreased risk for prostate cancer by 16% ($p$ for trend = 0.003). Significantly, consumption of tomato sauce, the primary source of bioavailable lycopene, was associated with an even greater reduction in risk of prostate cancer;

individuals who consumed more than eight servings per month had a 23% reduction of risk as compared to those having less than one serving over the same period. In a nested case-control study [50] embedded in the Physicians' Health Study (a randomized placebo-controlled trial of aspirin and β-carotene), the plasma carotenoid levels of 578 men who developed prostate cancer within 13 years of follow-up were evaluated and compared to 1294 age and smoking status-matched controls. Lycopene was the only antioxidant occurring at significantly lower mean levels in cancer cases than in matched controls. Individuals in the highest quintile of lycopene plasma concentrations had 25% lower prostate cancer risk than those with the lowest levels. The inverse association was even stronger for aggressive prostate cancers; a 44% reduction of risk being observed when comparing the highest quintile for lycopene plasma concentration with the lowest. A similar inverse association between higher lycopene plasma levels and lowered risk of aggressive prostate cancer was recently confirmed for Americans of differing ethnic origin [48].

Until now, no large human intervention trials addressing the effect of lycopene in prostate cancer prevention have been conducted; one small trial among prostate cancer patients afforded preliminary evidence of reduced cancer growth and decreases in associated biomarkers following intervention with a tomato extract containing 15 mg lycopene, fed twice daily [51]. Diffuse involvement of the prostate by high-grade prostatic intraepithelial neoplasia was found in 67% of subjects in the intervention compared with 100% in the control group ($p = 0.05$). Plasma prostate specific antigen decreased by 18% in the intervention group, whereas there was a 14% increase in the control population; however, the small sample size precludes too strong a conclusion being drawn. Tomato intake via daily consumption of a pasta dish containing 30 mg of lycopene for 3 weeks in patients scheduled for prostatectomy also reduced the levels of oxidized nucleotides in prostate tissue [52]. Prevention of DNA oxidation might, therefore, contribute to the overall cancer preventive function of lycopene.

Recently, Siler et al. (unpublished) investigated the effects of lycopene, alone or in combination with vitamin E, in the Dunning tumor model, an orthotopic rat model for aggressive prostate cancer. Rats were grouped and fed a diet supplemented with lycopene (200 ppm in feed), vitamin E (500 ppm) or lycopene (200 ppm) in combination with vitamin E (500 ppm) for 46 days; controls were either untreated or fed diets containing a placebo formulation as supplement. After 4 weeks of feeding, MatLyLu Dunning tumor cells, a highly aggressive and metastatic cell line, were injected into the ventral lobes of the prostate; 14 days after this injection, orthotopic prostate tumors were monitored for necrosis rates by intravenous magnetic resonance imaging (MRI). Tumor volumes and weights were measured after sacrifice on day 18. When supplemented alone, lycopene and vitamin E each increased tumor necrosis up to 36%, compared to 20 and 23%, respectively, in untreated or placebo treatment controls.

There was neither an additive/synergistic effect of lycopene and vitamin E nor any influence on tumor size. The finding that necrosis rates were enhanced in this model is consistent with observed effects of lycopene on aggressive cancers in humans (see above). A number of in vitro studies showing antiproliferative effects of lycopene in various cancer cell lines [53,54], including prostate [55,56] (see below), offer further support for a prostate cancer preventive function.

Recent research also indicates an association between tomato consumption (fresh and processed) and protection against cardiovascular disease. Coronary heart disease (CHD) has a high prevalence in western countries and is increasing in Asian countries due to changes in socioeconomic conditions, lifestyle, and diet. The link between LDL oxidation and atherosclerosis is hypothesized to be the basis for a beneficial effect of antioxidants in general on the incidence of subclinical and clinical CHD.

Two prospective studies from Finland demonstrate that low levels of plasma lycopene, resulting from a diet low in tomatoes, significantly increase the risk for heart attack, stroke, and early atherosclerosis among middle-aged men [57,58]. Similarly, a Dutch study revealed a lowered risk for aortic calcification in current and former smokers with higher lycopene plasma concentrations [59]. Finally, results from a European multicenter case control study (EURAMIC) are suggestive of a protective effect of lycopene against myocardial infarction [60].

## IV. THE PHYSIOLOGICAL FUNCTION OF LYCOPENE AND MECHANISMS OF PROTECTION

### A. Free Radical Scavenging

The high antioxidant properties of lycopene are suggested to be linked with its effect on prevention of cancer and chronic diseases, such as cardiovascular disease. Carotenoids have been known to be effective scavengers of singlet oxygen ($^1O_2^*$) and other excited species for many years [61,62], the efficiency of its quenching being linked to the number of conjugated double bonds in the polyenoid structure [63]. Singlet oxygen quenching is a result of energy being transferred from $^1O_2^*$ to the lycopene molecule, converting the latter to a highly energized triplet state that is readily returned to the ground state with dissipation of energy as heat or through physical quenching. Once returned to the ground state, lycopene can react with another $^1O_2^*$ molecule. Quenching of other free radicals can occur by electron transfer from the carotenoid to the radical, radical addition yielding a radical adduct, or hydrogen atom transfer from the radical.

Lycopene, together with β-carotene, is one of the most efficient singlet oxygen quenchers of all carotenoids in both organic solution [64] and biological membrane environments, such as liposomes [65,66]. In homogeneous organic solution, lycopene was the most rapidly destroyed carotenoid [67] upon reaction with hydroxyl radicals, providing clear evidence for its presence in the first line of defense. Lycopene may also provide protection against oxidative lipid, protein, and DNA damage in vivo; oxidation of nucleotides is implicated in DNA mutation, a crucial event in early carcinogenesis.

Several human studies report that consumption of tomatoes or tomato products reduces the content of 8-hydroxy-2′-deoxyguanosine (8-OHdG), a marker for oxidative DNA damage, in DNA isolated from whole white blood cells in subjects whose basal level of 8-OHdG were higher than the mean value [68]. Subjects ingesting either tomato juice or puree for 2 or 3 weeks increased their lycopene concentrations in plasma and blood lymphocytes [69]; moreover, lymphocyte DNA was significantly more resistant to oxidative damage by hydrogen peroxide [69–71]. Consumption of tomato sauce–based pasta dishes (30 mg lycopene/day) for 3 weeks led to a significant increase in serum and prostate tissue lycopene concentrations [52]. Compared to preintervention, leukocyte oxidative DNA damage was reduced in a statistically significant manner, as was oxidative DNA damage in prostate tissue.

Cigarette smoke is a significant source of oxidative stress, and smoking is known to deplete levels of plasma antioxidants. A variety of studies have demonstrated the negative influence of smoking on plasma lycopene level [72]. Some studies found plasma LDL oxidation to be reduced by dietary fruits/vegetables or lycopene itself [73,74], but other researchers were unable to confirm such an effect [75,76]. In a short-term intervention study, including a depletion period of 8 days and a supplementation period of 7 days, a dietary intake of >40 mg/day of lycopene through intake of food containing red vegetables by a group of nonsmoking individuals significantly reduced susceptibility of LDL to oxidation. However, an equivalent intake of lycopene by a group of smokers was without effect [77]. Lycopene can also efficiently quench nitrogen dioxide radical ($NO_2^*$), a damaging radical produced in tobacco smoke [78,79], and peroxynitrite ($ONOO^-$) [80], a powerful oxidant and nitrating agent. Losses of protein function due to nitro-tyrosine formation has been suggested to be involved in the pathogenesis of a variety of diseases. Lycopene is more rapidly consumed upon reaction with peroxynitrite than either β-carotene or tocopherols [81]. Exposure of human LDL to peroxynitrite in vitro, led to more rapid half-maximal losses of lycopene than other cartenoids or oxocarotenoids [80]. A lycopene-rich diet maintained for 2 weeks led to 10-fold increase in serum lycopene and was, moreover, found to confer in vitro protection of human lymphocytes against $NO_2^*$ and $^1O_2$ by factors of 17.6-fold and 6.3-fold, respectively [78].

Cell culture and animal experiments offer further evidence of the protective effect of lycopene against DNA oxidation. In cell culture, lycopene supplementation decreased levels of both 8-OHdG and lipid peroxidation induced by Fe-NTA/ascorbate [82]. Subsequently, this effect was also demonstrated in vivo. Rats treated with ferric nitrilotriacetate showed increased hepatic levels of 8-oxo-

7,8-dihydroxy-2′-deoxyguanosine and malondialdehyde, associated with histopathological changes in the liver [83]; pretreatment with lycopene almost completely prevented liver biomolecule oxidative damage and protected the tissue against liver damage [83].

Together, such studies demonstrate that lycopene can function as an antioxidant in vivo and in vitro and can reduce genetic damage in humans, animal models, and cell systems.

## B. Regulation of Phase II Enzymes

Tissues, mainly gastrointestinal and liver, involved in uptake and metabolism of external carcinogens and mutagens, such as heterocyclic amines, provide a first barrier to damage, and induction of phase II enzymes plays an elementary role in protecting against toxic compounds of low-molecular weight [84].

In male Wistar rats, lycopene significantly increased glutathione (GSH) levels, enhanced activities of glutathione peroxidase (GPx), glutathione-S-transferase (GST), and glutathione reductase (GR), and significantly reduced lipid peroxidation in stomach tissue [85]. In a dimethylbenz(a)anthracene (DMBA)-induced hamster buccal pouch (HBP) carcinogenesis model system, treatment with DMBA alone produced well-differentiated squamous cell carcinomas; diminished lipid peroxidation in the oral tumor tissue was accompanied by a significant increase in levels of GSH and increased activities of GPx, GST, and GR. Administration of lycopene significantly suppressed DMBA-induced oral carcinogenesis, as revealed by the absence of carcinomas [86]. In male Syrian hamsters, lycopene treatment significantly reduced hepatic lipid peroxidation and enhanced the activities of hepatic biotransformation enzymes such as GPx, GST, and GR, that use GSH as substrate [87]. Leal et al. [88] evaluated the GST, γ-glutamyltransferase (GGT) and GPx activities in liver homogenates from male broilers; dietary T2 toxin induced an increase in malondialdehyde, decrease in endogenous GSH, and marked increases in activities of GST, GGT, and GPx. Broilers fed lycopene plus T2 toxin showed some of these parameters to be diminished relative to levels in birds fed T2 toxin alone. The authors concluded that lycopene may act as an antioxidant and a protector of cellular GST. The common mechanism of phase II enzyme induction mediated by xenobiotics and antioxidants involves activation and binding of transcription factors, e.g., Nrf-Jun heterodimer to an antioxidant responsive element (ARE) within a common *cis* element of these genes [89].

While the mechanism of phase II enzyme induction by lycopene has not yet been characterized in detail, it is likely that it shares common elements with other antioxidants. Many carcinogens are activated and/or detoxified by a common set of oxidation-reduction reactions, and inhibitors of cancer initiation frequently exert their effect by modulating these processes [90]. Breinholt et al. [91]

have reported that lycopene increases hepatic ethoxyresorufin $O$-dealkylase and benzoxyresorufin $O$-dealkylase activities, suggesting an isozymic specificity towards CYP1A and CYP3A4. Furthermore, the same study revealed an increase in hepatic quinone reductase and UDP-glucuronosyltransferase activity within an inverted U-shaped dose-response curve [91]. Although unequivocal evidence for the mode of action of lycopene remains to be presented, an activating influence on the detoxification of carcinogens may be one explanation for its cancer-reducing properties.

## C. Interactions with Cell Cycle, Cell Communication, Gene Regulation, and Signal Tranduction Pathways

In addition to its direct or indirect antioxidant effects, described earlier, lycopene is also considered to act directly on regulation of cell function within specific target tissues. Among the crucial mechanisms reported to be affected by lycopene are the inhibition of cancer cell growth; regulation of the cell cycle; interference with signal transduction pathways of cytokines, hormones, and growth factors; increase in cell communication; and interaction with gene regulation [54,92–94]. However, specific lycopene-protein interactions have not yet been characterized. It should also be emphasized that the effects of lycopene may vary in differing target tissues, with obvious consequences for overall risk/benefit assessments.

Lycopene is reported to have an inhibitory influence on cell proliferation in a variety of cancer cells, including prostate [56], mammary [54], breast [54], and endometrium [94], and in promyelocytic leukemia cells [92]. In promyelocytic leukemia cells HL-60 [92], cell growth was reduced by inhibiting cell cycle progression through the G0/G1 phase and cell differentiation induced, as measured by phorbol ester-dependent reduction of nitroblue tetrazolium and expression of CD14. Lycopene and 1,25-dihydroxyvitamin $D_3$ showed a synergistic effect on cell proliferation and differentiation in this HL-60 model. In MCF-7 and T-47D breast cancer cells, and in ECC-1 endometrial cancer cells, lycopene delayed G1-S transition, mediated by lycopene-induced reduction of cyclin D1 and D3 protein levels [94]. The effect of lycopene alone, or in association with other antioxidants, has been studied on the growth of androgen-insensitive DU-145 and PC-3 cells, two human prostate carcinoma cell lines [55]. By itself, lycopene was not a potent inhibitor of prostate carcinoma cell proliferation, but together with α-tocopherol at physiological concentrations ($<1$ μM and 50 μM, respectively), a strong inhibitory effect of prostate carcinoma cell proliferation was observed, which reached values close to 90%. The effect of lycopene with α-tocopherol was synergistic but was not observed when α-tocopherol was replaced by its β-isomer, ascorbic acid or probucol.

Of further biological relevance is the finding that lycopene interfered with the IGF-1 signaling process. This and, in particular, the limitation of IGF-1 signaling is important for human health, since high serum concentrations of IGF-1 are associated with an increased risk of breast, prostate, colorectal, and lung cancers [95,96]. In a transgenic adenocarcinoma of mouse prostate (TRAMP) model that expresses the oncogene SV40 T antigen specifically in the epithelium of the prostate [97], prostate cancer progression is associated with increased IGF-1 expression [98]. Transgenic mice expressing IGF-1 in prostate basal epithelial cells develop intraepithelial neoplasia by 6–7 months of age [99].

Physiological concentrations of lycopene markedly reduced growth stimulation of MCF-7 cells by IGF-1, an inhibitory effect that was associated with delayed cell cycle progression through G1-S phase unaccompanied by either apoptotic or necrotic cell death. Moreover, the effect was due to neither changes in number nor affinity of IGF-1 receptors. In this model, lycopene increased the number of a cell surface–associated member of the IGF-binding protein family on the protein level, which had been previously shown to negatively regulate IGF-receptor activity. In this situation, inhibition of IGF-receptor activity resulted in reduced insulin receptor substrate-1 (IRS-1) tyrosine phosphorylation and, further downstream, to a reduced binding capacity in the AP-1 transcriptional complex [54]. Kucuk et al. [51] observed that blood levels of IGF-1 decreased in both lycopene intervention and control groups of patients with newly diagnosed, clinically localized prostate cancer, suggesting that lycopene, although able to influence cellular IGF downstream signaling, does not regulate circulating levels of IGF protein in vivo.

A number of recent studies provide evidence for an effect of lycopene on cell communication. The structural elements for cell communication are transmembrane proteins, the connexins, which constitute the gap junctions; decreased expression of connexins, including connexin 43, has been widely reported in human tumors, as compared with normal tissue [100]. In healthy prostate tissue, connexin 43 is immunolocalized in a punctuated manner along the border of basal epithelial cells, whereas in benign prostate hyperplasia there is a marked increase in both the incidence and intensity of connexins 43 and 32 within epithelial cells. Connexins are known to be decreased in prostate cancer [101], and in the human study carried out by Kucuk et al. [51], connexin 43 was increased in the prostate tissue from lycopene-treated subjects, although this effect was not statistically significant. Carruba et al. [102] recently reported increased intercellular communication to be associated with increased connexin 43 protein content and/or a reduction in connexin 32 and suggested that agents which increase the connexin 43:connexin 32 ratio could be used to restore gap junction–mediated intercellular communication in junction-deficient, nontumorigenic immortalized cells. In human fetal skin fibroblasts (HFFF2), gap junctional communication was stimulated by lycopene [93]. Lycopene significantly upregulated both the transcription and the expression of connexin 43 in KB-1 human oral tumor cells [103]. Furthermore, lycopene enhanced gap junctional communication between the cancer cells and strongly, and dose-dependently, inhibited proliferation of KB-1 human oral tumor cells, thereby pointing to a role in oral cancer.

In contrast to β-carotene, α-carotene, or β-cryptoxanthin, lycopene is not a provitamin A carotenoid; the absence of the β-ionone rings at the two ends of the molecule prevents its acceptance as a substrate by the 15,15′-central cleavage enzyme to generate retinal. A mammalian enzyme that catalyzes asymmetrical oxidative cleavage of β-carotene at the 9′,10′ position has recently been identified; this cleaved lycopene to form apolycopenales [104]. It is, thus, theoretically possible that, by analogy to the β-oxidation of β-apocarotenales to yield retinoic acid [105], acyclo-retinoic acid could be formed from lycopene in vivo. Kim et al. [106] identified acyclo-lycopenales and acyclo-retinal as in vitro oxidation products and suggested that, under the oxidative conditions in biological tissues, lycopene might be cleaved to a series of apolycopenales and short-chain carbonyl compounds and that acyclo-retinal may be further enzymatically converted to acyclo-retinoic acid. This raises the possibility that lycopene might act, in part at least, via formation of this oxidation product.

Stahl et al. [93] showed the effect of acyclo-retinoic acid or lycopene on retinoid signaling and gap junctional communication. Levels of acyclo-retinoic acid (50 μM) much higher than is present physiologically were necessary to transactivate the $RARB_2$ promoter [93]. Likewise, 500-fold greater amounts of acyclo-retinoic acid, as compared to all-E-retinoic acid, were required to exert a stabilizing effect on the mRNA of connexin 43 via elements located in the 3′-UTR. Moreover, although gap junctional communication was stimulated by both lycopene and acyclo-retinoic acid in human fetal skin fibroblasts, 10-fold higher amounts of the later (1 μM) were needed to produce a comparable effect. These data demonstrate that acyclo-retinoic acid is much less active than retinoic acid with respect to communication across gap junctions and retinoid-related signaling, leading the authors to conclude that lycopene affects gap junctional communication by mechanisms that are mainly independent of acyclo-retinoic acid formation. Lycopene, retinoic acid, and acyclo-retinoic acid all showed similar potency in reducing cell cycle progression from G1 to S phase in MCF-7 human mammary cancer cells [107], indicating that their antiproliferative effects are also not mediated through the retinoic acid receptor. Other investigators showed that acyclo-retinoic acid can inhibit cancer cell growth and induce apoptosis in several prostate cancer cells, except LNCAP [108]; however, these results were only obtained at relatively high, nonphysiological concentrations (up to 40 μM). Until now, acyclo-retinoic acid has not been reported in human or animal tissues, which may, of course, be due to tissue concentrations below detection limit and/or low stability during extraction. It still remains an open question as to whether lycopene exerts some of its effects via retinoid signaling, and further studies on these pathways is clearly required.

## V.  CONCLUSION

Observational, epidemiological studies relate tomato intake to decreased risk for cancers at various sites and heart disease. Mechanistic plausibility, animal, and in vitro studies offer support for a role of lycopene, the red carotenoid in tomatoes. Lycopene has been demonstrated to decrease markers for oxidative stress and to act as an antioxidant in vivo. Although several lycopene-induced cellular changes can be demonstrated at the molecular level, the mechanism(s) by which this hydrophobic, membrane-localized carotenoid might modulate gene expression remains to be determined. The possibility exists for direct lycopene-protein interactions with transmembrane proteins, i.e., growth factor receptors and others, thereby modulating their activity and subsequent upstream signalling pathways. Lycopene might, both as an antioxidant and through its presence in membranes, also interfere with reactive oxygen species/NO-regulated signaling cascades and/ or influence cellular membrane properties and/or redox potentials, resulting in altered gene expression; in this context it is noteworthy that several cellular redox sensors have already been described [109–112]. Further investigations are needed to examine the relationship between the membrane localization and antioxidant activity of lycopene, on the one hand, and its influence on gene expression level, on the other. It is important that additional, preclinical studies be carried out in appropriate animal models with respect to specific target tissues (e.g., prostate cancer) before larger human intervention trials are conducted. As elsewhere, the powerful tools of genomics and proteomics technologies will soon increase our knowledge, identify the role of lycopene in the pathogenesis of disease, and facilitate its use as a cancer-preventing and cancer-protecting compound.

## REFERENCES

1.  Khachik F, Carvalho L, Bernstein PS, Muir GJ, Zhao DY, Katz NB. Chemistry, distribution, and metabolism of tomato carotenoids and their impact on human health. Exp Biol Med (Maywood) 2002; 227:845–851.
2.  Giovannucci E, Ascherio A, Rimm EB, Stampfer MJ, Colditz GA, Willett WC. Intake of carotenoids and retinol in relation to risk of prostate cancer. J Natl Cancer Inst 1995; 87:1767–1776.
3.  Giovannucci E, Clinton SK. Tomatoes, lycopene, and prostate cancer. Proc Soc Exp Biol Med 1998; 218:129–139.
4.  Arab L, Steck S. Lycopene and cardiovascular disease. Am J Clin Nutr 2000; 71: 1691S–1697S.
5.  Rissanen T, Voutilainen S, Nyyssonen K, Salonen JT. Lycopene, atherosclerosis, and coronary heart disease. Exp Biol Med (Maywood) 2002; 227:900–907.
6.  Shi J, Le Maguer M. Lycopene in tomatoes: chemical and physical properties affected by food processing. Crit Rev Biotechnol 2000; 20:293–334.

7.  Tonucci LH, Holden JM, Beecher GR, Kachik F, Davis SD, Mulokozi G. Carotenoid content of thermally processed tomato-based food products. J Agric Food Chem 1995; 43:579–586.

8.  Schierle J, Bretzel W, Bühler I, Faccin N, Hess D, Steiner K, Schuep W. Content and isomeric ratio of lycopene in food and human blood plasma. Food Chem 1997; 59:459–465.

9.  Aoki H, Kieu NT, Kuze N, Tomisaka K, Van Chuyen N. Carotenoid pigments in GAC fruit (*Momordica cochinchinensis* Spreng.). Biosci Biotechnol Biochem 2002; 66:2479–2482.

10. van het Hof KH, de Boer BC, Tijburg LB, Lucius BR, Zijp I, West CE, Hautvast JG, Weststrate JA. Carotenoid bioavailability in humans from tomatoes processed in different ways determined from the carotenoid response in the triglyceride-rich lipoprotein fraction of plasma after a single consumption and in plasma after four days of consumption. J Nutr 2000; 130:1189–1196.

11. Isaacson T, Ronen G, Zamir D, Hirschberg J. Cloning of tangerine from tomato reveals a carotenoid isomerase essential for the production of *beta*-carotene and xanthophylls in plants. Plant Cell 2002; 14:333–342.

12. Giuliano G, Giliberto L, Rosati C. Carotenoid isomerase: a tale of light and isomers. Trends Plant Sci 2002; 7:427–429.

13. Eckardt NA. Tangerine dreams: cloning of carotenoid isomerase from *Arabidopsis* and tomato. Plant Cell 2002; 14:289–292.

14. Takeoka GR, Dao L, Flessa S, Gillespie DM, Jewell WT, Huebner B, Bertow D, Ebeler SE. Processing effects on lycopene content and antioxidant activity of tomatoes. J Agric Food Chem 2001; 49:3713–3717.

15. Dewanto V, Wu X, Adom KK, Liu RH. Thermal processing enhances the nutritional value of tomatoes by increasing total antioxidant activity. J Agric Food Chem 2002; 50:3010–3014.

16. Stahl W, Sies H. Uptake of lycopene and its geometrical isomers is greater from heat-processed than from unprocessed tomato juice in humans. J Nutr 1992; 122: 2161–2166.

17. Gartner C, Stahl W, Sies H. Lycopene is more bioavailable from tomato paste than from fresh tomatoes. Am J Clin Nutr 1997; 66:116–122.

18. Boileau TW, Boileau AC, Erdman JW. Bioavailability of all-*trans* and *cis*-isomers of lycopene. Exp Biol Med (Maywood) 2002; 227:914–919.

19. Boileau AC, Merchen NR, Wasson K, Atkinson CA, Erdman JW. *cis*-Lycopene is more bioavailable than *trans*-lycopene in vitro and in vivo in lymph-cannulated ferrets. J Nutr 1999; 129:1176–1181.

20. Holloway DE, Yang M, Paganga G, Rice-Evans CA, Bramley PM. Isomerization of dietary lycopene during assimilation and transport in plasma. Free Radic Res 2000; 32:93–102.

21. Cohn WTP, Aebischer C, Schierle J, Schalch W. Lycopene concentrations in plasma after consumption of tablets containing lycopene beadlets or lycopene from tomatoes. Ann Nutr Metab 2001; 45:37.

22. Johnson EJ. Human studies on bioavailability and plasma response of lycopene. Proc Soc Exp Biol Med 1998; 218:115–120.

23. Ford ES, Gillespie C, Ballew C, Sowell A, Mannino DM. Serum carotenoid concentrations in US children and adolescents. Am J Clin Nutr 2002; 76:818–827.

24. Van Kappel A. Carotenoids in blood: Biochemical markers of fruit and vegetable consumption and relation to the risk of breast cancer. Doctoral thesis presented to the University Claude Bernard, Lyon, 2000, Service Nutrition and Cancer, IARC, Lyon, 0.

25. Yuan JM, Ross RK, Chu XD, Gao YT, Yu MC. Prediagnostic levels of serum *beta*-cryptoxanthin and retinol predict smoking-related lung cancer risk in Shanghai, China. Cancer Epidemiol Biomarkers Prev 2001; 10:767–773.

26. van Breemen RB, Xu X, Viana MA, Chen L, Stacewicz-Sapuntzakis M, Duncan C, Bowen PE, Sharifi R. Liquid chromatography-mass spectrometry of *cis*- and all-*trans*-lycopene in human serum and prostate tissue after dietary supplementation with tomato sauce. J Agric Food Chem 2002; 50:2214–2219.

27. Carroll YL, Corridan BM, Morrissey PA. Lipoprotein carotenoid profiles and the susceptibility of low density lipoprotein to oxidative modification in healthy elderly volunteers. Eur J Clin Nutr 2000; 54:500–507.

28. Forman MR, Beecher GR, Muesing R, Lanza E, Olson B, Campbell WS, McAdam P, Raymond E, Schulman JD, Graubard BI. The fluctuation of plasma carotenoid concentrations by phase of the menstrual cycle: a controlled diet study. Am J Clin Nutr 1996; 64:559–565.

29. Lin S, Quaroni L, White WS, Cotton T, Chumanov G. Localization of carotenoids in plasma low-density lipoproteins studied by surface-enhanced resonance Raman spectroscopy. Biopolymers 2000; 57:249–256.

30. Khachik F, Beecher GR, Smith JC. Lutein, lycopene, and their oxidative metabolites in chemoprevention of cancer. J Cell Biochem Suppl 1995; 22:236–246.

31. Khachik F, Spangler CJ, Smith JC, Canfield LM, Steck A, Pfander H. Identification, quantification, and relative concentrations of carotenoids and their metabolites in human milk and serum. Anal Chem 1997; 69:1873–1881.

32. Khachik F SC, Smith JC. Identification, quantification, and relative concentrations of carotenoids and their metabolites in human milk and serum. Anal Chem 1997; 69:1873–1881.

33. Khachik F PH, Traber B. Proposed mechanisms for the formation of synthetic and naturally occuring metabolites of lycopene in tomato products and human serum. J Agric Food Chem 1998; 46:4885–4890.

34. Chen G, Djuric Z. Detection of 2,6-cyclolycopene-1,5-diol in breast nipple aspirate fluids and plasma: a potential marker of oxidative stress. Cancer Epidemiol Biomarkers Prev 2002; 11:1592–1596.

35. Stahl W, Schwarz W, Sundquist AR, Sies H. *cis-trans* Isomers of lycopene and *beta*-carotene in human serum and tissues. Arch Biochem Biophys 1992; 294: 173–177.

36. Clinton SK, Emenhiser C, Schwartz SJ, Bostwick DG, Williams AW, Moore BJ, Erdman JW. *cis-trans* Lycopene isomers, carotenoids, and retinol in the human prostate. Cancer Epidemiol Biomarkers Prev 1996; 5:823–833.

37. Clinton SK. Lycopene: chemistry, biology, and implications for human health and disease. Nutr Rev 1998; 56:35–51.

38. Freeman VL, Meydani M, Yong S, Pyle J, Wan Y, Arvizu-Durazo R, Liao Y. Prostatic levels of tocopherols, carotenoids, and retinol in relation to plasma levels and self-reported usual dietary intake. Am J Epidemiol 2000; 151:109–118.

39. Gerster H. The potential role of lycopene for human health. J Am Coll Nutr 1997; 16:109–126.

40. Weisburger JH. Evaluation of the evidence on the role of tomato products in disease prevention. Proc Soc Exp Biol Med 1998; 218:140–143.

41. Sengupta A, Das S. The anti-carcinogenic role of lycopene, abundantly present in tomato. Eur J Cancer Prev 1999; 8:325–330.

42. Agarwal S, Rao AV. Carotenoids and chronic diseases. Drug Metabol Drug Interact 2000; 17:189–210.

43. Agarwal S, Rao AV. Tomato lycopene and its role in human health and chronic diseases. Cmaj 2000; 163:739–744.

44. Bramley PM. Is lycopene beneficial to human health?. Phytochemistry 2000; 54: 233–236.

45. Arab L, Steck-Scott S, Bowen P. Participation of lycopene and beta-carotene in carcinogenesis: defenders, aggressors, or passive bystanders?. Epidemiol Rev 2001; 23:211–230.

46. Giovannucci E. Tomatoes, tomato-based products, lycopene, and cancer: review of the epidemiologic literature. J Natl Cancer Inst 1999; 91:317–331.

47. Miller EC, Giovannucci E, Erdman JW, Bahnson R, Schwartz SJ, Clinton SK. Tomato products, lycopene and prostate cancer risk. Urol Clin North Am 2002; 29:83–93.

48. Vogt TM, Mayne ST, Graubard BI, Swanson CA, Sowell AL, Schoenberg JB, Swanson GM, Greenberg RS, Hoover RN, Hayes RB, Ziegler RG. Serum lycopene, other serum carotenoids, and risk of prostate cancer in US blacks and whites. Am J Epidemiol 2002; 155:1023–1032.

49. Giovannucci E, Rimm EB, Liu Y, Stampfer MJ, Willett WC. A prospective study of tomato products, lycopene, and prostate cancer risk. J Natl Cancer Inst 2002; 94:391–398.

50. Gann PH, Ma J, Giovannucci E, Willett W, Sacks FM, Hennekens CH, Stampfer MJ. Lower prostate cancer risk in men with elevated plasma lycopene levels: results of a prospective analysis. Cancer Res 1999; 59:1225–1230.

51. Kucuk O, Sarkar FH, Sakr W, Djuric Z, Pollak MN, Khachik F, Li YW, Banerjee M, Grignon D, Bertram JS, Crissman JD, Pontes EJ, Wood DP. Phase II randomized clinical trial of lycopene supplementation before radical prostatectomy. Cancer Epidemiol Biomarkers Prev 2001; 10:861–868.

52. Chen L, Stacewicz-Sapuntzakis M, Duncan C, Sharifi R, Ghosh L, van Breemen R, Ashton D, Bowen PE. Oxidative DNA damage in prostate cancer patients consuming tomato sauce-based entrees as a whole-food intervention. J Natl Cancer Inst 2001; 93:1872–1879.

53. Levy J, Bosin E, Feldman B, Giat Y, Miinster A, Danilenko M, Sharoni Y. Lycopene is a more potent inhibitor of human cancer cell proliferation than either alpha-carotene or beta-carotene. Nutr Cancer 1995; 24:257–266.

54. Karas M, Amir H, Fishman D, Danilenko M, Segal S, Nahum A, Koifmann A, Giat Y, Levy J, Sharoni Y. Lycopene interferes with cell cycle progression and

insulin-like growth factor I signaling in mammary cancer cells. Nutr Cancer 2000; 36:101–111.

55. Pastori M, Pfander H, Boscoboinik D, Azzi A. Lycopene in association with *alpha*-tocopherol inhibits at physiological concentrations proliferation of prostate carcinoma cells. Biochem Biophys Res Commun 1998; 250:582–585.

56. Kotake-Nara E, Kushiro M, Zhang H, Sugawara T, Miyashita K, Nagao A. Carotenoids affect proliferation of human prostate cancer cells. J Nutr 2001; 131: 3303–3306.

57. Rissanen TH, Voutilainen S, Nyyssonen K, Lakka TA, Sivenius J, Salonen R, Kaplan GA, Salonen JT. Low serum lycopene concentration is associated with an excess incidence of acute coronary events and stroke: The Kuopio Ischaemic Heart Disease Risk Factor Study. Br J Nutr 2001; 85:749–754.

58. Rissanen T, Voutilainen S, Nyyssonen K, Salonen R, Salonen JT. Low plasma lycopene concentration is associated with increased intima-media thickness of the carotid artery wall. Arterioscler Thromb Vasc Biol 2000; 20:2677–2681.

59. Klipstein-Grobusch K, Launer LJ, Geleijnse JM, Boeing H, Hofman A, Witteman JC. Serum carotenoids and atherosclerosis. The Rotterdam Study. Atherosclerosis 2000; 148:49–56.

60. Kohlmeier L, Kark JD, Gomez-Gracia E, Martin BC, Steck SE, Kardinaal AF, Ringstad J, Thamm M, Masaev V, Riemersma R, Martin-Moreno JM, Huttunen JK, Kok FJ. Lycopene and myocardial infarction risk in the EURAMIC Study. Am J Epidemiol 1997; 146:618–626.

61. Truscott TG, Land EJ, Sykes A. The in vitro photochemistry of biological molecules III. Absorption spectra, lifetimes and rates of oxygen quenching of the triplet states of *beta*-carotene, retinal and related polyenes. Photochem Photobiol 1973; 17: 43–51.

62. Conn PF, Schalch W, Truscott TG. The singlet oxygen and carotenoid interaction. J Photochem Photobiol B 1991; 11:41–47.

63. Hirayama O, Nakamura K, Hamada S, Kobayasi Y. Singlet oxygen quenching ability of naturally occurring carotenoids. Lipids 1994; 29:149–150.

64. Di Mascio P, Devasagayam TP, Kaiser S, Sies H. Carotenoids, tocopherols and thiols as biological singlet molecular oxygen quenchers. Biochem Soc Trans 1990; 18:1054–1056.

65. Cantrell A, McGarvey DJ, Truscott TG, Rancan F, Bohm F. Singlet oxygen quenching by dietary carotenoids in a model membrane environment. Arch Biochem Biophys 2003; 412:47–54.

66. Stahl W, Junghans A, de Boer B, Driomina ES, Briviba K, Sies H. Carotenoid mixtures protect multilamellar liposomes against oxidative damage: synergistic effects of lycopene and lutein. FEBS Lett 1998; 427:305–308.

67. Woodall AA, Britton G, Jackson MJ. Carotenoids and protection of phospholipids in solution or in liposomes against oxidation by peroxyl radicals: relationship between carotenoid structure and protective ability. Biochim Biophys Acta 1997; 1336: 575–586.

68. Rehman AB, Halliwell B, Rice-Evans CA. Tomato consumption modulates oxidative DNA damage in humans. Biochem Biophys Res Commun 1999; 262:828–831.

69. Porrini M, Riso P. Lymphocyte lycopene concentration and DNA protection from oxidative damage is increased in women after a short period of tomato consumption. J Nutr 2000; 130:189–192.

70. Pool-Zobel BL, Bub A, Muller H, Wollowski I, Rechkemmer G. Consumption of vegetables reduces genetic damage in humans: first results of a human intervention trial with carotenoid-rich foods. Carcinogenesis 1997; 18:1847–1850.

71. Torbergsen AC, Collins AR. Recovery of human lymphocytes from oxidative DNA damage; the apparent enhancement of DNA repair by carotenoids is probably simply an antioxidant effect. Eur J Nutr 2000; 39:80–85.

72. Alberg A. The influence of cigarette smoking on circulating concentrations of antioxidant micronutrients. Toxicology 2002; 180:121–137.

73. Agarwal S, Rao AV. Tomato lycopene and low density lipoprotein oxidation: a human dietary intervention study. Lipids 1998; 33:981–984.

74. Hininger I, Chopra M, Thurnham DI, Laporte F, Richard MJ, Favier A, Roussel AM. Effect of increased fruit and vegetable intake on the susceptibility of lipoprotein to oxidation in smokers. Eur J Clin Nutr 1997; 51:601–606.

75. Hininger IA, Meyer-Wenger A, Moser U, Wright A, Southon S, Thurnham D, Chopra M, Van Den Berg H, Olmedilla B, Favier AE, Roussel AM. No significant effects of lutein, lycopene or *beta*-carotene supplementation on biological markers of oxidative stress and LDL oxidizability in healthy adult subjects. J Am Coll Nutr 2001; 20:232–238.

76. Dugas TR, Morel DW, Harrison EH. Dietary supplementation with *beta*-carotene, but not with lycopene, inhibits endothelial cell-mediated oxidation of low-density lipoprotein. Free Radic Biol Med 1999; 26:1238–1244.

77. Chopra M, O'Neill ME, Keogh N, Wortley G, Southon S, Thurnham DI. Influence of increased fruit and vegetable intake on plasma and lipoprotein carotenoids and LDL oxidation in smokers and nonsmokers. Clin Chem 2000; 46:1818–1829.

78. Bohm F, Edge R, Burke M, Truscott TG. Dietary uptake of lycopene protects human cells from singlet oxygen and nitrogen dioxide – ROS components from cigarette smoke. J Photochem Photobiol B 2001; 64:176–178.

79. Bohm F, Tinkler JH, Truscott TG. Carotenoids protect against cell membrane damage by the nitrogen dioxide radical. Nat Med 1995; 1:98–99.

80. Panasenko OM, Sharov VS, Briviba K, Sies H. Interaction of peroxynitrite with carotenoids in human low density lipoproteins. Arch Biochem Biophys 2000; 373:302–305.

81. Pannala AS, Rice-Evans C, Sampson J, Singh S. Interaction of peroxynitrite with carotenoids and tocopherols within low density lipoprotein. FEBS Lett 1998; 423:297–301.

82. Matos HR, Di Mascio P, Medeiros MH. Protective effect of lycopene on lipid peroxidation and oxidative DNA damage in cell culture. Arch Biochem Biophys 2000; 383:56–59.

83. Matos HR, Capelozzi VL, Gomes OF, Mascio PD, Medeiros MH. Lycopene inhibits DNA damage and liver necrosis in rats treated with ferric nitrilotriacetate. Arch Biochem Biophys 2001; 396:171–177.

84. Kwak MK EP, Dolan PM, Ramos-Gomez M, Groopman JD, Itoh K, Yamamoto M, Kensler TW. Role of Phase 2 enzyme induction in chemoprotection by dithiolethiones. Mutat Res 2001; 480–481:305–315.

85. Velmurugan B, Bhuvaneswari V, Burra UK, Nagini S. Prevention of N-methyl-N'-nitro-N-nitrosoguanidine and saturated sodium chloride-induced gastric carcinogenesis in Wistar rats by lycopene. Eur J Cancer Prev 2002; 11:19–26.

86. Bhuvaneswari V, Velmurugan B, Balasenthil S, Ramachandran CR, Nagini S. Chemopreventive efficacy of lycopene on 7,12-dimethylbenz[a]anthracene-induced hamster buccal pouch carcinogenesis. Fitoterapia 2001; 72:865–874.

87. Bhuvaneswari V, Velmurugan B, Nagini S. Induction of glutathione-dependent hepatic biotransformation enzymes by lycopene in the hamster cheek pouch carcinogenesis model. J Biochem Mol Biol Biophys 2002; 6:257–260.

88. Leal M, Shimada A, Ruiz F, Gonzalez de Mejia E. Effect of lycopene on lipid peroxidation and glutathione-dependent enzymes induced by T-2 toxin in vivo. Toxicol Lett 1999; 109:1–10.

89. Dhakshinamoorthy S LDn, Jaiswal AK. Antioxidant regulation of genes encoding enzymes that detoxify xenobiotics and carcinogens. Curr Top Cell Regul 2000; 36:201–216.

90. Cantelli-Forti G HP, Paolini M. The pitfall of detoxifying enzymes. Mutat Res 1998; 402:179–183.

91. Breinholt V, Lauridsen ST, Daneshvar B, Jakobsen J. Dose-response effects of lycopene on selected drug-metabolizing and antioxidant enzymes in the rat. Cancer Lett 2000; 154:201–210.

92. Amir H, Karas M, Giat J, Danilenko M, Levy R, Yermiahu T, Levy J, Sharoni Y. Lycopene and 1,25-dihydroxyvitamin D3 cooperate in the inhibition of cell cycle progression and induction of differentiation in HL-60 leukemic cells. Nutr Cancer 1999; 33:105–112.

93. Stahl W, von Laar J, Martin HD, Emmerich T, Sies H. Stimulation of gap junctional communication: comparison of acyclo-retinoic acid and lycopene. Arch Biochem Biophys 2000; 373:271–274.

94. Nahum A, Hirsch K, Danilenko M, Watts CK, Prall OW, Levy J, Sharoni Y. Lycopene inhibition of cell cycle progression in breast and endometrial cancer cells is associated with reduction in cyclin D levels and retention of p27(Kip1) in the cyclin E-cdk2 complexes. Oncogene 2001; 20:3428–3436.

95. Furstenberger G, Senn HJ. Insulin-like growth factors and cancer. Lancet Oncol 2002; 3:298–302.

96. Pollak M. Insulin-like growth factors and prostate cancer. Epidem Rev 2001; 23:59–66.

97. Foster BA, Kaplan PJ, Greenberg NM. Peptide growth factors and prostate cancer: new models, new opportunities. Cancer Metastasis Rev 1998–99; 17:317–324.

98. Kaplan PJ, Mohan S, Cohen P, Foster BA, Greenberg NM. The insulin-like growth factor axis and prostate cancer: lessons from the transgenic adenocarcinoma of mouse prostate (TRAMP) model. Cancer Res 1999; 59:2203–2209.

99. DiGiovanni J, Kiguchi K, Frijhoff A, Wilker E, Bol DK, Beltran L, Moats S, Ramirez A, Jorcano J, Conti C. Deregulated expression of insulin-like growth factor 1 in prostate epithelium leads to neoplasia in transgenic mice. Proc Natl Acad Sci USA 2000; 97:3455–3460.

100. Neveu M, Bertram JS. Gap Junctions and Neoplasia. Greenwich. CT: JAI Press, 2000.

101. Habermann H, Ray V, Habermann W, Prins GS. Alterations in gap junction protein expression in human benign prostatic hyperplasia and prostate cancer. J Urol 2002; 167:655–660.
102. Carruba G, Webber MM, Quader ST, Amoroso M, Cocciadiferro L, Saladino F, Trosko JE, Castagnetta JE. Regulation of cell-to-cell communication in non-tumorigenic and malignant human prostate epithelial cells. Prostate 2002; 50:73–82.
103. Livny O, Kaplan I, Reifen R, Polak-Charcon S, Madar Z, Schwartz B. Lycopene inhibits proliferation and enhances gap-junction communication of KB-1 human oral tumor cells. J Nutr 2002; 132:3754–3759.
104. Kiefer C, Hessel S, Lampert JM, Vogt K, Lederer MO, Breithaupt DE, von Lintig J. Identification and characterization of a mammalian enzyme catalyzing the asymmetric oxidative cleavage of provitamin A. J Biol Chem 2001; 276:14110–14116.
105. Wang XD, Russell RM, Liu C, Stickel F, Smith DE, Krinsky NI. *Beta*-oxidation in rabbit liver in vitro and in the perfused ferret liver contributes to retinoic acid biosynthesis from *beta*-apocarotenoic acids. J Biol Chem 1996; 271:26490–26498.
106. Kim SJ, Nara E, Kobayashi H, Terao J, Nagao A. Formation of cleavage products by autoxidation of lycopene. Lipids 2001; 36:191–199.
107. Ben-Dor A, Nahum A, Danilenko M, Giat Y, Stahl W, Martin HD, Emmerich T, Noy N, Levy J, Sharoni Y. Effects of acyclo-retinoic acid and lycopene on activation of the retinoic acid receptor and proliferation of mammary cancer cells. Arch Biochem Biophys 2001; 391:295–302.
108. Kotake-Nara E, Kim SJ, Kobori M, Miyashita K, Nagao A. Acyclo-retinoic acid induces apoptosis in human prostate cancer cells. Anticancer Res 2002; 22: 689–695.
109. Georgiou G. How to flip the (redox) switch. Cell 2002; 111:607–610.
110. Fan C, Li Q, Ross D, Engelhardt JF. Tyrosine phosphorylation of IkappaBalpha activates NF-kappaB through a redox-regulated and c-Src-dependent mechanism following hypoxia/reoxygenation. J Biol Chem 2002; 278:2072–2080.
111. Helmann JD. OxyR: a molecular code for redox sensing?. Sci STKE 2002; 2002: PE46.
112. Delaunay A, Pflieger D, Barrault MB, Vinh J, Toledano MB. A thiol peroxidase is an h(2)o(2) receptor and redox-transducer in gene activation. Cell 2002; 111: 471–481.
113. van Kappel AL, Martinez-Garcia C, Elmstahl S, Steghens JP, Chajes V, Bianchini F, Kaaks R, Riboli E. Plasma carotenoids in relation to food consumption in Granada (southern Spain) and Malmo (southern Sweden). Int J Vitam Nutr Res 2001; 71: 97–102.
114. Olmedilla B, Granado F, Southon S, Wright AJ, Blanco I, Gil-Martinez E, Berg H, Corridan B, Roussel AM, Chopra M, Thurnham DI. Serum concentrations of carotenoids and vitamins A, E and C in control subjects from five European countries. Br J Nutr 2001; 85:227–238.
115. Ruiz Rejon F, Martin-Pena G, Granado F, Ruiz-Galiana J, Blanco I, Olmedilla B. Plasma status of retinol, *alpha*- and *gamma*-tocopherols, and main carotenoids to first myocardial infarction: case control and follow-up study. Nutrition 2002; 18: 26–31.

116. Polidori MC, Mecocci P, Stahl W, Parente B, Cecchetti R, Cherubini A, Cao P, Sies H, Senin U. Plasma levels of lipophilic antioxidants in very old patients with type 2 diabetes. Diabetes Metab Res Rev 2000; 16:15–19.

117. Nagao T, Ikeda N, Warnakulasuriya S, Fukano H, Yuasa H, Yano M, Miyazaki H, Ito Y. Serum antioxidant micronutrients and the risk of oral leukoplakia among Japanese. Oral Oncol 2000; 36:466–470.

118. Irwig MS, El-Sohemy A, Baylin A, Rifai N, Campos H. Frequent intake of tropical fruits that are rich in *beta*-cryptoxanthin is associated with higher plasma *beta*-cryptoxanthin concentrations in Costa Rican adolescents. J Nutr 2002; 132: 3161–3167.

119. Mayne ST, Cartmel B. Antioxidant vitamin supplementation and lipid peroxidation in smokers. Am J Clin Nutr 1999; 69:1292.

120. Ford ES, Giles WH. Serum vitamins, carotenoids and angina pectoris: findings from the National Health and Nutrition Examination Survey III. Ann Epidemiol 2000; 10:106–116.

121. Ribaya-Mercado JD, Garmyn M, Gilchrest BA, Russell RM. Skin lycopene is destroyed preferentially over *beta*-carotene during ultraviolet irradiation in humans. J Nutr 1995; 125:1854–1859.

122. Rao AV, Agarwal S. Role of lycopene as antioxidant carotenoid in the prevention of chronic diseases: a review. Nutr Res 1999; 19:305–323.

# 15

# Effects of Oltipraz on Phase 1 and Phase 2 Xenobiotic-Metabolizing Enzymes

**Sophie Langouët, Fabrice Morel, and André Guillouzo**
*University of Rennes, Rennes, France*

## I. INTRODUCTION

Many chemicals require initial metabolic activation to electrophilic intermediates in order to exert carcinogenic effects. This phase 1 reaction is mostly catalyzed by the cytochrome P450 (CYP) superfamily. If not subsequently inactivated by phase 2 conjugation reactions, electrophiles can interact with DNA, form DNA adducts, and cause genetic lesions. The phase 2 reactions include those catalyzed by various enzymes, particularly glutathione transferases (GSTs), which are recognized to play a major role in inactivation of carcinogens. Activation of chemicals can also lead to generation of reactive oxygen species (ROS) that will oxidize DNA. If not repaired, these different DNA lesions can be carcinogenic.

Carcinogenesis is a multistep process. Diverse natural products and chemical agents can prevent the formation of tumors; these are classified according to their level of intervention, either at the initiation stage or during the promotion and/or progression of tumorigenesis. These chemoprotective agents may act by shifting the balance between activation and inactivation of carcinogens through their action on phase 1 and phase 2 xenobiotic-metabolizing enzymes. Brassica vegetables—particularly cabbage, brussels sprouts, broccoli, and cauliflower—contain substantial amounts of compounds such as indoles, isothiocyanates, and dithiolethiones, each of which possesses chemoprotective properties.

Oltipraz [4-methyl-5-(2-pyrazinyl)-1,2-dithiole-3-thione] (OPZ), a synthetic derivative of 1,2-dithiole-3-thione, has been extensively investigated for its chemopreventive properties; first developed as an antischistosomal agent, this compound was found to increase cellular thiol levels and to induce enzymes involved in the maintenance of reduced glutathione (GSH) pools and in detoxification of electrophilic intermediates. Induction of phase 2 enzymes such as GSTs, epoxide hydrolases, and NADP(H):quinone reductases was considered to represent the key mechanism of chemoprotection of OPZ. More recently this compound was also found to modulate phase 1 enzymes, antioxidant enzymes, and other inactivating proteins, thereby indicating that its chemoprevention mechanism(s) are more complex than first envisaged. In this chapter, current information about the effects of OPZ on phase 1 and phase 2 enzymes is presented.

## II. SOURCE AND PHARMACEUTICAL HISTORY OF DITHIOLETHIONES

Dithiolethiones belong to one of the most important class of chemopreventive agents. The five-membered cyclic compound containing sulfur, 1,2-dithiole-3-thione, was long considered to be a natural product present in cabbage and having potent anticarcinogenic properties. However in 1991 Marks et al. [1] demonstrated that this compound was actually formed by thermal reactions during isolation. A number of substituted derivatives have, however, been isolated from natural sources [2], while several substituted 1,2-dithiole-3-thiones have been synthesized and medicinally used as antischistosomal, anticholeretic, and salivary secretion–stimulating agents. The synthetic derivative OPZ (Fig. 1), developed by the Rhône-Poulenc Pharmaceutical Company in the 1980s, was the focus of at least 18 studies involving 1284 patients conducted to evaluate its antischistosomal effectiveness [3]. Several side effects, centered mainly in the digestive tract, as well as photosensitivity led to arrest of its commercial development. The research indicated that the antischistosomal properties of OPZ were probably due to its powerful inhibition of GSH levels in *Schistosoma mansoni*–infected mice, as

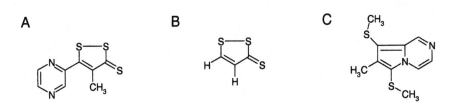

**Figure 1**   Chemical structure of OPZ (A), 1,2-dithiole-3-thione (D3T) (B), and OPZ M3 (C).

well as of GST and glutathione reductase (GR) activities in the host [4,5]; together these resulted in a reduced ability of the parasite to defend itself against toxic metabolites. Surprisingly, OPZ administration had the opposite effect in the host, where induction of GSH levels and activities of enzymes catalyzing the inactivation of toxic compounds were observed [3,6]. Such differential effects between the worm and its host suggested that OPZ might possess chemotherapeutic properties, and several groups conducted research projects to examine this possibility.

## III. PHARMACOKINETICS AND METABOLISM

In order to predict the optimal dosing and schedules for human treatment, pharmacokinetics, metabolism, and plasma concentrations of OPZ must be determined. OPZ is highly liposoluble so that its absorption depends on the fat content of the diet [7–9]. Recently, Dimitrov et al. [10] showed large variations in the distribution of OPZ within the lipid fraction, albumin, and plasma depending on the administration schedule; in certain individuals the compound was undetectable. OPZ is actively metabolized in various species; when orally administered to animals (e.g., mouse, rat and monkey) and humans, large amounts of free and conjugated metabolites were found in urine and the parent compound represented less than 1% of the original dose [11]. Metabolism of OPZ occurs by molecular rearrangement to yield a pyrolopyrazine derivative, OPZ M3 (Fig. 1), which is further metabolized to at least 10 products depending on the species studied. By measuring OPZ and OPZ M3 in human plasma, O'Dwyer et al. [12] concluded that the bioavailability of OPZ decreases with increasing dose while OPZ M3 is saturable. Therefore, the questions arise as to whether the effects are due to OPZ itself or to its metabolites and whether these are dependent on the species and tissues studied. Several observations suggest that OPZ is, in fact, a pro-drug and that its biological effects are correlated with its metabolism [13].

## IV. CHEMOPROTECTIVE PROPERTIES

The protective effects OPZ were found to be dependent on the species and tissues investigated, the starting time and duration of treatment, and the dosing schedule. The most potent protective effects were associated with simultaneous exposure of animals to OPZ and the carcinogen, treatment by OPZ being started 1 or 2 days before the first administration of the carcinogen [14]. When added concomitantly in the diet, OPZ was reported to significantly decrease tumor incidence and multiplicity in various rodent tissues, including lung, trachea, mammary glands, forestomach, small intestine, colon, skin, liver, and bladder, exposed to a range of carcinogens including benzo[a]pyrene (BaP), aflatoxin $B_1$ ($AFB_1$), diethylnitrosamine, uracil mustard, azoxymethane, methylazoxymethanol acetate, 4-[methylnitroso-amino-1-(3-pyridyl)]-1-butanone (NNK), and 7,12-dimethylbenz[a]anthracene (DMBA) [15–20].

A number of studies conducted in animals has demonstrated the efficacy of OPZ as a chemoprotective agent in experimental carcinogenesis. Thus, when administered to rats several days before azoxymethane, the incidence and multiplicity of adenocarcinomas in the small intestine and colon decreased by 66% [16]; moreover, the formation of aberrant foci was inhibited, supporting the view that OPZ can act during the initiation phase of colon carcinogenesis [21]. When administered to mice one week before N-butyl-N(4-hydroxybutyl)nitrosamine, incidence of bladder carcinomas was reduced by 30% and the tumors were less invasive [18]. When administered to female mice 48 hours prior to BaP, OPZ (500 mg/kg) inhibited occurrence of pulmonary adenomas and tumors of the forestomach by up to 62%. When female mice were exposed to diethylnitrosamine or uracil mustard, the incidence of pulmonary adenomas diminished by 30% [15]. At 600 mg/kg, OPZ prevented ductal adenocarcinomas induced by N-nitrosobis(2-oxopropyl)-amine in Syrian golden hamsters [22]. When added concurrently to the diet, OPZ caused suppression of lymphomas induced by oral administration of 2-amino-1-methyl-6-phenylimidazo[4,5-b]pyridine (PhIP) in rats [23]. OPZ has, in addition, radioprotective effects; for example, an increased survival rate from 48% to 91% was observed following the exposure of mice to lethal doses of radiations in the presence of OPZ [24].

The protective effects of OPZ have also been demonstrated in experiments using cell lines. Thus, OPZ was very effective in a human epidermal cell assay in which propane sultone–induced growth inhibition and increased differentiation were reduced by 90% in the presence of OPZ [25]. OPZ acts as an anti-initiator in mammary breast cancer organ cultures and was shown to inhibit formation of dibenz(a,l)pyrene-derived DNA adducts in human breast tumor epithelial cells by 80% [26].

The best studied example of the protective role of OPZ was conducted by Kensler et al. [27] using the potent hepatocarcinogen $AFB_1$. In an acute toxicity study, pretreatment with OPZ reduced the mortality caused by 10 mg/kg $AFB_1$ in rats from 83% to 36% [28]. γ-Glutamyl transpeptidase–positive foci in $AFB_1$-treated rat liver were inhibited in the presence of 0.075% OPZ in the diet and complete suppression of formation of hyperplasic nodules, and hepatocarcinoma did not result in any increased incidence of tumors in extrahepatic tissues [18,14].

From experiments conducted with OPZ administered before, during, and after $AFB_1$, it was concluded that OPZ acts as a "blocking" rather than a "suppressive" chemoprotective agent [29]. Such an anti-initiating agent can, in principle, act at several levels, including inhibition or induction of CYPs, induction of phase 2 enzymes, scavenging of electrophilic metabolites and ROS, and induction of DNA repair [30].

In order to determine the exact mechanism, structure-activity studies were performed; OPZ and 5-(4-methoxyphenyl)-3H-1,2-dithiole-3-thione were shown to be similarly effective on the inhibition of preneoplastic lesions induced by $AFB_1$, while treatment with 1,3-dithiole-2-thione was without effect [31]. The efficacy of OPZ compared to other dithiolethiones was further studied by evaluating the effects of 17 dithiolethiones on acute hepatotoxicity of $AFB_1$ in rats [32]. The results demonstrated that, although several dithiolethiones were more active than OPZ in inhibiting toxicity, no structural motif linked to these chemoprevention properties could be identified.

## V. EFFECTS ON PHASE 2 ENZYMES

The ability of OPZ to block the carcinogenic effects of several chemical agents was initially attributed to its effectiveness in inducing a battery of phase 2 detoxification enzymes: glutathione transferases (GSTs), epoxide hydrolase, NADP(H): quinone oxidoreductase (NQO), aflatoxin $B_1$–aldehyde reductase, and UDP-glucuronosyl transferases (UDP-GT); OPZ was, hence, classified as a monofunctional inducer since it advantageously elevated phase 2 detoxification enzymes [27]. It effectively induced NQO in Hepa 1c1c7 cells defective in the aryl hydrocarbon receptor (AhR) function required by bifunctional inducers [33]. Structure/ function analyses indicate that specificity for phase 2 enzyme induction was associated with the unsubstituted 1,2-dithiole-3-thione nucleus [14].

Data from numerous animal studies support the use of GST activity as a biomarker of the chemopreventive effect of OPZ. The GSTs are a superfamily of proteins that detoxify chemical carcinogens; they have been divided into eight families, alpha, mu, pi, theta, kappa, sigma, zeta, and omega [34–38], and are highly conserved between species. The reduced hepatotoxicity of acetaminophen and carbon tetrachloride following OPZ administration was accompanied by significant elevations in the total GST activity of mouse liver [39]. Western blotting and high-pressure liquid chromatography (HPLC) analyses indicated that OPZ induced at least three classes of GST ($\alpha$, $\mu$, and $\pi$) both in vivo and in vitro [40–43], and sex-specific variability in GST isozyme induction has been reported [44]. The level of induction ranged from 2.5- to 6-fold, depending on the dose administered, the duration of exposure and the species examined [6,14,27,39,45]. The ability of OPZ to inhibit DNA damage via GST induction is suggested by an inverse correlation between the amount of $AFB_1$ bound to DNA following treatment with the carcinogen plus OPZ and the level of GST activity within the liver [27]. Lately, this has been associated with induction of GST$\alpha$, particularly GSTA5, an enzyme that is not expressed in adult rat liver [46]. The effect on human GSTs was investigated using human hepatocytes; in these experiments, induction of GSTA1 and GSTA2 and, to a lesser extent, GSTM1 mRNA and protein levels was observed following OPZ treatment [40,47].

A single administration of OPZ to mice produced coordinate elevation of transcripts encoding GSTα, μ, and π 2 days after treatment [45]. In this study and that conducted by Davidson et al. [48] on rats, transcriptional activation, rather than RNA stability, of the GST isozymes was confirmed using nuclear run-on assays. In both studies, maintenance of elevated enzymatic activity after RNA levels returned to baseline suggested that additional molecular mechanisms are required to regulate GST expression. Other genes such as UDP-GT, NQO, and γ-glutamylcysteine synthetase are also regulated by OPZ at a transcriptional level [6,14,49,50]. Several cis-acting regulatory elements with binding sites in flanking regulatory region of phase 2 genes control the expression and inducibility of these genes. However, the antioxidant responsive element (ARE) was the first regulatory element shown to be involved in regulation of both basal and inducible expression of cytoprotective genes. This cis-acting sequence is located in the 5'-flanking region of genes encoding many phase 2 enzymes, including mouse and rat GSTA1, human NQO1, and human γ-glutamylcysteine ligase [51,52]. Recent findings suggest an increasingly central role of the ARE sequence in regulating expression of GST and also UDP-GT and NQO1 by OPZ. The consensus sequence of the ARE in the rat GSTA2 promoter was initially defined as TGACNNNGC by mutational analysis [53]. Recent work has defined the consensus of the ARE, or electrophile responsive element, from mouse GSTA1 as A/GTGAC/TNNNGCA/G [54]. Most significantly, this element shares similarity with the NF-erythroid 2–binding site A/GTGAC/GTCAGCA/G and the recognition sequence TGCTGACTCAGCA of v-Maf (a viral oncoprotein associated with musculoaponeurotic fibrosarcoma), otherwise called the 12-O-tetradecanoylphorbol-13-acetate–responsive element-Maf recognition element (T-MARE). Recognition of the ARE sequence by CNC basic-region leucine-zipper proteins NF-erythroid 2–related factor (Nrf)1 and 2 has been shown using an ARE-driven gene reporter construct transfected into human HepG2 cells [55].

The most persuasive evidence that Nrf2 mediates regulation of ARE-driven genes comes from mouse gene knockout experiments that have demonstrated both a repression of the basal expression of several genes and a failure of OPZ to induce many genes [56]. Other basic leucine zipper transcription factors typically form heterodimers, Nrf1/Nrf2, and are able to dimerize with other factors in order to activate the ARE [57]. Nrf2 transcription factor is localized mainly in the cytosol bound to a chaperone, Keap1, which suppresses Nrf2 transcriptional activity [58]. Keap1 is able to completely repress the activity of Nrf2, as demonstrated by co-transfection experiments with increasing concentrations of Keap1-expression plasmid with a Nrf2 vector [58]. Such observations support the hypothesis that Keap1 is a key regulatory molecule of Nrf2 and that the Keap1-Nrf2 complex acts as a sensor to oxidative or electrophilic stresses to induce protective genes promoting cell survival.

Exposure of cells to electrophilic agents, including OPZ, disrupts the Keap1-Nrf2 complex. Nrf2 migrates to the nucleus, binds to the enhancer regions of phase 2 genes, and stimulates their transcription (Fig. 2). Nuclear accumulation of Nrf2 is both rapid and persistent and can be mediated by inducers of distinct chemical classes; however, this accumulation does not appear to be due simply to translocation of preexistent Nrf2 from the cytoplasm. For example, dissociation of Nrf2 from Keap1 leads to an initial elevation of nuclear Nrf2 within 20–60 minutes: given the short half-life of Nrf2 (around 20 min), the predominant factor

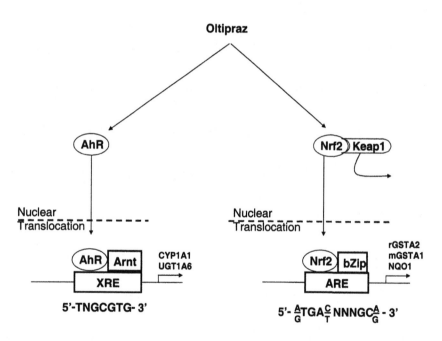

**Figure 2** Regulation of XRE- and ARE-driven gene expression by OPZ. This diagram illustrates the two main molecular mechanisms, transcription factors and *cis*-acting elements, involved in transcriptional activation of genes by OPZ. This compound increases CYP1A1 and UGT1A6 transcription through interaction of AhR with the xenobiotic responsive element (XRE). The other pathway involves the antioxidant responsive element (ARE). This sequence recruits either Nrf1 or Nrf2 as heterodimers with small Maf proteins, FosB, c-Jun, JunD, ATF2, or ATF4; the complex above ARE in the diagram is simply Nrf-bZIP and is involved in induction of phase II enzymes such as GST and NQO1.

driving the sustained accumulation and transactivation of phase 2 genes must result from enhanced de novo synthesis [56].

Although the involvement of the ARE in phase 2 gene induction by OPZ has been extensively studied, at least three other regulatory elements are implicated in the process of OPZ induction: the CCAAT/enhancer binding (C/EBP), the xenobiotic responsive element (XRE), and the NF-κB sequences [59].

Recently, Kang et al. [60] demonstrated that in OPZ-treated cells, C/EBP translocated to the nucleus and bound to the consensus sequence of C/EBP (TTGCGCAA). OPZ treatment increased luciferase reporter-gene activity in rat hepatoma H4IIE cells transfected with the C/EBP-containing regulatory region of the GSTA2 gene. Deletion of the C/EBP binding site or overexpression of a dominant-negative mutant form of C/EBP (AC/EBP) abolished the reporter gene activity. It was further found that PI3-kinase was required for C/EBP β-dependent induction of GSTA2 by OPZ.

Although the xenobiotic-responsive element (XRE)/aromatic hydrocarbon responsive element (AhRE) is mostly involved in the induction of CYP1A and 1B genes, it also plays a role in some limited phase 2 genes [61]. The involvement of the AhR and aryl hydrocarbon nuclear translocator (ARNT) in mediating the effects of OPZ on the XRE was supported by electrophoretic mobility supershift data, transient transfection, and AhR/ARNT overexpression studies. Recently, the XRE/AhR pathway was demonstrated to be involved in UDP-GT 1A6 induction by OPZ [62]. Overexpression of NQO1 induced by OPZ through a mechanism involving the NF-κB response element in the NQO1 5′-flanking region was first reported by Yao and O'Dwyer [59], thereby indicating a role for this element in the control of detoxification responses to environmental changes.

Taken together, these studies demonstrate that the pleiotropic effects of OPZ on phase 2 enzymes might be mediated by at least four distinct pathways; this might well explain the broad range of genes regulated by this compound.

## VI.  EFFECTS ON PHASE I ENZYMES

More recent studies have established that OPZ and other dithiolethiones are also effective on phase 1 enzymes. Dual effects have been demonstrated on CYPs; the use of primary human hepatocytes has shown that overall metabolism of $AFB_1$ and formation of $AFM_1$ were strongly inhibited when cells were co-treated with the carcinogen and the chemoprotective agent (Fig. 3) [40]. 7-Ethoxyresorufin O-deethylation and nifedipine oxidation catalyzed by CYP1A2 and CYP3A4, respectively, the two major CYPs involved in $AFB_1$ metabolism in humans, were rapidly and strongly reduced following the addition of the carcinogen in the presence of OPZ. Such inhibition was confirmed in OPZ-treated rat liver by measuring caffeine $N^3$-demethylation catalyzed by CYP1A2 [63]. Using human

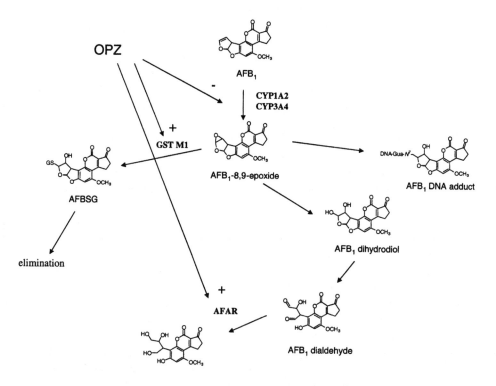

**Figure 3** Effects of OPZ on $AFB_1$ metabolism in humans. AFBSG: GSH conjugate of $AFB_1$-8,9-epoxide.

recombinant CYPs expressed in *Escherichia coli* membranes, OPZ was shown to behave as both a competitive- and mechanism-based inhibitor of CYP1A2, exhibiting $K_i$ of 1.5 μM and $k_{inactivation}$ of 0.19 min$^{-1}$ [64], this inhibition being, in part, due to heme inactivation. OPZ also inhibits, in descending order, CYP3A4, 1A1, 1B1, and 2E1; its metabolite, OPZ M3, is equally effective, leading to the conclusion that the parent molecule does not require transformation to exert its inhibitory effect (Fig. 4). Inhibition of CYPs by OPZ has been confirmed in vivo in humans during the phase IIb clinical trial: weekly administration of 500 mg OPZ led to a significant decrease of $AFM_1$ and completely masked its effect on GST since $AFB_1$ mercapturic acid formation was also strongly inhibited [65]. OPZ is also effective at lower doses; thus, 8 days after a single dose of 125 mg OPZ, caffeine *N*-demethylation was inhibited by up to 80% [66].

OPZ is also an inducer of CYPs; a 3- and 11-fold increase of CYP1A1/2 and 2B1/2 mRNA levels, respectively, was observed in OPZ-treated rat liver

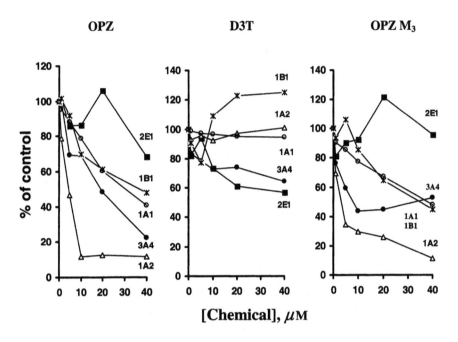

**Figure 4** Effects of OPZ, 1,2-dithiole-3-thione (D3T), and OPZ M3 on CYP-catalyzed activities in *Escherichia coli* membranes. CYP1A1, 1A2, and 1B1 activities were estimated by 7-ethoxyresorufin *O*-deethylation, CYP3A4 activity by testosterone 6β-hydroxylation, and CYP2E1 by chlorzoxazone 6-hydroxylation. (From Ref. 64.)

[63,67]. CYP1A expression and activity were augmented in rat kidney and lung after a 48-hour treatment; in contrast, CYP2B1 mRNA levels were unchanged in lung and the amount of the corresponding apoprotein and activity were dramatically decreased, due to a rapid catabolism of the protein via the proteasome-dependent pathway [68]. Le Ferrec et al. [69] have recently shown that CYP1A1 transcripts were also augmented in rat intestine and in the human Caco-2 cell line, derived from a colon adenocarcinoma. Both the CYP1A protein content and its corresponding activity, estimated by ethoxyresorufin *O*-deethylation, were increased, reaching a maximum of 30-fold; such induction occurred at the transcriptional level and was mediated by the AhR. In addition, a rapid increase in intracellular calcium concentration was required to observe CYP1A induction. It is likely that CYP induction by OPZ depends on the species, tissues, and CYP enzymes.

## VII. EFFECTS ON OTHER DETOXICATION FUNCTIONS

In addition to phase 1 and phase 2 enzymes, several other nonenzymatic and enzymatic proteins more or less involved in xenobiotic metabolism have been shown to be induced by OPZ; these include $\gamma$-glutamylcysteine synthetase (a rate-limiting enzyme involved in GSH synthesis), glutathione conjugate efflux pumps [50,70], heme oxygenase 1, heavy and light subunits of ferritin, catalase, $AFB_1$ aldehyde reductase (AFAR) [71], leukotriene B4 dehydrogenase [72,73], and manganese superoxide dismutase. No effect was observed on two other antioxidant enzymes: copper/zinc superoxide dismutase and glutathione peroxidase [74].

AFAR is highly induced by OPZ; this enzyme, first purified from ethoxyquin-treated rat liver, has been shown to play a role in $AFB_1$ detoxication [75,76]. CYP-mediated $AFB_1$-8,9-epoxides are mostly spontaneously hydrolyzed to $AFB_1$-8,9-dihydrodiol (Fig. 3) that, under physiological conditions, can rearrange to form a reactive dialdehyde; this metabolite does not bind to DNA but can react with primary amine groups in intracellular proteins [77]. AFAR catalyzes the NADPH-dependent reduction of $AFB_1$-dialdehyde to $AFB_1$-dialcohol, thereby reducing $AFB_1$ cytotoxicity [78,79].

Several studies favored the idea that OPZ was also capable of increasing nucleotide excision repair, the major pathway of elimination of chemical carcinogen–derived DNA adducts. OPZ was shown to protect against colon carcinogenesis when administered following azoxymethane [17] and to increase removal of platinium-derived DNA adducts in colon carcinoma cells [80]. However, approaches involving measurement of whole DNA repair in cisplatin-damaged DNA plasmid in the presence of cell-free extracts from OPZ-treated liver cells or monitoring the removal of $AFB_1$-derived DNA adducts formed in primary human hepatocytes exposed to the chemoprotective agent both failed to provide evidence of an influence of OPZ on nucleotide excision repair activity [81].

A moderate increase in the production of reactive oxygen species was also found in primary rat hepatocytes exposed to OPZ [74]; it may be postulated that these molecules act as secondary messengers that trigger transcription of the many genes induced by this chemoprotective agent.

## VIII. PROTECTION AGAINST $AFB_1$ IN HUMANS

Several studies have been designed to evaluate the chemopreventive potential of OPZ in humans; most have been conducted by Kensler and coworkers and have centered on Chinese populations exposed to high levels of $AFB_1$ and having a high incidence of hepatocellular carcinoma. The effects of the chemoprotective agent have been estimated by measuring biomarkers that reflect exposure of individuals to the carcinogen, i.e., $AFB_1$-albumin adducts, $AFM_1$, a phase 1 inactive metabolite, and $AFB_1$-mercapturic acid, a derivative of the $AFB_1$-GSH conjugate. $AFB_1$-albumin adducts in peripheral blood are assayed by ELISA or radioimmunoassay, while the other biomarkers are measured in urine using sequential immunoaffinity approaches and liquid chromatography with fluorescence detection or coupling to a mass spectrometer [65,82].

A phase I clinical trial showed that oral doses of 125, 250, 375, and 500 mg OPZ had similar half-lives, with no significant differences between clearances at the 125 and 250 mg levels. Peak plasma levels ranged from 0.35 to 4.92 μg/mL, with marked intersubject variability being noted. Administration of a daily dose of 125 mg OPZ for 6 months was well tolerated [8]. The rapid clearance of OPZ from the body can explain the low steady-state concentrations; however, these concentrations are sufficient to induce phase 2 enzymes in cultured liver cells [8,9,83].

A phase IIa clinical trial was conducted with 230 residents of Daxin, Qidong County of Jiangsu Province, China [65]. Daily administration of 125 mg of OPZ led to a 2.6-fold increase in excretion of aflatoxin-mercapturic acid without any effect on $AFM_1$ level; in contrast, weekly administration of 500 mg over one month led to a 51% decrease of $AFM_1$ in urine without any effect on aflatoxin-mercapturic acid excretion. The decreased $AFM_1$ excretion can be interpreted as an inhibition of CYP1A2 activity and the increase of $AFB_1$-mercapturic acid as an increased conjugation of $AFB_1$-8,9-epoxides with GSH.

This study was prolonged by a 12-month phase IIb study in which modulation of $AFB_1$ biomarkers by weekly doses of 250 or 500 mg OPZ was evaluated. Although this study was conducted over the period 1999–2000, the results have not yet been published. This study should determine a safe and effective dose for a phase III trial that will establish the efficacy of OPZ in chemoprevention and determine a relationship between the drug and cancer incidence.

The population of Qidong has also been used to estimate the mutagenic potency of urine from smokers. A significant correlation was found with the number of cigarettes smoked per day and even with urinary cotinine [84]. By contrast, no increased mutagenicity was detected in the urines of individuals treated by OPZ either daily (125 mg) or weekly (500 mg) [85]. These data provide an additional argument in favor of the efficacy of OPZ as a chemopreventive agent.

## IX. CONCLUSIONS

Among the diverse natural products and synthetic chemicals known to have chemopreventive properties against cancer, OPZ has been one of the most investigated compounds as a result of its well-recognized anticarcinogenic effects in animals and its good tolerance in humans. It can modulate both phase 1 and phase 2 xenobiotic-metabolizing enzymes as well as other genes involved in metabolism of electrophiles and reactive oxygen species through multiple regulatory mechanisms. It is a strong phase 2 inducer as well as both an inhibitor and an inducer of certain CYPs involved in activation of various carcinogens. However, large species and organ differences have been observed, probably due to bioavailability of the drug.

A recent phase IIa clinical trial in China supports the claim that OPZ may be a promising chemopreventive agent in humans, based on its capacity to modulate formation of both phase 1– and phase 2–mediated $AFB_1$ metabolism. However, this finding, though interesting, does not allow a conclusion that long-term treatment by OPZ will markedly reduce development of cancer. Only ongoing phase III clinical studies will provide the information necessary to enable such a conclusion to be reached. Moreover, it will be critical to determine whether OPZ is suitable for prevention against cancer induced by diverse carcinogens or whether its use should be limited to specific compounds, depending on chemical structure, target organ(s), and/or metabolic pathways.

## ACKNOWLEDGMENTS

Our personal work was supported in part by the Institut National de la Santé et de la Recherche Médicale, the Association pour la Recherche contre le Cancer, La Ligue contre le Cancer (comité d'Ille et Vilaine), and the Fondation Langlois.

## REFERENCES

1. Marks HS. Analysis of a reported organosulfur, carcinogenesis inhibitor: 1,2-dithiole-3-thione in cabbage. J Agric Food Chem 1991; 39:893–895.
2. Kensler TW, Groopman JD, Roebuck BD, Curphey TJ. Chemoprotection by 1,2-dithiole-3-thiones. In: Huang MT, Osawa T, Ho CT, Rosen RT, Eds. Food Phytochemicals for Cancer Prevention I: Fruits and Vegetables. Washington, DC: American Chemical Society, 1994:154–163.
3. Archer S. The chemotherapy of schistosomiasis. Annu Rev Pharmacol Toxicol 1985; 25:485–508.
4. Mkoji GM, Smith JM, Prichard RK. Glutathione redox state, lipid peroxide levels, and activities of glutathione enzymes in oltipraz-treated adult *Schistosoma mansoni*. Biochem Pharmacol 1989; 38:4307–4313.
5. Nare B, Smith JM, Prichard RK. Oltipraz-induced decrease in the activity of cytosolic glutathione S-transferase in *Schistosoma mansoni*. Int J Parasitol 1991; 21:919–925.

6. Ansher SS, Dolan P, Bueding E. Biochemical effects of dithiolthiones. Food Chem Toxicol 1986; 24:405–415.
7. Wattenberg LW. Chemoprevention of cancer. Cancer Res 1985; 45:1–8.
8. Benson AB. Oltipraz: a laboratory and clinical review. J Cell Biochem 1993; 17F(suppl):278–291.
9. Gupta E, Olopade OI, Ratain MJ, Mick R, Baker TM, Berezin FK, Benson AB, Dolan ME. Pharmacokinetics and pharmacodynamics of oltipraz as a chemopreventive agent. Clin Cancer Res 1995; 1:1133–1138.
10. Dimitrov NV, Leece CM, Tompkins ER, Seymour E, Bennink M, Gardiner J, Crowell J, Hawk E, Nashawaty M, Bennett JL. Oltipraz concentrations in plasma, buccal mucosa cells, and lipids: pharmacological studies. Cancer Epidemiol Biomarkers Prev 2001; 10:201–207.
11. Bieder A, Decouvelaere B, Gaillard C, Depaire H, Heusse D, Ledoux C, Lemar M, J P, Raynaud L, Snozzi C. Comparison of the metabolism of oltipraz in the mouse, rat and monkey and in man. Distribution of the metabolites in each species. Arzneimittelforschung 1983; 33:1289–1297.
12. O'Dwyer PJ, Szarka C, Brennan JM, Laub PB, Gallo JM. Pharmacokinetics of the chemopreventive agent oltipraz and of its metabolite M3 in human subjects after a single oral dose. Clin Cancer Res 2000; 6:4692–4696.
13. Moreau N, Martens T, Fleury MB, Leroy JP. Metabolism of oltipraz and glutathione reductase inhibition. Biochem Pharmacol 1990; 40:1299–1305.
14. Kensler TW, Egner PA, Dolan PM, Groopman JD, Roebuck BD. Mechanism of protection against aflatoxin tumorigenicity in rats fed 5- (2-pyrazinyl)-4-methyl-1,2-dithiol-3-thione (oltipraz) and related 1,2- dithiol-3-thiones and 1,2-dithiol-3-ones. Cancer Res 1987; 47:4271–4277.
15. Wattenberg LW, Bueding E. Inhibitory effects of 5-(2-pyrazinyl)-4-methyl-1,2-dithiol-3-thione (oltipraz) on carcinogenesis induced by benzo[a]pyrene, diethylnitrosamine and uracil mustard. Carcinogenesis 1986; 7:1379–1381.
16. Rao CV, Tokomo K, Kelloff G, Reddy BS. Inhibition by dietary oltipraz of experimental intestinal carcinogenesis induced by azoxymethane in male F344 rats. Carcinogenesis 1991; 12:1051–1055.
17. Rao CV, Rivenson A, Katiwalla M, Kelloff GJ, Reddy BS. Chemopreventive effect of oltipraz during different stages of experimental colon carcinogenesis induced by azoxymethane in male F344 rats. Cancer Res 1993; 53:2502–2506.
18. Moon RC, Kelloff GJ, Detrisac CJ, Steele VE, Thomas CF, Sigman CC. Chemoprevention of OH-BBN-induced bladder cancer in mice by oltipraz, alone and in combination with 4-HPR and DFMO. Anticancer Res 1994; 14:5–11.
19. Roebuck BD, Liu YL, Rogers AE, Groopman JD, Kensler TW. Protection against aflatoxin B1-induced hepatocarcinogenesis in F344 rats by 5-(2-pyrazinyl)-4-methyl-1,2-dithiole-3-thione (oltipraz): predictive role for short-term molecular dosimetry. Cancer Res 1991; 51:5501–5506.
20. Kensler TW, Helzlsouer KJ. Oltipraz: clinical opportunities for cancer chemoprevention. J Cell Biochem 1995; 22(suppl):101–107.
21. Wargovich MJ, Chen CD, Jimenez A, Steele VE, Velasco M, Stephens LC, Price R, Gray K, Kelloff GJ. Aberrant crypts as a biomarker for colon cancer: evaluation

of potential chemopreventive agents in the rat. Cancer Epidemiol Biomarkers Prev 1996; 5:355–360.

22. Clapper ML, Wood M, Leahy K, Lang D, Miknyoczki S, Ruggeri BA. Chemopreventive activity of oltipraz against N-nitroso-bis(2-oxopropyl)amine (BOP)-induced ductal pancreatic carcinoma development and effects on survival of Syrian golden hamsters. Carcinogenesis 1995; 16:2159–2165.

23. Rao CV, Rivenson A, Zang E, Steele V, Kelloff G, Reddy BS. Inhibition of 2-amino-1-methyl-6-phenylimidazo[4,5]pyridine-induced lymphoma formation by oltipraz. Cancer Res 1996; 56:3395–3398.

24. Kim ND, Kwak MK, Kim SG. Inhibition of cytochrome P450 2E1 expression by 2-(allylthio)pyrazine, a potential chemoprotective agent: hepatoprotective effects. Biochem Pharmacol 1997; 53:261–269.

25. Elmore E, Luc TT, Li HR, Buckmeier JA, Steele VE, Kelloff GJ, Redpath JL. Correlation of chemopreventive efficacy data from the human epidermal cell assay with in vivo data. Anticancer Res 2000; 20:27–32.

26. Mehta RG, Moon RC. Characterization of effective chemopreventive agents in mammary gland in vitro using an initiation-promotion protocol. Anticancer Res 1991; 11:593–596.

27. Kensler TW, Egner PA, Trush MA, Bueding E, Groopman JD. Modification of aflatoxin B1 binding to DNA in vivo in rats fed phenolic antioxidants, ethoxyquin and a dithiothione. Carcinogenesis 1985; 6:759–763.

28. Liu YL, Roebuck BD, Yager JD, Groopman JD, Kensler TW. Protection by 5-(2-pyrazinyl)-4-methyl-1,2-dithiol-3-thione (oltipraz) against the hepatotoxicity of aflatoxin B1 in the rat. Toxicol Appl Pharmacol 1988; 93:442–451.

29. Wattenberg LW, Hanley AB, Barany G, Sparnins VL, Lam LK, Fenwick GR. Inhibition of carcinogenesis by some minor dietary constituents. Princess Takamatsu Symp 1985; 16:193–203.

30. Morel F, Langouët S, Maheo K, Guillouzo A. The use of primary hepatocyte cultures for the evaluation of chemoprotective agents. Cell Biol Toxicol 1997; 13:323–329.

31. Maxuitenko YY, Curphey TJ, Kensler TW, Roebuck BD. Protection against aflatoxin B1-induced hepatic toxicity as short-term screen of cancer chemopreventive dithiolethiones. Fundam Appl Toxicol 1996; 32:250–259.

32. Maxuitenko YY, Libby AH, Joyner HH, Curphey TJ, MacMillan DL, Kensler TW, Roebuck BD. Identification of dithiolethiones with better chemopreventive properties than oltipraz. Carcinogenesis 1998; 19:1609–1615.

33. Prochaska HJ, Talalay P. Regulatory mechanisms of monofunctional and bifunctional anticarcinogenic enzyme inducers in murine liver. Cancer Res 1988; 48:4776–4782.

34. Kanaoka Y, Fujimori K, Kikuno R, Sakaguchi Y, Urade Y, Hayaishi O. Structure and chromosomal localization of human and mouse genes for hematopoietic prostaglandin D synthase. Conservation of the ancestral genomic structure of sigma-class glutathione S-transferase. Eur J Biochem 2000; 267:3315–3322.

35. Hayes JD, Pulford DJ. The glutathione S-transferase supergene family: regulation of GST and the contribution of the isoenzymes to cancer chemoprotection and drug resistance. Crit Rev Biochem Mol Biol 1995; 30:445–600.

36. Board PG, Coggan M, Chelvanayagam G, Easteal S, Jermiin LS, Schulte GK, Danley DE, Hoth LR, Griffor MC, Kamath AV, Rosner MH, Chrunyk BA, Perregaux DE,

Gabel CA, Geoghegan KF, Pandit J. Identification, characterization, and crystal structure of the omega class glutathione transferases. J Biol Chem 2000; 275: 24798–24806.

37. Blackburn AC, Woollatt E, Sutherland GR, Board PG. Characterization and chromosome location of the gene GSTZ1 encoding the human zeta class glutathione transferase and maleylacetoacetate isomerase. Cytogenet Cell Genet 1998; 83:109–114.

38. Pemble SE, Wardle AF, Taylor JB. Glutathione S-transferase class kappa: characterization by the cloning of rat mitochondrial GST and identification of a human homologue. Biochem J 1996; 319:749–754.

39. Ansher SS, Dolan P, Bueding E. Chemoprotective effects of two dithiolthiones and of butylhydroxyanisole against carbon tetrachloride and acetaminophen toxicity. Hepatology 1983; 3:932–935.

40. Langouët S, Coles B, Morel F, Becquemont L, Beaune P, Guengerich FP, Ketterer B, Guillouzo A. Inhibition of CYP1A2 and CYP3A4 by oltipraz results in reduction of aflatoxin B1 metabolism in human hepatocytes in primary culture. Cancer Res 1995; 55:5574–5579.

41. Langouët S, Morel F, Meyer DJ, Fardel O, Corcos L, Ketterer B, Guillouzo A. A comparison of the effect of inducers on the expression of glutathione-S-transferases in the liver of the intact rat and in hepatocytes in primary culture. Hepatology 1996; 23:881–887.

42. Maheo K, Antras-Ferry J, Morel F, Langouët S, Guillouzo A. Modulation of glutathione S-transferase subunits A2, M1, and P1 expression by interleukin-1 β in rat hepatocytes in primary culture. J Biol Chem 1997; 272:16125–16132.

43. Maheo K, Morel F, Antras-Ferry J, Langouët S, Desmots F, Corcos L, Guillouzo A. Endotoxin suppresses the oltipraz-mediated induction of major hepatic glutathione transferases and cytochromes P450 in the rat. Hepatology 1998; 28:1655–1662.

44. Meyer DJ, Harris JM, Gilmore KS, Coles B, Kensler TW, Ketterer B. Quantitation of tissue- and sex-specific induction of rat GSH transferase subunits by dietary 1,2-dithiole-3-thiones. Carcinogenesis 1993; 14:567–572.

45. Clapper ML, Everley LC, Strobel LA, Townsend AJ, Engstrom PF. Coordinate induction of glutathione S-transferase alpha, mu, and pi expression in murine liver after a single administration of oltipraz. Mol Pharmacol 1994; 45:469–474.

46. Hayes JD, Judah DJ, McLellan LI, Neal GE. Contribution of the glutathione S-transferases to the mechanisms of resistance to aflatoxin B1. Pharmacol Ther 1991; 50:443–472.

47. Morel F, Fardel O, Meyer DJ, Langouët S, Gilmore KS, Meunier B, Tu CP, Kensler TW, Ketterer B, Guillouzo A. Preferential increase of glutathione S-transferase class alpha transcripts in cultured human hepatocytes by phenobarbital, 3-methylcholanthrene, and dithiolethiones. Cancer Res 1993; 53:231–234.

48. Davidson NE, Egner PA, Kensler TW. Transcriptional control of glutathione S-transferase gene expression by the chemoprotective agent 5-(2-pyrazinyl)-4-methyl-1,2-dithiole-3-thione (oltipraz) in rat liver. Cancer Res 1990; 50:2251–2255.

49. Egner PA, Kensler TW, Prestera T, Talalay P, Libby AH, Joyner HH, Curphey TJ. Regulation of phase 2 enzyme induction by oltipraz and other dithiolethiones. Carcinogenesis 1994; 15:177–181.

50. Yamane Y, Furuichi M, Song R, Van NT, Mulcahy RT, Ishikawa T, Kuo MT. Expression of multidrug resistance protein/GS-X pump and gamma-glutamylcysteine synthetase genes is regulated by oxidative stress. J Biol Chem 1998; 273: 31075–31085.

51. Rushmore TH, King RG, Paulson KE, Pickett CB. Regulation of glutathione S-transferase Ya subunit gene expression: identification of a unique xenobiotic-responsive element controlling inducible expression by planar aromatic compounds. Proc Natl Acad Sci USA 1990; 87:3826–3830.

52. Favreau LV, Pickett CB. The rat quinone reductase antioxidant response element. Identification of the nucleotide sequence required for basal and inducible activity and detection of antioxidant response element-binding proteins in hepatoma and non-hepatoma cell lines. J Biol Chem 1995; 270:24468–24474.

53. Rushmore TH, Morton MR, Pickett CB. The antioxidant responsive element. Activation by oxidative stress and identification of the DNA consensus sequence required for functional activity. J Biol Chem 1991; 266:11632–11639.

54. Wasserman WW, Fahl WE. Functional antioxidant responsive elements. Proc Natl Acad Sci USA 1997; 94:5361–5366.

55. Venugopal R, Jaiswal AK. Nrf1 and Nrf2 positively and c-Fos and Fra1 negatively regulate the human antioxidant response element-mediated expression of NAD(P)H: quinone oxidoreductase1 gene. Proc Natl Acad Sci USA 1996; 93:14960–14965.

56. Kwak MK, Itoh K, Yamamoto M, Kensler TW. Enhanced expression of the transcription factor Nrf2 by cancer chemopreventive agents: role of antioxidant response element-like sequences in the nrf2 promoter. Mol Cell Biol 2002; 22:2883–2892.

57. Kang KW, Cho IJ, Lee CH, Kim SG. An Nrf2/small Maf heterodimer mediates the induction of phase II detoxifying enzyme genes through antioxidant response elements. Biochem Biophys Res Commun 1997; 236:313–322.

58. Itoh K, Wakabayashi N, Katoh Y, Ishii T, Igarashi K, Engel JD, Yamamoto M. Keap1 represses nuclear activation of antioxidant responsive elements by Nrf2 through binding to the amino-terminal Neh2 domain. Genes Dev 1999; 13:76–86.

59. Yao KS, O'Dwyer PJ. Involvement of NF-kappa B in the induction of NAD(P)H: quinone oxidoreductase (DT-diaphorase) by hypoxia, oltipraz and mitomycin C. Biochem Pharmacol 1995; 49:275–282.

60. Kang KW, Cho IJ, Lee CH, Kim SG. Essential role of phosphatidylinositol 3-kinase-dependent CCAAT/enhancer binding protein beta activation in the induction of glutathione S-transferase by oltipraz. J Natl Cancer Inst 2003; 95:53–66.

61. Whitlock JP. Induction of cytochrome P4501A1. Annu Rev Pharmacol Toxicol 1999; 39:103–125.

62. Auyeung DJ, Kessler FK, Ritter JK. Mechanism of rat UDP-glucuronosyltransferase 1A6 induction by oltipraz: evidence for a contribution of the aryl hydrocarbon receptor pathway. Mol Pharmacol 2003; 63:119–127.

63. Langouët S, Maheo K, Berthou F, Morel F, Lagadic-Gossman D, Glaise D, Coles B, Ketterer B, Guillouzo A. Effects of administration of the chemoprotective agent oltipraz on CYP1A and CYP2B in rat liver and rat hepatocytes in culture. Carcinogenesis 1997; 18:1343–1349.

64. Langouët S, Furge LL, Kerriguy N, Nakamura K, Guillouzo A, Guengerich FP. Inhibition of human cytochrome P450 enzymes by 1,2-dithiole-3-thione, oltipraz and its derivatives, and sulforaphane. Chem Res Toxicol 2000; 13:245–252.

65. Wang JS, Shen X, He X, Zhu YR, Zhang BC, Wang JB, Qian GS, Kuang SY, Zarba A, Egner PA, Jacobson LP, Munoz A, Helzlsouer KJ, Groopman JD, Kensler TW. Protective alterations in phase 1 and 2 metabolism of aflatoxin B1 by oltipraz in residents of Qidong, People's Republic of China. J Natl Cancer Inst 1999; 91: 347–354.

66. Sofowora GG, Choo EF, Mayo G, Shyr Y, Wilkinson GR. In vivo inhibition of human CYP1A2 activity by oltipraz. Cancer Chemother Pharmacol 2001; 47: 505–510.

67. Buetler TM, Gallagher EP, Wang C, Stahl DL, Hayes JD, Eaton DL. Induction of phase I and phase II drug-metabolizing enzyme mRNA, protein, and activity by BHA, ethoxyquin, and oltipraz. Toxicol Appl Pharmacol 1995; 135:45–57.

68. Le Ferrec E, Ilyin G, Maheo K, Bardiau C, Courtois A, Guillouzo A, Morel F. Differential effects of oltipraz on CYP1A and CYP2B in rat lung. Carcinogenesis 2001; 22:49–55.

69. Le Ferrec E, Lagadic-Gossmann D, Rauch C, Bardiau C, Maheo K, Massiere F, Le Vee M, Guillouzo A, Morel F. Transcriptional induction of CYP1A1 by oltipraz in human Caco-2 cells is aryl hydrocarbon receptor- and calcium-dependent. J Biol Chem 2002; 277:24780–24787.

70. Courtois A, Payen L, Vernhet L, Morel F, Guillouzo A, Fardel O. Differential regulation of canalicular multispecific organic anion transporter (cMOAT) expression by the chemopreventive agent oltipraz in primary rat hepatocytes and in rat liver. Carcinogenesis 1999; 20:2327–2330.

71. Hayes JD, McLellan LI. Glutathione and glutathione-dependent enzymes represent a co-ordinately regulated defence against oxidative stress. Free Radic Res 1999; 31: 273–300.

72. Prestera T, Talalay P, Alam J, Ahn YI, Lee PJ, Choi AM. Parallel induction of heme oxygenase-1 and chemoprotective phase 2 enzymes by electrophiles and antioxidants: regulation by upstream antioxidant-responsive elements (ARE). Mol Med 1995; 1: 827–837.

73. Primiano T, Li Y, Kensler TW, Trush MA, Sutter TR. Identification of dithiolethione-inducible gene-1 as a leukotriene B4 12-hydroxydehydrogenase: implications for chemoprevention. Carcinogenesis 1998; 19:999–1005.

74. Antras-Ferry J, Maheo K, Chevanne M, Dubos MP, Morel F, Guillouzo A, Cillard P, Cillard J. Oltipraz stimulates the transcription of the manganese superoxide dismutase gene in rat hepatocytes. Carcinogenesis 1997; 18:2113–2117.

75. Judah DJ, Hayes JD, Yang JC, Lian LY, Roberts GC, Farmer PB, Lamb JH, Neal GE. A novel aldehyde reductase with activity towards a metabolite of aflatoxin B1 is expressed in rat liver during carcinogenesis and following the administration of an anti-oxidant. Biochem J 1993; 292:13–18.

76. Hayes JD, Judah DJ, Neal GE. Resistance to aflatoxin B1 is associated with the expression of a novel aldo-keto reductase which has catalytic activity towards a cytotoxic aldehyde-containing metabolite of the toxin. Cancer Res 1993; 53: 3887–3894.

77. Guengerich FP, Cai H, McMahon M, Hayes JD, Sutter TR, Groopman JD, Deng Z, Harris TM. Reduction of aflatoxin B1 dialdehyde by rat and human aldo-keto reductases. Chem Res Toxicol 2001; 14:727–737.

78. Primiano T, Gastel JA, Kensler TW, Sutter TR. Isolation of cDNAs representing dithiolethione-responsive genes. Carcinogenesis 1996; 17:2297–2303.

79. Manson MM, Ball HW, Barrett MC, Clark HL, Judah DJ, Williamson G, Neal GE. Mechanism of action of dietary chemoprotective agents in rat liver: induction of phase I and II drug metabolizing enzymes and aflatoxin B1 metabolism. Carcinogenesis 1997; 18:1729–1738.

80. O'Dwyer PJ, Johnson SW, Khater C, Krueger A, Matsumoto Y, Hamilton TC, Yao KS. The chemopreventive agent oltipraz stimulates repair of damaged DNA. Cancer Res 1997; 57:1050–1053.

81. Sparfel L, Langouët S, Fautrel A, Salles B, Guillouzo A. Investigations on the effects of oltipraz on the nucleotide excision repair in the liver. Biochem Pharmacol 2002; 63:745–749.

82. Jacobson LP, Zhang BC, Zhu YR, Wang JB, Wu Y, Zhang QN, Yu LY, Qian GS, Kuang SY, Li YF, Fang X, Zarba A, Chen B, Enger C, Davidson NE, Gorman MB, Gordon GB, Prochaska HJ, Egner PA, J D, Munoz A, Helzlsouer KJ, Kensler TW. Oltipraz chemoprevention trial in Qidong, People's Republic of China: study design and clinical outcomes. Cancer Epidemiol Biomarkers Prev 1997; 6:257–265.

83. Dimitrov NV, Bennett JL, McMillan J, Perloff M, Leece CM, Malone W. Clinical pharmacology studies of oltipraz—a potential chemopreventive agent. Invest New Drugs 1992; 10:289–298.

84. Camoirano A, Bagnasco M, Bennicelli C, Cartiglia C, Wang JB, Zhang BC, Zhu YR, Qian GS, Egner PA, Jacobson LP, Kensler TW, De Flora S. Oltipraz chemoprevention trial in Qidong, People's Republic of China: results of urine genotoxicity assays as related to smoking habits. Cancer Epidemiol Biomarkers Prev 2001; 10: 775–783.

85. De Flora S, D'Agostini F, Balansky R, Camoirano A, Bennicelli C, Bagnasco M, Cartiglia C, Tampa E, Longobardi MG, Lubet RA, Izzotti A. Modulation of cigarette smoke-related end-points in mutagenesis and carcinogenesis. Mutat Res 2003; 523–524:237–252.

# 16
# Future Perspectives in Phytochemical and Health Research

**Yongping Bao and Roger Fenwick**
*Institute of Food Research, Colney, Norwich, United Kingdom*

## I. INTRODUCTION

Consideration of the foregoing chapters demonstrates that the pace of development of research addressing phytochemicals and their potential for improving human health and well-being is increasing. In the face of such rapid change, the wisdom of highlighting key areas for future research may be questioned; first, because it is to be expected that major advances in some areas will be achieved within an unexpectedly short time, and second, because the significance of others will become more obvious with hindsight. In addition, serendipity will continue to provide unexpected opportunities, and each new advance will present the research community with additional challenges.

Milner et al. [1] have pointed out that the development and implementation of a successful interdisciplinary effort that integrates and exploits a molecular approach to nutrition-related health and disease research will require both time and patience, to which might be added the availability of adequate resources. The quantity of data that such studies will generate over the next few years is difficult to overestimate, and, in science as elsewhere, quantity does not necessarily equate with quality. As pointed out by the authors of the opening chapter, the onus will be on scientists to adhere to well-established good scientific principles of study design and best practice if the anticipated research is to be exploited to the maximum.

The detailed knowledge, technologies, best practice, guidance, and expertise that have been gained and exploited in pharmacological research should be judiciously applied to phytochemicals, with necessary attention being paid to the essential differences between a drug and a food-derived chemical. The variety, complementarities, advantages, and disadvantages of existing methodologies is particularly well illustrated in the second chapter, addressing bioavailability, and this also highlights the need to gain as complete as possible an understanding of the parameters involved before individual experimental procedures are finalized. The time and attention to detail expended at the outset of proposed transdisciplinary investigations will ultimately yield considerable dividends.

One comment that must be made at the outset is the battery of innovative molecular biology-based techniques and approaches now being developed and exploited are but one means (albeit a powerful one) to address the key questions raised within this volume, and should not be seen as an end in themselves.

The following points, while not intended to be comprehensive, are put forward as a means to initiate discussion in and around the topic of phytochemicals, health, and disease and in the hope that they will contribute to the effective development of this area for the mutual benefit of the scientific community and its stakeholders.


## II. MECHANISMS OF ACTION

The information presented in this volume emphasizes the diverse opportunities that are becoming available to apply "trinomics" (a platform of technologies embracing genomics, proteomics, and metabonomics) to further elucidate mechanisms of action. This knowledge may facilitate the design of even more effective agents or serve to identify and exploit combinations of phytochemicals to target multiple pathways. Such combination is analogous to the use of multiagent cytotoxic therapies for cancer treatment [2] and may be seen as similar to hurdle technology in food safety. According to Brenner [3], the approach is challenging and high risk, but also holds out possibilities of high gain. It may offer a particular advantage when, for example, the potential of a single compound is limited by toxicity over the relevant dose range.

The exploitation of anticarcinogenic targeting agents in a sequence aimed at reducing cellular proliferative signals while enhancing apoptotic function may be an optimal approach to the design and testing of combinations of chemopreventive agents [4]. However, such an approach highlights the need to understand, and ideally predict, the nature and extent of interactions between the individual chemicals involved. In this context, it is interesting that the idea of a single dietary components acting best at a certain optimal level and in combination with other compounds (enshrined in the recommendation to enjoy a mixed diet) has recently been shown to have a scientific basis [5].

The complexities of phytochemical metabolism outlined in Chapter 3 and elsewhere and the observations that individual metabolites elicit large differences in biological activity emphasize the crucial importance of determining, and focusing on, the appropriate metabolites from the individual tissues under investigation. Such studies are an indispensable element of phytochemical research even if they lack some of the glamour associated with much trinomic activity.

This complexity of biological activity at the metabolic level is overlaid on the biological activities of the many thousands of naturally occurring chemicals in the food chain. The enthusiasm of growing numbers of consumers for plant-based medicines and tonics will add considerably to this number. It will be important for these compounds, as well as those formed by secondary reactions during postharvest and processing stages, to be classified and categorized according to structural and biological criteria so that the assessment of their nature and effect in humans can be made. Application of quantitative structure-activity relationships (QSARs) and predictive modeling techniques will have a significant role to play in simplifying and prioritizing the necessary screening activities required.

The outcome of mechanistic studies will also help to determine whether future efforts should be concentrated on manipulating levels and defining bio-availabilities in foods or whether an approach involving purified supplements is more appropriate. The requirements for establishing safety and efficacy of such products have been recently set out by Talalay and Talalay [6]. Finally, in common with practice in pharmacology, questions need to be asked about the possible influences that age, ethnicity, gender, and existing disease states may exert on mechanisms of action.

## III. EARLY BIOMARKERS: GENE EXPRESSION, NOVEL TARGETS

Human intervention studies, characterized by their long-term, double-blind, placebo-controlled protocol, are necessary to determine whether specific dietary factors are effective against chronic disease. Such studies require enormous resources that can only generally be provided by commercial organizations or government bodies. The financial benefit to industry from development and exploitation of a naturally occurring phytochemical are clearly very different from those associated with a widely marketed drug, thereby limiting even further the likelihood of such funding. It is, therefore, of very great importance to identify biochemical and cellular biomarkers that may be confidently used to assess the early events in disease initiation and progression.

This is an area of very considerable activity and developments will undoubtedly serve to facilitate and promote research broadly addressing the biological effects of dietary phytochemicals. The aim should be to have available a battery, or suite, of cost-effective biomarkers capable of being exploited across a range of organisms. Such biomarkers will be based upon key proteins and enzymes, appropriate metabolites, DNA alterations, and immunological and histopathological indicators. The importance of understanding any limitations in the interpretation of the resulting data cannot be overstressed.

Among molecular biomarkers, DNA adducts such as 8-hydroxy-dG, M1-dG, and PhIP-DNA, discussed in Chapter 7, represent potential candidates for the assessment of the efficacy of an intervention study; however, as pointed out recently by Halliwell [7], there is considerable circumstantial evidence rather than proof beyond reasonable doubt as to the validity of oxidative DNA as a biomarker of subsequent cancer development. As well as validated biomarkers, interlaboratory agreement is urgently needed on their protocols and use. Developments in functional trinomics will facilitate characterization of molecular and genomic biomarkers that can be used to determine risk in prospective cohorts as well as surrogate endpoints in clinical studies.

## IV. PHYTOCHEMICAL ENHANCEMENT: FROM TRADITIONAL BREEDING TO GMO MANIPULATION

A primary focus of nutritional research over the last century has been on the limited number of minerals and vitamins required to overcome possible nutritional disorders. According to Harborne [8], there are some 80,000 phytochemicals—representing 80% of the total secondary metabolites on the planet. While the majority of these may be assigned to specific plant species or genera, many are confined to individual plant species where they have evolved in response to, for example, specific predation and stress responses. Phytochemicals such as cyanogenic glycosides and pyrrolizidine alkaloids are associated with severe, even fatal, consequences if consumed. Such situations represent a particular problem in areas of the world where food security is uncertain; in the Indian subcontinent, for example, lathyrism and anemia still occur widely despite their causes being well understood.

Assuming that there is sufficient evidence to support such action, levels of individual phytochemicals may be enhanced by choosing the appropriate cultivar and/or plant part and selecting an appropriate means of processing, as indicated in Chapter 6. In the last century, the primary aims of conventional breeding programs were, necessarily, yield and productivity, although in the area of animal feeding stuffs attention was paid to reducing levels of glucosinolates in rapeseed, as a result of which new "zero" glucosinolate canola cultivars were introduced.

The attention of breeders is now increasingly turning to protein quality and micronutrient content and density. As an example, so-called "superbroccoli" was produced by crossing a normal broccoli variety, Marathon, with a wild-type cabbage from Sicily. The level of the isothiocyanate sulforaphane in the cross was 10-fold higher than in conventional broccoli, and the induction of quinone reductase, a phase 2 enzyme, was increased almost 100-fold [9]. Acceptance trials have revealed no detrimental taste or sensory characteristics, and the product is expected to reach the marketplace within the next few years. Developments such as this will, hopefully, serve to make the consumer aware of the health benefits of such foods as part of an overall balanced diet and may go some way to prepare the ground for a broader acceptance of genetically manipulated produce.

In all cases the expenditure of resources can only be justified if the basis of the association between the compound in question and the health of the consumer is well founded (and takes into account overall dietary variability). This places a barrier in the path of selection for individual phytochemical content. The same is, of course, true for approaches that exploit the new molecular biological technologies.

Although most plant secondary metabolic pathways have now been mapped [10], the absence of suitable gene probes to elucidate the mechanisms regulating synthesis and accumulation of phytochemicals made it difficult to develop effective approaches to manipulate their levels. Trinomics offers the possibility of an integrative approach to address these challenges. Nevertheless, some difficulties do remain. Phytochemicals are generally present in the diet at relatively low amounts, and they are frequently of limited distribution in the plant kingdom; these two factors make it more difficult to successfully identify the relevant genes.

Unlike the situation for conventional nutrients such as vitamins and minerals, where the role in disease prevention and recommended levels are both understood, the situation for phytochemicals remains obscure. Crucial decisions need to be made as to which secondary metabolites, individual or group, should be manipulated, to what level(s), in which crops to facilitate optimum nutritional and health benefits. Given that attempts to increase, for example, brassica consumption generally founder on the rock of well-established food preferences, would it be better to enhance sulforaphane levels in brassicas or to introduce the relevant genes and controls into other, more widely consumed species? Of course, one option does not exclude the other.

The mere enhancement of phytochemical levels is not an answer in itself. Such higher levels may not be effectively absorbed and may pass through the body and be excreted without any biological benefit. Again, it must be remembered that the "improved" plants must first be consumed if their benefits are to be achieved. Many phytochemicals, especially those having a role to play in plant protection, are bitter, acrid, astringent, or pungent. Since taste is the most important influence on food choice [11] and sensitivity towards bitterness is a heritable trait [12], it is important to ensure that levels of phytochemicals are manipulated only within the boundaries of consumer preference (which may vary with the geographical or ethnic origins of the consumer). In this context it is also important to consider the extended food chain, so that achievements in enhancing phytochemical levels in the crop are not obviated in subsequent postharvest or processing (e.g., debittering) stages.

In the same manner, the use of genetic manipulation to optimize levels of nutrients and phytochemicals will provide many members of society with a dilemma. Scientists, food producers, and regulators must make every effort to enter into a dialogue with consumers to explain the benefits and risks associated with these new products. Ultimately the consumer alone will determine how successful an improved, phytochemically enhanced food product will be. Experience has shown that the concerns and distrust of consumers over one genetically manipulated foodstuff may, however "unscientific," be extended to all.

Given that levels of phytochemicals are affected by variety, agronomic, and cultural conditions, plant part, postharvest, and processing conditions it will be important to develop robust methods for characterization of the compound(s) of interest and for the cultivation and processing technological conditions to ensure reproducible preparations. Much early work on the effects of phytochemicals was hampered by researchers inadequately characterizing the compounds, extracts, or foods administered, thereby compromising effective interlaboratory comparison.

## V. GENETIC DEFECTS: POLYMORPHISM, PERSONALIZED AND INDIVIDUAL NUTRITION

Zhao et al. [13] reported an interaction between the GST genotype and reduction in risk of lung cancer among a Chinese female population, including many non-smokers, following ingestion of a variety of isothiocyanates derived from nine cruciferous vegetables. The inverse association observed was modified by *GSTM1* and *GSTT1* genotypes; those with the null genotype for either or both enzymes experienced a significant reduction in risk with higher isothiocyanate intake, while the effect was smaller and not statistically different if either or both genes were present. Earlier, London et al. [14] had reported a similar result in a Chinese male population in Shanghai: individuals with detectable urinary isothiocyanate levels were found to have a significantly reduced risk of lung cancer, and this subgroup was primarily comprised of individuals with *GSTM1*, *GSTT1*, or both null genotypes. Following an investigation of a U.S. population, Spitz et al. [15] reported that a combination of low intake of isothiocyanates and *GSTM1* and *GSTT1* null genotypes conferred the highest risk of lung cancer among smokers.

GSTs are important biotransformation enzymes that detoxify many xenobiotics. In individuals with null genotype of certain GST isoforms, the biological activity of isothiocyanates is more marked. These data emphasize the importance of the fine balance between the metabolizing enzymes if phytochemicals and/or their metabolites are to exert their optimal biological activities.

Ratnasinghe et al. [16] and Hu and Diamond [17] have demonstrated that the leucine-containing allele of the glutathione peroxidase codon 198 was more frequently associated with breast and lung cancers that the corresponding proline-containing allele. Significantly, the polymorphism involving leucine-containing hGPX1 germline codon 198 is confined to Caucasians and is not found among ethnic Chinese. This GPX gene polymorphism also affects enzyme activity following selenium supplementation, demonstrating the differences in an individual's selenium requirements that are necessary for optimum activity. This is a challenge to traditional concepts of recommended daily allowances, RDA, and Daily Recommended Intakes (DRI), which fail to address individual variations in nutrient requirements. These data emphasize the importance of taking polymorphisms into account when considering possible dietary recommendations and advice. Future dietary advice should take into account individual genetic polymorphism, environmental exposure to the risk factors, and a relevant spectrum of both phytochemicals and nutrients.

In summary, with the decoding of all of the human genomes, the time and opportunity will arise for nutritionists and health professionals to prescribe the appropriate diets/foods (comprising the optimum combination of nutrients and phytochemicals) to the appropriate people (according to their individual polymorphisms in certain key genes related to nutrients and phytochemical metabolism) at the appropriate time (according to age, gender, ethnicity, and overall health status). Such an approach to diet is not intended to imply that science seeks to remove the pleasure from eating; we should certainly live to eat rather than eat to live.

## VI.  PHYTOCHEMICAL INTERACTIONS: SYNERGISM, ADDITIVE, OR ANTAGONISTIC EFFECT

Together with the intractable topic of bioavailability, which is addressed in Chapter 2, and the development and exploitation of reliable and robust biomarkers, "interactions" pose one of the biggest challenges and, therefore, opportunities of the next decade. These interactions are manifold: between phytochemicals; between phytochemicals and the food matrix; between phytochemicals, nutrients, and prescribed drugs. An understanding of the natures, strengths, and factors affecting such interactions is fundamental to determining the true nature of the relationship between phytochemicals and health because, in general, isolated phytochemicals and nutrients do not form part of the diet, and even when supplements are consumed these may be extracts or individual compounds capable of subsequently interacting with other dietary ingredients. In addition to the effects brought about by the interactions of individual compounds, which are dependent on their chemical structures and mechanisms of action, attention must also be paid to how such interactions may affect overall in vivo redox balance [18].

As has been mentioned earlier, combinations of phytochemicals may prove optimal in targeting multiple pathways; thus, Zhang et al. [19] have demonstrated that selenium and sulforaphane have a synergistic effect in inducing thioredoxin reductase 1 and in protecting against free radical–mediated cell death. The results demonstrate that synergy can result from a combination of induction at the transcriptional (by sulforaphane) and translational (by selenium) levels. Very recently, several phytochemicals including curcumin, caffeic acid phenethyl ester, and sulforaphane [20] have been shown to disrupt the Nrf2-Keap1 complex, leading to increased Nrf2 binding to the antioxidant responsive element (ARE) and induced transcription of phase II enzymes. Oltipraz has been shown to elicit similar effects (Chapter 15). Such knowledge may allow optimal synergies to be predicted, thereby directing the formulation of future food products.

In the long term, sufficient knowledge should be generated and analyzed to allow predications of the nature and degree of phytochemical binding to the various components of the food matrix, the aim being to predict how the overall composition of a foodstuff, or meal, would affect the biological properties of the ingested phytochemical(s). Such information would facilitate the development of personalized foods suited to the needs of the individual.

The reactivity of many biologically active compounds, the organosulfur compounds of brassicas and alliums being but two examples, means that their levels and biological effects may be altered during food processing and domestic preparation and following ingestion. Little or nothing is known about the effects of domestic food preparation on the chemical reaction and breakdown of phytochemicals. This is a dynamic situation; the biologically active organosulfur compounds, for example, are formed by enzyme-substrate interactions during processing/cooking and, thereafter, decline as a result of volatility, leaching, and secondary reactions. The "flavor profile" of the volatile fraction of onions, garlic, cabbage, and broccoli are highly complex and greatly affected by processing conditions. In general, the higher the processing temperature, the more complex the chemical reactions undergone by biologically active compounds. The collection and analysis of reliable data in this area provides a particular challenge for the chemist.

## VII. RISK/BENEFIT ASSESSMENT

Indole-3-carbinol, capsaicin, and allium-derived organosulfur compounds have all been studied with respect to their anticancer properties and mechanisms have been proposed. However, as pointed out by Lee and Park [21], evidence has also been put forward that each of these compounds possess properties that may promote or assist cancer development.

Indole-3-carbinol has cancer-promoting activities in the colon, thyroid, pancreas, and liver, while capsaicin alters the metabolism of chemical carcinogens and may also promote carcinogenesis. While the effects of organosulfur compounds are less severe, there are some adverse reports, and selenium-enriched garlic has been shown to inhibit the early stage, but not the late stage, of mammary carcinogenesis [22]. However, more recent work [23] has not confirmed this latter claim, emphasizing both the importance and complexity of chemical interactions. Many of these effects may be associated with the levels of compound administered.

As described elsewhere in this volume, phase 1 enzymes are able to covert precarcinogens into ultimate carcinogens that can damage biomolecules, including DNA, while phase 2 enzymes are responsible for the removal of such toxic species. DNA damage may also be brought about by reactive oxygen species. It is known that some chemopreventive agents may act as antioxidants and prooxidants according to their concentration. The dose of the chemical concerned may thus be a crucial determinant of benefit/risk.

Walle et al. [24] have recently reported that quercetin, perhaps following bioactivation to quinine/quinine methide metabolites, can to bind to DNA and protein in human intestinal Caco-2 cells and hepatoma HepG2 cells. This report fuels the longstanding debate about the adverse biological effects of quercetin and other flavonoids. It is, therefore, obvious that the overall toxicological effects of potential chemo-protective agents should be thoroughly evaluated under as broad a range of conditions as possible, and that the focus should not be just primarily on the beneficial effects, however attractive these may seem. Such consideration applies both to supplements and foods. Unless such a balanced approach is undertaken, significant resources will be wasted and the legitimate concerns and interests of society frustrated. Very recently, Conney [25] has pointed out that some compounds may be extremely effective cancer chemopreventive agents in one experimental setting but enhance carcinogenesis in another. In many instances, it may be necessary to tailor cancer chemopreventive agents to individual subjects with known carcinogen exposures or to individuals at high risk for cancer with mechanistically understood pathways of carcinogenesis so that chemopreventive regimens can be targeted at the individual and selected on a more rational basis.

## VIII.   COMMUNICATION AND COOPERATION

Science is becoming increasingly transdisciplinary, and phytochemical and health research is no exception to this trend. The development and effective exploitation of plant secondary metabolites will require the actions of plant scientists, human nutritionists, food scientists and technologists, medical and clinical practitioners, consumer scientists, and experts in many other areas.

Communication between such disciplines is frequently difficult, with a need to agree on common terms and language. A conference in Norwich a few years ago brought plant, food, and nutritional scientists together to discuss ''quality'';

each group perceived and defined the word from their own narrow perspective. Narrow, discipline-related "quality improvement" could not always be related to overall quality and fitness for use. Interdisciplinary studies require effective management, communication, and dissemination of results if they are to be successful. The management and administration of such interdisciplinary research projects is already emerging as a career opportunity for highly qualified researchers.

Communication with society, consumers, patient groups, and their families is not a luxury that scientists can afford to ignore. If we fail in this area, then members of the public will seek information (and certainty) from less well-informed and less independent sources. Since many of these individuals will have personal or family reasons for seeking and acting upon such information, the consequences could be both serious and dangerous.

Communication, gaining and sharing knowledge, is also important between the practitioners of conventional and alternative and complementary medicine. In recent years, much has come to be known about the nature, structures, and mechanisms of action of the active ingredients from traditional Chinese, Indian, and other plant-based medicines. The application of the new genomics and other technologies in these areas and effective exploitation of the data will undoubtedly afford considerable benefits to society.

To cite but one example, the peoples of China have an extensive and well-documented history of understanding the links between food and well-being and have long believed that foods and drugs come from the same source. The proposition that food could be used instead of drugs to treat diseases was discussed in a book entitled "*Effective Emergency Treatments*" from the Tang Dynasty (618–907 A.D.). Easily the most comprehensive work describing the functionality of food is that of Shizhen Li, published in 1578. The author devoted 27 years to studying Material Medica and presented information on 1892 different kinds of medicinal materials within 25 volumes. In Chinese medicine, foodstuffs having highly active healing effects are categorized as drugs, and those having milder effects are categorized as foods. The Chinese Health Authorities have listed 99 items in a special category called "Items with both food and drug properties," mostly of plant origin (flowers, fruits, peels, seeds, leaves, whole plant, roots). The scientific examination of the modes of action of such national and regional plant drugs is underway and will undoubtedly reap considerable rewards.

It has been mentioned by several authors that the advanced genomic, analytical technologies (such as Accelerator Mass Spectrometry, LC-NMR-MS), and clinical cancer prevention approaches (high-resolution endoscopy, laser-capture microdissection, and small interfering RNA) [26] that are now necessary for state-of-the-art studies of phytochemicals and health are limited by cost and expertise to a number of advanced centers nationally and internationally. This poses the question of whether this area of natural science will become the confine of researchers from a limited number of advanced centers, with all the associated problems that such constraints on scientific endeavor produce, or whether such centers can be opened up to a wider group of researchers through proactive training and mobility programs and networking. This is especially important when the plant material under investigation derives from a less developed region (see above) where subsequent commercial development should provide an appropriate local return.

Even in the comparatively resource-rich North American, European Union, and Pacific Rim countries, not every center, or even country, can possess all of the necessary technologies, ideas, and expertise to address the kinds of interdisciplinary challenges described in this volume. Transnational cooperation is, therefore, necessary. Among other organisations, the European Union, through its 4-year Framework Programmes, has been at the forefront of facilitating and encouraging such interactions. Intra- and international cooperation and communication are also vital if methodologies are to be reviewed and amended with the aim of establishing and promoting optimal protocols.

Of course, such transnational activities place even more responsibilities on good management, transparent communication, trust and teamwork, but it is undoubtedly the direction in which many areas of science is moving. Phytochemical researchers share the general responsibility of the scientific community to nurture and encourage the next generation of inquisitive and responsible scientists.

## REFERENCES

1. Milner JA, McDonald SS, Anderson DE, Greenwald P. Molecular targets for nutrients involved with cancer prevention. Nutr Cancer 2001; 41:1–16.
2. DeVita V. Principles of cancer management: chemotherapy. In: DeVita V, Hellman A, Rosenberg S, Eds. Cancer: Principles and Practice of Oncology. Philadelphia: Lippincott-Raven, 1997:333–347.
3. Brenner DE. Multiagent chemopreventive agent combinations. J Cell Pharmacol 2000; 34(suppl):121–124.
4. Sporn MB, Hong KW. Recent advances in chemoprevention of cancer. Science 1997; 278:1073–1077.
5. Nyberg F, Hou SM, Perdhagen G, Lambert B. Dietary fruit and vegetables protect against somatic mutation in vivo, but low or high intake of carotenoids does not. Carcinogenesis 2003; 24:689–696.
6. Talalay P, Talalay P. The importance of using scientific principles in the development of medicinal agents from plants. Acad Med 2001; 76:238–247.
7. Halliwell B. Effect of diet on cancer development: is oxidative DNA damage a biomarker. Free Radic Biol Med 2002; 32:968–974.
8. Harborne JB. Introduction to Ecological Biochemistry. 4th ed.. San Diego: Academic, 1993.
9. Faulkner K, Mithen R, Williamson G. Selective increase of the potential anticarcinogen 4-methylsulphinylbutyl glucosinolate in broccoli. Carcinogenesis 1998; 19: 605–609.

10. DellaPenna D. Nutritional genomics: manipulating plant micronutrients to improve human health. Science 1999; 285:375–379.

11. Glanz K, Basil M, Maibach E, Goldberg J, Snyder D. Why Americans eat what they do: taste, nutrition, cost, convenience and weight control concerns as influences on food consumption. J Am Diet Assoc 1998; 98:1118–1126.

12. Drewnowski A, Gomez-Carneros C. Bitter taste, phytonutrients and the consumer: a review Amer. J Clin Nutr 2000; 72:1424–1435.

13. Zhao B, Seow A, Lee EJD, Poh WT, Teh M, Wang YT, Tan WC, Yu MC, Lee HP. Dietary isothiocyanates, glutathione-S—transferase—M1, -T1 polymorhisms and lung cancer risk among Chinese women in Singapore. Cancer Epidemiol Biomark Prev 2001; 10:1063–1067.

14. London SJ, Yuan JM, Chung FL, Gao YT, Coetzee GA, Ross RK, Yu MC. Isothiocyanates, glutathione-S- transferase M1 and T1 polymorphisms and lung cancer risk;: a prospective study of men in Shanghai, China. Lancet 2000; 356:724–729.

15. Spitz MR, Duphorne CM, Detry MA, Pillow PC, Amos CI, Lei L, de Andrade M, Gu Z, Hong WK, Lee HP. Dietary intake of isothiocyanates: evidence of a joint effect with glutathione-S-transferase polymorphisms in lung cancer risk.. Cancer Epidemiol Biomark Prev 2000; 9:1017–1020.

16. Ratnasinghe D, Tangrea JA, Anderson MA, Barrett MJ, Virtamo J, Taylor PR, Albanes D. Glutathione peroxidase codon 198 polymorphism variant increases lung cancer risk. Cancer Res 2000; 60:6381–6383.

17. Hu YJ, Diamond AM. Role of glutathione peroxidase 1 in breast cancer: loss of heterozygosity and allelic difference in the response to selenium. Cancer Res 2003; 63:3347–3351.

18. Seifried HE, McDonald SS, Anderson DE, Greenwald P, Milner JA. The antioxidant conundrum in cancer. Cancer Res 2003; 63:4295–4298.

19. Zhang JS, Svehlikova V, Bao YP, Howie AF, Beckett GJ, Williamson G. Synergy between sulforaphane and selenium in the induction of thioredoxin reductase 1 requires both transcriptional and translational modulation. Carcinogenesis 2003; 24: 497–503.

20. Surh YJ. Cancer chemoprevention with dietary phytochemicals. Nat Rev Cancer 2003; 3:768–780.

21. Lee BM, Park KK. Beneficial and adverse effects of chemopreventive agents. Mutat Res 2003; 523–4:265–273.

22. Lisk DJ, Thompson J. Selenium-enriched garlic inhibits the early stage but not the late stage of mammary carcinogenesis. Carcinogenesis 1996; 17:1979–1982.

23. Milner JA. A historical perspective on garlic and cancer. J Nutr 2001; 131(suppl): 1027S–1031S.

24. Walle T, Vincent TS, Walle UK. Evidence of covalent binding of dietary flavonoid quercetin to DNA and protein in human intestinal and hepatic cells. Biochemical Pharmacology 2003; 65:1603–1610.

25. Conney AH. Enzyme induction and dietary chemicals as approaches to cancer chemoprevention: The Seventh DeWith S. Goodman Lecture. Cancer Res 2003; 63: 7005–7031.

26. Sabichi AL, Demierre MF, Hawk ET, Lerman CE, Lippman SM. Frontiers in cancer prevention research. Cancer Res 2003; 63:5649–5655.

# Index